Human Genome Informatics

Translational and Applied Genomics Series

Human Genome Informatics

Translating Genes into Health

Edited by

Christophe G. Lambert

Center for Global Health, Division of Translational Informatics,
Department of Internal Medicine, University of New Mexico
Health Sciences Center, Albuquerque, NM, United States

Darrol J. Baker

The Golden Helix Foundation, London, United Kingdom

George P. Patrinos

Department of Pharmacy, School of Health Sciences, University of Patras,
Patras, Greece

Department of Pathology, College of medicine and Health Sciences,
United Arab Emirates University, Al-Ain, United Arab Emirates

Department of Pathology—Bioinformatics Unit, Faculty of Medicine and Health
Sciences, Erasmus University Medical Center, Rotterdam, The Netherlands

ACADEMIC PRESS

An imprint of Elsevier

Academic Press is an imprint of Elsevier
125 London Wall, London EC2Y 5AS, United Kingdom
525 B Street, Suite 1650, San Diego, CA 92101, United States
50 Hampshire Street, 5th Floor, Cambridge, MA 02139, United States
The Boulevard, Langford Lane, Kidlington, Oxford OX5 1GB, United Kingdom

Notices
Knowledge and best practice in this field are constantly changing. As new research and experience
broaden our understanding, changes in research methods, professional practices, or medical treatment
may become necessary.

Practitioners and researchers must always rely on their own experience and knowledge in evaluating
and using any information, methods, compounds, or experiments described herein. In using such
information or methods they should be mindful of their own safety and the safety of others, including
parties for whom they have a professional responsibility.

To the fullest extent of the law, neither the Publisher nor the authors, contributors, or editors, assume
any liability for any injury and/or damage to persons or property as a matter of products liability,
negligence or otherwise, or from any use or operation of any methods, products, instructions, or ideas
contained in the material herein.

Library of Congress Cataloging-in-Publication Data
A catalog record for this book is available from the Library of Congress

British Library Cataloguing-in-Publication Data
A catalogue record for this book is available from the British Library

ISBN 978-0-12-809414-3

For information on all Academic Press publications visit our website
at https://www.elsevier.com/books-and-journals

Working together
to grow libraries in
developing countries

www.elsevier.com • www.bookaid.org

Publisher: John Fedor
Acquisition Editor: Rafael E. Teixeira
Editorial Project Manager: Mariana L. Kuhl
Production Project Manager: Poulouse Joseph
Cover Designer: Mark Rogers

Typeset by SPi Global, India

Contents

PART 2   Human Genome Informatics Tools and Related Resources

CHAPTER 7   A Review of Tools to Automatically Infer Chromosomal Positions From dbSNP and HGVS Genetic Variants

Alexandros Kanterakis, Theodora Katsila, George Potamias, George P. Patrinos, Morris A. Swertz

CHAPTER 8   Translating Genomic Information to Rationalize Drug Use

Alexandros Kanterakis, Theodora Katsila, George P. Patrinos

CHAPTER 9 Minimum Information Required for Pharmacogenomics Experiments 179

J. Kumuthini, L. Zass, Emile R. Chimusa, Melek Chaouch, Collen Masimiremwa

CHAPTER 10 Human Genomic Databases in Translational Medicine 195

Theodora Katsila, Emmanouil Viennas, Marina Bartsakoulia, Aggeliki Komianou, Konstantinos Sarris, Giannis Tzimas, George P. Patrinos

Contributors

Darrol J. Baker The Golden Helix Foundation, London, United Kingdom

Marina Bartsakoulia Department of Pharmacy, School of Health Sciences, University of Patras, Patras, Greece

Melek Chaouch Institute of Pasteur of Tunisia, Tunis, Tunisia

Emile R. Chimusa University of Cape Town, Cape Town, South Africa

Alexandros Kanterakis Institute of Computer Science, Foundation for Research and Technology Hellas (FORTH), Heraklion; Department of Pharmacy, School of Health Sciences, University of Patras, Patras, Greece

Theodora Katsila Department of Pharmacy, School of Health Sciences, University of Patras, Patras, Greece

Aggeliki Komianou Department of Pharmacy, School of Health Sciences, University of Patras, Patras, Greece

J. Kumuthini Centre for Proteomic and Genomic Research, Cape Town, South Africa

Christophe G. Lambert Center for Global Health, Division of Translational Informatics, Department of Internal Medicine, University of New Mexico Health Sciences Center, Albuquerque, NM, United States

Collen Masimiremwa African Institute of Biomedical Science and Technology, Harare, Zimbabwe

George P. Patrinos Department of Pharmacy, School of Health Sciences, University of Patras, Patras, Greece; Department of Pathology, College of medicine and Health Sciences, United Arab Emirates University, Al-Ain, United Arab Emirates; Department of Pathology—Bioinformatics Unit, Faculty of Medicine and Health Sciences, Erasmus University Medical Center, Rotterdam, The Netherlands

George Potamias Institute of Computer Science, Foundation for Research and Technology Hellas (FORTH), Heraklion, Greece

Konstantinos Sarris Department of Pharmacy, School of Health Sciences, University of Patras, Patras, Greece

Morris A. Swertz Genomics Coordination Center, University Medical Center Groningen, University of Groningen, Groningen, The Netherlands

Giannis Tzimas Department of Computer and Informatics Engineering, Technological Educational Institute of Western Greece, Patras, Greece

Emmanouil Viennas University of Patras, Faculty of Engineering, Department of Computer Engineering and Informatics, Patras, Greece

L. Zass Centre for Proteomic and Genomic Research, Cape Town, South Africa

Preface

We are delighted to offer this textbook to the scientific community, entitled Human Genome Informatics, which covers a timely topic in the rapidly evolving discipline of bioinformatics.

In the postgenomic era, the development of electronic tools to translate genomic information into a clinically meaningful format is of utmost importance to expedite the transition of genomic medicine into mainstream clinical practice. There are several well-established textbooks that cover the bioinformatics field, and there are also numerous protocols for bioinformatics analysis that one can retrieve from the Internet. However, the field of human genome informatics is a relatively new one that emerged in the postgenomic era, constituting a niche research discipline. As such, there are hardly any books that discuss this important new discipline, despite its broad implications for human health.

We therefore decided to deliver a textbook focused on human genome informatics, in order to first define the field and some of its history, and then provide an overview of the most commonly used electronic tools to analyze and interpret human genomic information into a clinically meaningful format, hence expediting the integration of genomic medicine into the mainstream clinical practice. At the same time, the book will provide an update on related topics, such as genomic data sharing, human genomic databases, and informatics tools in pharmacogenomics. To our knowledge, no other existing book deals exclusively with this topic.

We envision that this textbook will be of particular benefit to graduate and doctoral students, postdoctoral researchers in the field of genome informatics and bioinformatics, and representatives from bioinformatics companies and diagnostic laboratories interested in establishing such tools to translate/interpret the findings from their analyses. Also, this textbook will be ultimately useful as the main course material or supplementary reading in related graduate courses.

Our effort to compile most of the chapters included in this textbook has been assisted by many internationally renowned experts in their field, who have kindly accepted our invitation to share with us and our readers their expertise, experience, and results through contributed chapters. In addition, we made efforts to formulate the book contents using simple language and terminology, along with self-explanatory illustrations, in order that the book be useful not only to experienced professionals and academics, but also to undergraduate medical and life science students.

We are grateful to the publishing editors, Drs. Mariana Kuhl, Rafael Texeira, and Peter Linsley at Elsevier, who helped us in close collaboration to overcome encountered difficulties. We also express our gratitude to all contributors for delivering outstanding compilations that summarize their experience and many years of hard work in their field of research and to those colleagues who provided constructive comments and criticisms on the chapters. We are indebted to the copy editor, Jude Fernando, who has refined the final manuscript prior to letting it into production. Also, we owe special thanks to the academic reviewers for their constructive criticisms of the chapters and their positive evaluation of our proposal for this compilation.

We feel certain that some points in this textbook can be further improved. Therefore, we would welcome comments and criticism from attentive readers, which will contribute to improve the contents of this book even further in its future editions.

Christophe G. Lambert
Center for Global Health, Division of Translational Informatics, Department of Internal Medicine, University of New Mexico Health Sciences Center, Albuquerque, NM, United States

Darrol J. Baker
The Golden Helix Foundation, London, United Kingdom

George P. Patrinos
Department of Pharmacy, School of Health Sciences, University of Patras, Patras, Greece; Department of Pathology, College of medicine and Health Sciences, United Arab Emirates University, Al-Ain, United Arab Emirates; Department of Pathology—Bioinformatics Unit, Faculty of Medicine and Health Sciences, Erasmus University Medical Center, Rotterdam, The Netherlands

CHAPTER 1

Human Genome Informatics: Coming of Age

Christophe G. Lambert*, Darrol J. Baker[†], George P. Patrinos[‡,§,¶]

*Center for Global Health, Division of Translational Informatics, Department of Internal Medicine, University of New Mexico Health Sciences Center, Albuquerque, NM, United States, [†]The Golden Helix Foundation, London, United Kingdom, [‡]Department of Pharmacy, School of Health Sciences, University of Patras, Patras, Greece, [§]Department of Pathology, College of medicine and Health Sciences, United Arab Emirates University, Al-Ain, United Arab Emirates, [¶]Department of Pathology—Bioinformatics Unit, Faculty of Medicine and Health Sciences, Erasmus University Medical Center, Rotterdam, The Netherlands

1.1 INTRODUCTION

Human genome informatics is the application of information theory, including computer science and statistics, to the field of human genomics. Informatics enlists computation to augment our capacity to form models of reality with diverse sources of information. When forming a model of reality, one engages in a process of abstraction. The word *"abstraction"* comes from the Latin *abstrahere*, which means to "draw away," which is a metaphor, based in human vision, that as we back away from something, the details fall away and we form mental constructs about what we can discern from the more distant vantage point. That more distant vantage point both encompasses a greater portion of reality and yet holds in mind a smaller amount of detail about that larger space.

Given the human mind's limit on the number of variables it can manage, as we form our mental models of reality, we pay attention to certain facets of reality and ignore others, perhaps leaving them to subconscious or unconscious processing mechanisms. When we form models of reality, we have a field of perception that encompasses a subset of reality at a particular scale and a particular time horizon and that includes a subset of the variables at that spatio-temporal scale. Those variables are recursively composed using abstractive processes, for instance, by scale: an atom, a base pair, a gene, a chromosome, a strand of DNA, the nucleus, a cell, a tissue, an organ, an organ system, the human body, a family, a racial group defined by geography and heredity, or all of humanity. Note this abstraction sequence was only spatial and ignored time. Because our perceivable universe is seen through the lens of three spatial and one apparently

1

Human Genome Informatics. https://doi.org/10.1016/B978-0-12-809414-3.00001-2

nonreversible temporal dimension, the mental models we compose describe the transformations of matter-energy forwards through space-time. Let us relate this to information theory and computer science, then bring it back to genomics.

In the 1930s, Alan Turing introduced an abstract model of computation, called the Turing machine (Turing, 1937). The machine is comprised of an infinite linear blank tape with a tape head that can read/write/erase only the current symbol and can move one space to the left or right or remain stationary. This tape head is controlled by a controller that contains a finite set of states and contains the rules for operating the tape head, based only on the current state and the current symbol on the tape (the algorithm or program). Despite the simplicity of this model, it turns out that it can represent the full power of every algorithm that a computer can perform and is thus a universal model of computation.

Suppose we wanted an algorithm to write down the first billion digits of the irrational number π. We could create a Turing machine that had the billion digits embedded in the finite controller (the program) and we could run that program to write the digits to the tape one at a time. In this case, the length of the program would be proportional to the billion digits of output. This might be coded in a language like C++ as: printf("3.1415926[…]7,504,551"), with "[…]" filled in with the remaining digits. If a billion-digit number was truly random and had no regularity, this would approach being the shortest program that we could write (the information-theoretic definition of randomness). However, π is not a random number, but can be computed to an arbitrary number of digits via a truncated infinite series. An algorithm to perform a series approximation of π could thus be represented as a much shorter set of instructions.

In algorithmic information theory, the *Kolmogorov complexity* or descriptive complexity of a string is the length of the shortest Turing machine instruction set (i.e., shortest computer program) that can produce that string (Kolmogorov, 1963). We can think of the problem of modeling a subset of reality as generating a parsimonious algorithm that prints out a representation of the trajectory of a set of variables representing an abstraction of that subset of reality to some level of approximation. That is, we say, "under such and such conditions, thus and such will happen over a prescribed time period". The idea of Kolmogorov complexity motivates the use of *Occam's razor*, where, given two alternate explanations of reality that explain it comparably well, we will choose the simpler one.

In our modeling of reality, we are not generally trying to express the state space transitions of the universe down to the level of every individual atom or quark in time intervals measured by Planck time units, but rather at some level of abstraction that is useful with respect to the outcomes we value in a particular context. Also, because reality has constraints (i.e., laws), and thus regularity, we can observe a small spatial-temporal subset of reality from models that not only describe that observed behavior, but also that generalize to predict the behavior of a broader subset of reality. That is, we don't just model specific concrete observables in the here and now, but we model abstract notions of observables that can be applied beyond the here and now.

The most powerful models are the most universal, such as laws of physics, which are hypothesized to hold over all of reality and can thus be falsified if any part of reality fails to behave according to those laws, and yet, cannot be proven because all reality would have to be observed over all time. This then forms the basis of the scientific method where we form and falsify hypotheses but can never prove them. Unlike with hydrogen atoms or billiard balls where the units of observation may be considered in most contexts as near-identical, when we operate on abstractions such as cells, or people, we create units of observation that may have enormous differences.

1.2 FROM INFORMATICS TO BIOINFORMATICS AND GENOME INFORMATICS

In biology, we often blithely assume that the notion of *ceteris paribus* (all things being equal) holds, but it can lead us astray (Lambert and Black, 2012; Meehl, 1990). For instance, while genetics exists at a scale where *ceteris paribus* generally holds, we are nevertheless trying to draw relations with genetic variations at the molecular scale, with fuzzy phenotypes at the level of populations of nonidentical people.

So unlike our previous example of writing a program to generate the first billion digits of π, which has a very precise answer, our use of abstraction to model biology involves leaving out variables of small effect, which nevertheless, when left unaccounted for, may result in error when we extrapolate our projections of the future with abstract models. We would do well to mind the words of George Box, "all models are wrong, but some are useful":

> Since all models are wrong, the scientist cannot obtain a "correct" one by excessive elaboration. On the contrary, following William of Occam, he should seek an economical description of natural phenomena. Just as the ability to devise simple but evocative models is the signature of the great scientist, overelaboration and overparameterization is often the mark of mediocrity (Box, 1976).

How then do we choose what variables to study at what level of abstraction over what time scale? To begin to answer this question, it is useful to talk about control in the context of goal-directedness and to turn to a field that preceded and contributed to the development of computer science, namely *Cybernetics*. In 1958, Ross Ashby introduced the *Law of Requisite Variety* (Ashby, 1958). *Variety* is measured as the logarithm of the number of states available to a system. *Control*, when stripped of its negative connotations of coercion, can be defined as restricting the variety of a system to a subset of states that are valued and preventing the other states from being visited. For instance, an organism will seek to restrict its state space to healthy and alive ones. For every disturbance that can move a system from its current state to an undesirable one, the system must have a means of acting upon or regulating that disturbance. Ashby's example of a fencer staving off attack is helpful:

> Again, if a fencer faces an opponent who has various modes of attack available, the fencer must be provided with at least an equal number of modes of defense if the outcome is to have the single value: attack parried. (Ashby, 1958)

The law of requisite variety says that "variety absorbs variety," and thus that the number of states of the regulator or control mechanism whose job is to keep a system in desirable states (i.e., absorb or reduce the variety of outcomes) must be at least as large as the number of disturbances that could put the system in an undesirable state. All organisms engage in goal-directed activity, the primary one being sustaining existence or survival. The fact that humanity has dominated as a species reflects our capacity to control our environment—to both absorb and enlist the variety of our environment in the service of sustaining health and life.

In computing, a *universal Turing machine* is a Turing machine that can simulate any Turing machine on arbitrary input. If DNA is the computer program for the "Turing machine of life," the field of human genome informatics is metaphorically moving towards the goal of a universal Turing machine that can answer "what-if" questions about modifying the governing variables of life. Note, the computer science concept of self-modifying code also enriches this metaphor. In particular, cancer genomics addresses the situation where the DNA program goes haywire, creating cancer cells with distorted copies where portions of the genome are deleted, copied extra times, and/or rearranged. Self-modifying code in computer science is enormously difficult to debug and is usually discouraged. Similarly, in cancer, we acknowledge that it is too difficult to repair rapidly replicating agents of chaos, and thus, most treatments involve killing or removing the offending cancer cells. Also, with the advent of emerging technologies such as CRISPR genome editing, humanity is now poised on the threshold of directly modifying our genome (Cong et al., 2013). Such technologies, guided by understanding of the genome, have the potential to recode portions of the program of life in order to cure genetic diseases.

With the human genome having a state space of three billion base pairs times two sets of chromosomes, compounded by epigenetic modifiers that can vary by tissue, compounded by replication errors, compounded by a microbiome living in synergy with its host, compounded by effects of the external environment, the complete modeling of the time evolution of the state space of a human organism at a molecular level appears intractable. Suppose we wanted to perform molecular dynamics simulations of the human body at a femtosecond time scale and do so for an hour. A human body contains approximately 7×10^{27} atoms that we would want to simulate over 3.6×10^{18} timesteps. Such a simulation might require 10,000 floating point operations per atom per timestep to account for various molecular forces, taking on the order of 10^{50} floating point operations. The world's fastest supercomputer at the time of this writing approaches 100 petaflops, or 10^{17} floating point operations per second (Fu et al., 2016). Such a simulation would take 10^{33} s or 3.17×10^{25} years on such a computer. However, if Moore's Law (Moore, 2006) holds and computation capacity is able to double indefinitely every 2 years, in 220 years we could perform such a simulation in 1 s.

By the law of requisite variety, it would seem that our goal of controlling the complete "computer program" of human life is doomed until long after the hypothesized *technological singularity* when human computation capacity is exceeded by computers (Vinge, 1993). We don't appear to have the capacity to model all possible deviations from health at a molecular level and form appropriate responses. However, the successes in modern treatment of numerous diseases suggest that there is enough constraint and regularity in how the building blocks of life assemble to form recognizable and treatable categories of processes and outcomes that we may hope that the set of variables we need to control may not be ultimately intractable.

The evolution of a field of knowledge towards becoming a science begins with classification (e.g., taxonomies), followed by searching for correlations (e.g., genome-wide association studies), followed by forming cause and effect models (e.g., well-characterized molecular mechanisms), and theories (e.g., Darwinian evolution). Other than the successes of understanding monogenic disease processes, much of the past 15–20 years of molecular genetic research has been in the classification and correlation stage. We are still figuring out the relevant variables in the field of genomic health, and only baby steps have been taken to form dynamic causal models of complex systems.

To better understand causation, we need to measure and model the time evolution of systems. This means that, in addition to understanding the germline DNA, we need to understand the time evolution of epigenetic modifications, gene expression, and the microbiome and how these all function together at many orders of magnitude of temporal-spatial scale, including the variation we observe in human populations. Unfortunately, the field has barely broached studies of this kind and will thus not be covered in this book.

1.3   INFORMATICS IN GENOMICS RESEARCH AND CLINICAL APPLICATIONS

As indicated above, informatics plays a vital role not only in genomics research by interpreting high-throughput genotyping and gene expression and deep DNA sequencing analysis data, but also in clinical applications of these new technologies. This not only include informatics tools for genomics, and reciprocally proteomics and metabolomics, analysis, but also tools for the proper annotation, including but not limited to variant nomenclature. It is also of equal importance to establish incentives for openly sharing genomics research results with the scientific community.

1.3.1   Genome Informatics Analysis

Creating transparent and reproducible pipelines is essential for developing best practices for tools, data, and workflow management systems. In Chapter 2, we present a deeper dive into the software tools for managing genomic analysis pipelines, including coverage of such systems as Galaxy and TAVERNA, emphasizing the importance of creating a reliable reproducible workflow upon which others can build and contribute. Recommendations are made on coding standards, code testing and quality control, project organization, documentation, data repositories, data ontologies, virtualization, data visualization, crowdsourcing, and the support of metastudies. Discussion is made of the tradeoffs between the modularity and maintainability of command line tools versus the usability of graphical user interfaces and how modern workflow management systems combine the two.

Similarly, in this book, we present examples on how cytogenetics paradigms shape decision making in translational genomics (Chapter 3). In particular, we describe how early genomic technologies, such as low-resolution cytogenetic testing, shaped worldviews which continue to influence our mental models of genomic medicine. In particular, as our capacity to sequence the human genome has grown exponentially, our capacity to turn this data into understanding has not kept pace. As a result, in medical decision making, the two poles of the central conflict may be verbalized as: "give me only the information I am sure about so that I don't make errors of omission" versus "give me as much information as possible so that I don't make errors of omission." The advent of high-resolution genome-wide microarrays, followed by whole-exome and whole-genome sequencing, has only exacerbated this conflict, as cytogenetics gives way to cytogenomics and as physicians attempt to perform decision making under ever more uncertainty.

Ultimately, next-generation sequencing is gaining momentum in all aspects of genomics research as well as clinical applications with several different platforms existing today, making data resulting from deep DNA sequencing impossible to interpret without dedicated tools and databases. As such, we opted to present a range of available tools and databases accompanied by practical guidelines for next-generation sequencing analysis (Chapter 4). In particular, we present the modular steps involved in the processing and secondary analysis of next-generation sequencing pipelines, including both DNA- and RNA-seq, while touching upon the follow-on tertiary analysis that may be applied once genomic variants have been identified. We describe common formats for storing raw sequencing data, such as FASTQ and SAM/BAM, as well as some popular online data repositories for sequencing data. We describe the basic building blocks of next-generation sequencing pipelines, including sequence alignment as well as approaches to annotation. We close reviewing some of the limitless applications of next-generation sequencing as a modular technology.

Apart from genomics application, proteomics and metabolomics are also coming of age in the postgenomic era, and as such, proteomics and metabolomics data analysis is also key for translational medicine. In Chapter 5, we underscore the importance and pitfalls of large-scale proteomics and metabolomics measures in the clinic for characterizing biological processes and objectively characterizing phenotypes in translational medicine. Discussion is made of ways and means for managing the complexity of these datasets as genome/proteome/metabolome interactions are considered and what challenges remain for broader adoption of these technologies in clinical practice.

1.3.2 Genomics Data Sharing

The continued deposition of genomic data in the public domain is essential to maximize both its scientific and clinical utility. However, rewards for data sharing are currently very few, representing a serious practical impediment to data submission. Moreover, a law of diminishing returns currently operates both in terms of genomic data publication and submission since manuscripts describing a single or few genomic variants cannot be published alone. To date, two main strategies have been adopted as a means to encourage the submission of human genomic variant data: (a) database journal linkups involving the affiliation of a scientific journal with a publicly available database and (b) microattribution, involving the unambiguous linkage of data to their contributors via a unique identifier. The latter could, in principle, lead to the establishment of a microcitation-tracking system that acknowledges individual endeavor and achievement (Giardine et al., 2011; Patrinos et al., 2012).

In Chapter 6, we discuss an important trend in science that started early in the field of genomics, as an outgrowth of public funding. That is, data generated for one research purpose by one organization can be shared with the entire field to augment our collective capacity to model the complexity of the human genome and genomic processes. We also discuss both social and technical challenges of realizing the full potential of collaborative science, with an emphasis on reward systems that enable credit and attribution to be made to genomic data contributors, which could eventually become more widely adopted as novel scientific publication modalities.

1.3.3   Genomic Variant Reporting and Annotation Tools

Advances in bioinformatics required to annotate human genomic variants and to place them in public data repositories uniformly have not kept pace with their discovery. At present, there are a handful of tools that are used to annotate genomic variants so that they are reported with a constant nomenclature. In Chapter 7, we discuss the Human Genome Variation Society (HGVS) nomenclature system for variant reporting and its challenges with ambiguous reporting of variants. We then describe a new tool, MutationInfo, to automatically infer chromosomal positions from dbSNP and HGVS genetic variants. This tool combines existing tools with a BLAST-like alignment tool (BLAT) search in order to successfully locate a much larger fraction of genomic positions for HGVS variants. Finally, we compare the available tools for checking the quality of variants documented in HGVS resources and dbSNP and we highlight the challenge of consistently representing genomic mutations across databases due to multiple versions of different coordinate systems in use.

1.4   PHARMACOGENOMICS AND GENOME INFORMATICS

Pharmacogenomics aim to rationalize drug use by delineating adverse reactions and lack of drug response with the underlying genetic profile of an individual. Since there is a documented lack of (pharmaco)genomics knowledge from clinicians (Mai et al., 2014), there is an urgent need to develop informatics solutions and tools to translate genomic information into a clinically meaningful format, especially in the case of individualization of drug treatment modalities. In other words, a tool that would be able to translate genotyping information from a few or more pharmacogenomic biomarkers into recommendations for drug use.

Chapter 8 touches upon the challenges in translating pharmacogenomics knowledge—the relationship between genetic factors and drug safety and efficacy—to decision making in a clinical setting. We present pharmacogenomics information system approaches to managing both the complexity and uncertainty of pharmacogenomic decision making towards achieving better health outcomes through personalized drug treatment, in the context of a rapidly changing heterogeneous distributed body of knowledge where data sharing and preservation of patient privacy must coexist. Similarly, we also explain how standardization of reporting of pharmacogenomics investigations is crucial in order to enhance the findability, accessibility, interoperability, and reusability of pharmacogenomics studies, promoting study quality and multidisciplinary scientific collaboration, by establishing the concept of the minimum information required for pharmacogenomics experiments (Chapter 9). Special emphasis is placed on experiments using the Drug Metabolism Enzymes and Transporters (DMET+) platform (Affymetrix Inc., Santa Clara, CA, United States) and the Minimum Information required for a DMET experiment (MIDE) to support variant reporting with microarray technologies in support of personalized medicine.

1.5 DATABASES, ARTIFICIAL INTELLIGENCE, AND BIG-DATA IN GENOMICS

In recent years, human genomics research has progressed at a rapid pace, resulting in very high amounts of data production in many laboratories. Therefore, it is imperative to efficiently integrate all of this information in order to establish a detailed understanding of how variants in the human genome sequence affect human health. This fact dictated the development of several genomic databases, i.e., online repositories of genomic variants, aimed to document the vast genomic data production for a single (*locus-specific*) or more (*general*) genes or specifically for a population or ethnic group (*national/ethnic*). Chapter 10 deals with human genomic databases in translational medicine. It describes and summarizes some of the key existing genomic databases that are currently in use in the field of genomic research and medicine, focusing in particular on general locus-specific and national/ethnic genomic databases, as well as an assessment of their strengths and limitations.

A discipline of equal importance in genomics research and genomic medicine, affecting the future landscape of genomic medical diagnosis, is in-silico artificial intelligent clinical information and machine learning systems. In Chapter 11, we describe some of the leading edge approaches towards genomic healthcare medicine that integrate genomic interpretation software, large databases, and approaches to capturing the genomics of large patient populations to ever increase our understanding of genomics and translate it to clinical utility. We highlight how the ability to collaborate and share findings within the wider global research community accelerates the development of global capacity for the diagnosis of genetic disorders.

Lastly, in Chapter 12, we present Genomics England as a paradigm of the future of genomic medical diagnosis, especially in the context of fruitful integration of genomics, informatics, and clinical decision support systems. In particular, we examine the Genomics England approach to genome healthcare medicine and how this pioneering project illustrates the challenges of data sharing within the global scientific community. We also raise concerns about how the erecting of national borders on data sharing (see also Chapter 6) is hampering international disease prevention.

1.6   CONCLUSIONS

The story of the field of human genome informatics to date is really the massive individual and collective effort to understand and thus control the computer program of life with the goal of health. We clearly have far to go, and no one human mind can hope to encompass all of this variety except at the most abstract levels. Nevertheless, with computation and collective endeavor, enormous progress has been made and we chronicle in this book some of the heroic efforts that have been made to date and how they are being translated into clinical use.

Acknowledgments

We wish to cordially thank the authors of all chapters who have contributed significantly in putting together this unique textbook, dealing exclusively with the use of informatics in human genomics research and genomics medicine.

References

Ashby, W.R., 1958. Requisite variety and its implications for the control of complex systems. Cybernetica 1:2, 83–99.

Box, G.E.P., 1976. Science and statistics. J. Am. Stat. Assoc. 71, 791–799.

Cong, L., Ran, F.A., Cox, D., Lin, S., Barretto, R., Habib, N., Hsu, P.D., Wu, X., Jiang, W., Marraffini, L.A., Zhang, F., 2013. Multiplex genome engineering using CRISPR/Cas systems. Science 339, 819–823.

Fu, H., Liao, J., Yang, J., Wang, L., Song, Z., Huang, X., Yang, C., Xue, W., Liu, F., Qiao, F., Zhao, W., Yin, X., Hou, C., Zhang, C., Ge, W., Zhang, J., Wang, Y., Zhou, C., Yang, G., 2016. The Sunway TaihuLight supercomputer: system and applications. Sci. China Inf. Sci. 59. 072001.

Giardine, B., Borg, J., Higgs, D.R., Peterson, K.R., Philipsen, S., Maglott, D., Singleton, B.K., Anstee, D.J., Basak, A.N., Clark, B., Costa, F.C., Faustino, P., Fedosyuk, H., Felice, A.E., Francina, A., Galanello, R., Gallivan, M.V., Georgitsi, M., Gibbons, R.J., Giordano, P.C., Harteveld, C.L., Hoyer, J.D., Jarvis, M., Joly, P., Kanavakis, E., Kollia, P., Menzel, S., Miller, W., Moradkhani, K., Old, J., Papachatzopoulou, A., Papadakis, M.N., Papadopoulos, P., Pavlovic, S., Perseu, L., Radmilovic, M., Riemer, C., Satta, S., Schrijver, I., Stojiljkovic, M., Thein, S.L., Traeger-Synodinos, J., Tully, R., Wada, T., Waye, J.S., Wiemann, C., Zukic, B., Chui, D.H., Wajcman, H., Hardison, R.C., Patrinos, G.P., 2011. Systematic documentation and analysis of human genetic variation in hemoglobinopathies using the microattribution approach. Nat. Genet. 43, 295–301.

Kolmogorov, A.N., 1963. On Tables of Random Numbers. Sankhyā: Indian J. Stat. Ser. A (1961–2002) 25, 369–376.

Lambert, C.G., Black, L.J., 2012. Learning from our GWAS mistakes: from experimental design to scientific method. Biostatistics 13, 195–203.

Mai, Y., Mitropoulou, C., Papadopoulou, X.E., Vozikis, A., Cooper, D.N., van Schaik, R.H., Patrinos, G.P., 2014. Critical appraisal of the views of healthcare professionals with respect to pharmacogenomics and personalized medicine in Greece. Perinat. Med. 11, 15–26.

Meehl, P.E., 1990. Why summaries of research on psychological theories are often uninterpretable. Psychol. Rep. 66, 195–244.

Moore, G.E., 2006. Cramming more components onto integrated circuits, reprinted from electronics, volume 38, number 8, April 19, 1965, pp. 114 ff. IEEE Solid-State Circuits Soc. Newslett. 11, 33–35.

Patrinos, G.P., Cooper, D.N., van Mulligen, E., Gkantouna, V., Tzimas, G., Tatum, Z., Schultes, E., Roos, M., Mons, B., 2012. Microattribution and nanopublication as means to incentivize the placement of human genome variation data into the public domain. Hum. Mutat. 33, 1503–1512.

Turing, A.M., 1937. On computable numbers, with an application to the Entscheidungsproblem. Proc. Lond. Math. Soc. 2, 230–265.

Vinge, V., 1993. The coming technological singularity: How to survive in the post-human era. In: Vision-21: Interdisciplinary Science and Engineering in the Era of Cyberspace. Presented at the Proceedings of a symposium cosponsored by the NASA Lewis Research Center and the Ohio Aerospace Institute, NASA, pp. 11–22.

PART

1

Human Genome
Informatics Applications

Creating Transparent and Reproducible Pipelines: Best Practices for Tools, Data, and Workflow Management Systems

Alexandros Kanterakis*,‡, George Potamias*, Morris A. Swertz†, George P. Patrinos‡

*Institute of Computer Science, Foundation for Research and Technology Hellas (FORTH), Heraklion, Greece, †Genomics Coordination Center, University Medical Center Groningen, University of Groningen, Groningen, The Netherlands, ‡Department of Pharmacy, School of Health Sciences, University of Patras, Patras, Greece

2.1 INTRODUCTION

Recently, publishing the source code, data, and other implementation details of a research project serves two basic purposes. The first is to allow the community to scrutinize, validate, and confirm the correctness and soundness of a research methodology. The second is to allow the community to properly utilize the methodology in novel data, or to adjust it to test new research hypotheses. This is a natural process that pertains to practically every new invention or tool and can be reduced down to the over-simplistic sequence: create a tool, test the tool, and use the tool. Yet it is surprising that in bioinformatics this natural sequence was not standard practice until the 1990s when specific groups and initiatives like BOSC (Harris et al., 2016) started advocating its use. Fortunately, today we can be assured that this process has become common practice, although there are still grounds for improvement. Many researchers state that ideally, the "Materials and Methods" part of a scientific report should be an object that encapsulates the complete methodology, is directly available and executable, and should accompany (rather than supplement) any biologic investigation (Hettne et al., 2012). This highlights the need for another set of tools that automate the sharing, replication and validation of results, and the conclusions from published studies. In addition, there is a need to include external tools and data easily (Goodman et al., 2014) and to be able to generate more complex or integrated research pipelines. This new family of tools is referred to as workflow management systems (WMSs) or data workflow systems (Karim et al., 2017).

The tight integration of open source policies and WMSs is a highly anticipated milestone that promises to advance many aspects of bioinformatics. In Section 2.6 we present some of these aspects, with fighting the replication

Human Genome Informatics. https://doi.org/10.1016/B978-0-12-809414-3.00002-4

crisis and advancing health care through clinical genetics being the most prominent. One critical question is where do we stand now on our way to making this milestone a reality. Estimating the number of tools that are part of reproducible analysis pipelines is not a trivial task. Galaxy (Giardine et al., 2005) has a publicly available repository of available tools called "toolshed,"[1] which lists 3356 different tools. myExperiment (Goble et al., 2010) is a social website where researchers can share Research Objects like scientific workflows; it contains approximately 2700 workflows. For comparison bio.tools (Ison et al., 2016), a community-based effort to list and document bioinformatics resources, lists 6842 items.[2] Bioinformatics.ca (Brazas et al., 2010) curates a list of 1548 tools and 641 databases, OMICtools (Henry et al., 2014) contains approximately 5000 tools and databases, and the *Nucleic Acids Research* journal curates a list of 1700 databases (Galperin et al., 2017). Finally, it is estimated that there are approximately 100 different WMSs[3] with very diverse design principles.

Despite the plethora of available repositories, tools, databases, and WMSs, the task of combining multiple tools in a research pipeline is still considered a cumbersome procedure that requires above-average IT skills (Leipzig, 2016). This task becomes even more difficult when the aim is to combine multiple existing pipelines, use more than one WMS, or to submit the analysis to a highly customized, high performance computing (HPC) environment. Since the progress and innovation in bioinformatics lies to a great extent in the correct combination of existing solutions, we should expect significant progress on the automation of pipeline building in the future (Perkel, 2017). In the meantime, today's developers of bioinformatics tools and services can follow certain software development guidelines that will help future researchers tremendously in building complex pipelines with these tools. In addition, data providers and curators can follow clear directions in order to improve the availability, reusability, and semantic description of these resources. Finally, WMS engineers should follow certain guidelines that can augment the inclusiveness, expressiveness, and user-friendliness of these environments. Here we present a set of easy-to-follow guidelines for each of these groups.

2.2   EXISTING WORKFLOW ENVIRONMENT

Thorough reviews on existing workflow environments in bioinformatics can be found in many relevant studies (Leipzig, 2016; Karim et al., 2017; Spjuth et al., 2015). Nevertheless, it is worthwhile to take a brief look at some of the most well-known environments and frequently used techniques.

---

[1] Galaxy Tool Shed: https://toolshed.g2.bx.psu.edu/.

[2] See https://bio.tools/stats.

[3] Awesome Pipeline: https://github.com/pditommaso/awesome-pipeline.

The most prominent workflow environment and perhaps the only success story in the area—with more than a decade of continuous development and use—is Galaxy (Giardine et al., 2005). Galaxy has managed to build a lively community that includes biologists and IT experts. It also acts as an analysis frontend in many research institutes. Other features that have contributed significantly to Galaxy's success are: (1) capability to build independent repositories of workflows and tools and allow the inclusion of these from one installation to another, (2) having a very basic and simple wrapping mechanism for ambiguous tools, (3) native support for a large set of HPC environments (like TORQUE, cloud, grid) (Krieger et al., 2016), and (4) offering a simple, interactive web-based environment for graph-like workflow creation. Despite Galaxy's success, it still lacks many of the qualitative criteria that we will present later. Perhaps most important is the final criterion that discusses enabling users to collaborate in analyses; share, discuss, and rate results; and exchange ideas and coauthor scientific reports.

The second most well-known environment is perhaps TAVERNA (Wolstencroft et al., 2013). TAVERNA is built around the integration of many web services and has limited adoption in the area of genetics. The reason for this is that it is a more complex framework, it is not a web environment and forces users to adhere to specific standards (i.e., BioMoby, WSDL). We believe that the reasons why TAVERNA is lagging GALAXY, despite having a professional development team and extensive funding, should be studied more deeply in order to generate valuable lessons for future reference.

In this thesis we use a new workflow environment, MOLGENIS compute (Byelas et al., 2011, 2012, 2013). This environment offers tool integration and workflow description that is even simpler than GALAXY and it is built around MOLGENIS, another successful tool used in the field (Swertz et al., 2010). Moreover, it comes with "batteries included" tools and workflows for genotype imputation and RNA-sequencing analysis.

Other environments with promising features include Omics Pipe (Fisch et al., 2015) and EDGE (Li et al., 2017) for next generation sequencing (NGS) data, and Chipster (Kallio et al., 2011) for microarray data. All these environments claim to be decentralized and community-based although their wide adoption by the community still needs to be proven.

Besides integrated environments, it is worth noting some workflow solutions at the level of programming languages, for example, Python packages like Luigi and bcbio-nextgen (Guimera, 2012), and Java packages like bpipe (Sadedin et al., 2012). Other solutions are Snakemake,[4] Nextflow,[5] and BigDataScript

---

[4] Similar to "make" tool, with a targeted repository of workflows https://bitbucket.org/snakemake/snakemake/wiki/Home.

[5] See http://www.nextflow.io/.

(Cingolani et al., 2015), which are novel programming languages dedicated solely to scientific pipelines.[6] Describing a workflow in these packages gives some researchers the ability to execute their analyses easily in a plethora of environments, but unfortunately the packages are targeted at skilled IT users.

Finally, existing workflows and languages for workflow description and exchange are YAWL (van der Aalst and ter Hofstede, 2005), Common Workflow Language[7] and Workflow Description Language.[8] The support of one or more workflow languages by an environment is essential and comprises one of the most important qualitative criteria.

2.3   WHAT SOFTWARE SHOULD BE PART OF A SCIENTIFIC WORKFLOW?

Even in the early phases of development of a bioinformatics tool, one should take into consideration that it will become part of a yet-to-be-designed workflow at some point. Since the needs of that workflow are unknown at the moment, all the design principles used for the tool should focus on making it easy to configure, adapt, and extend. This is on top of other quality criteria that bioinformatics methods should possess (Wilson et al., 2014; Leprevost et al., 2014). After combining tens of different tools together for the scope of this thesis, we can present a checklist of qualitative criteria for modern bioinformatics tools to cover coding, quality checks, project organization, and documentation.

- *On coding*:
  - Is the code readable or else written in an extrovert way, assuming that there is a community ready to read, correct, and extend it?
  - Can people, unfamiliar with the project's codebase get a good grasp of the architecture and module organization of the project?
  - Is the code written according to the idiom used by the majority of programmers of the chosen language (i.e., camel case for Java or underscores for Python?)
  - Is the code style (i.e., formatting, indentation, comment style) consistent throughout the project?

- *On quality checks*:
  - Do you provide test cases and unit tests that cover the entirety of the code base?
  - When you correct bugs, do you make the previously failing input a new test case?

---

[6] For more detail, see https://www.biostars.org/p/91301/.

[7] See https://github.com/common-workflow-language/common-workflow-language.

[8] See https://github.com/broadinstitute/wdl.

- Do you use assertions?
  - Does the test data come from "a real-world problem"?
  - Does your tool generate logs that could easily trace errors on: input parameters, input data, user's actions, or implementation?

- *On project organization*
  - Is there a build tool that automates compilation, setup, and installation?
  - Does the build tool check for necessary-dependent libraries and software?
  - Have you tested the build tool in some basic commonly used environments?
  - Is the code hosted in an open repository? (i.e., github, bitbucket, gitlab)
  - Do you make incremental changes, with sufficient commit messages?
  - Do you "re-invent the wheel"? Or else, is any part of the project already implemented in a mature and easily embedded way that you don't use?

- *On documentation*
  - Do you describe the tool sufficiently well?
  - Is some part of the text more targeted to novice users?
  - Do you provide tutorials or step-by-step guides?
  - Do you document interfaces, basic logic, and project structure?
  - Do you document installation, setup, and configuration?
  - Do you document memory, CPU, disk space, or bandwidth requirements?
  - Do you provide execution time for average use?
  - Is the documentation self-generated? (i.e., Sphinx)
  - Do you provide means of providing feedback or contacting the main developer(s)?

- Having a user interface is always a nice choice, but does it always support command line execution? Command line tools are far more easily adapted to pipelines.
- Recommendations for command line tools in bioinformatics include (Seemann, 2013): always have help screens, use stdout, stdin, and stderr when applicable, check for as many errors as possible and raise exceptions when they occur, validate parameters, and do not hard code paths.
- Create a script (e.g., with BASH) or a program that takes care of the installation of the tool in a number of known operating systems.
- Adopt open and widely used standards and formats for input and output data.
- If the tool will most likely be at the end of an analysis pipeline and will create a large list of findings (e.g., thousands of variants) that require

human inspection, consider generating an online database view. Excellent choices for this are MOLGENIS (Swertz et al., 2010) and BioMart. These systems are easily installed and populated with data while they allow other researchers to explore the data with intuitive user interfaces.

- Finally choose light web interfaces rather than local GUIs. Web interfaces allow easy automatic interaction in contrast to local GUIs. Each tool can include a modular web interface that in turn can be a component of a larger website. Web frameworks like Django offer very simple solutions for this. An example of an integrated environment in Django is given by Cobb (2015). Tools that include web interfaces are Mutalyzer (Wildeman et al., 2008) and MutationInfo.[9]

Before checking these lists a novice bioinformatician might wonder what minimum IT knowledge is required in the area of bioinformatics. Apart from basic abstract notions in computer programming, new comers should get experience in BASH scripting, the Linux operating system and modern interpreted languages like Python (Perkel, 2015). They should also get accustomed to working with basic software development techniques (Merali, 2010) and tools that promote openness and reproducibility (Ince et al., 2012; Liu and Pounds, 2014) like GIT (Ram, 2013).

These guidelines are not only targeted to novice but also to experienced users. The bio-star project was formed to counteract the complexity, bad quality, poor documentation, and organization of bioinformatics software, which was discouraging potential new contributors. The purpose of this project was to create "low barrier entry" environments (Bonnal et al., 2012) that enforce stability and scale-up community creation. Moreover, even if some of these guidelines are violated, publishing the code is still a very good idea (Barnes, 2010). It is important to note here that at one point the community was calling for early open publication of code, regardless of the quality, while at the same time part of the community was being judgmental and rejecting this step (Corpas et al., 2012). So having guidelines for qualitative software should not mean that the community does not support users who do not follow them mainly due to inexperience.

Another question is whether there should be guidelines for good practices on scientific software usage (apart from creating new software). Nevertheless, since software is actually an experimental tool in a scientific experiment, the abuse of the tool might actually be a good idea! For this reason the only guideline that is crucial when using scientific software regard tracking results and data provenance. This guideline urges researchers to be extremely cautious and

---

[9] See https://github.com/kantale/MutationInfo and http://www.epga.gr/MutationInfo/.

responsible when monitoring, tracking, and managing scientific data and logs that are generated from scientific software. In particular, software logs are no difference to wet-lab notebook logs and should always be kept (and archived) for further validation and confirmation.[10] Here we argue that this responsibility is so essential to the scientific ethic that it should not be delegated lightheartedly to automatic workflow systems without careful human inspection.

2.4 PREPARING DATA FOR AUTOMATIC WORKFLOW ANALYSIS

Scientific analysis workflows are autonomous research objects in the same fashion that as independent tools and data. This means that workflows need to be decoupled from specific data and should be able to analyze data from sources beyond the reach of the bioinformatician author. Being able to locate open, self-described and directly accessible data is one of the most sensitive areas of life sciences research. There are two reasons for this. The first covers law, ethical, and privacy concerns regarding the individuals participating in a study. The second is the reluctance of researchers to release data that only they have the benefit of accessing, thus giving them an advantage in scientific exploitations. This line of thought has placed scientific data management as a secondary concern, often given low priority in the scope of large projects and consortia. It is interesting in this regard to take a look at the conservative views of the medical community on open data policies (Longo and Drazen, 2016). These views consider the latest trends for openness as a side effect of the technology from which they should be protected rather than as a revolutionary chance for progress. Instead of choosing sides in this controversy, we argue that technology itself can be used to resolve the issue. This is feasible by making data repositories capable of providing research data while protecting private sensitive information and ensuring proper citation.

Recently it is evident that the only way to derive clinically applicable results from the analysis of genetic data is through the collective analysis of diverse datasets that are open not only for analysis, but also for quality control and validation scrutiny (Manrai et al., 2016). For this reason we believe that significant progress has to be made in order to establish the right legal frameworks that will empower political and decision-making boards to create mechanisms that enforce and control the adaptation of open data policies. Even then, the most difficult change remains the paradigm shift that is needed: from the deep-rooted philosophy of researchers who treat data as personal or institutional

---

[10] See http://www.nature.com/gim/journal/v18/n2/full/gim2015163a.html.

property toward more open policies (Axton, 2011). Nevertheless, we are optimistic—some changes have already started to take place due to the open data movement. Here, we present a checklist for open data management guidelines within a research project (Tenopir et al., 2011).

- Make long-term data storage plans early on. Most research data will remain valuable even decades after publication.
- Release the consent forms, or other legal documents and policies under which the data were collected and released.
- Make sure that the data are discoverable. Direct links should not be more distant than two or three clicks away from relevant search in a search engine.
- Consider submitting the data to an open repository, for example, Gene Expression Omnibus or the European Genome-phenome Archive (EGA).
- Provide meta-data. Try to adopt open and widely used ontologies and formats, and make the datasets self-explanatory. Make sure all data are described uniformly according to the meta-data and make the meta-data specifications available.
- Provide the "full story" of the data. Experimental protocols, equipment, preprocessing steps, data quality checks, and so on.
- Provide links to software that has been used for preprocessing and links to tools that are directly applicable for downstream analysis.
- Make the data citable and suggest a proper citation for researchers use. Also show the discoveries already made using the data and suggest further experiments.
- Provide a manager/supervisor's email and contact information for potential feedback from the community.

2.5 QUALITY CRITERIA FOR MODERN WORKFLOW ENVIRONMENTS

Although the notion of a scientific workflow is a simple concept with a historic presence in the bioinformatics community there are many factors that affect their overall quality that are still being overlooked today. In this section we present some of these factors.

2.5.1 Being Able to Embed and to Be Embedded

Any workflow should have the ability to embed other workflows as components regardless of their in-between unfamiliarity. Modern workflow environments tend to suffer from the "lock-in" syndrome. Namely, they demand their users invest considerable time and resources to wrap an external component with the required meta-data, libraries, and configuration files into a workflow.

Workflows should be agnostic regarding the possible components supported and should provide efficient mechanisms to wrap and include them with minimum effort.

Similarly, a workflow environment should not assume that it will be the primarily tool with which the researcher make the complete analysis. This behavior is selfish and reveals a desire to dominate a market rather than to contribute to a community. That being said, workflow systems should offer researchers the ability to export their complete analysis in formats that are easily digested by other systems. Examples include simple BASH scripts with meta-data described in well-known and easily parsed formats like XML, JSON, or YAML. Another option is to directly export meta-types, scripts, and analysis code in serialized objects like PICKLE and CAMEL[11] that can be easily loaded as a whole from other tools.

2.5.2 Support Ontologies

Workflow systems tend to focus more on the analysis part of the tool and neglect the semantic part. The semantic enrichment of a workflow can be achieved by adhering and conforming to the correct ontologies. Analysis pipelines that focus on high throughput genomics, like NGS, or proteomics have indeed a limited need for semantic integration mainly because a large part of this research landscape is uncharted so far. Nevertheless, when a pipeline approaches findings closer to the clinical level, the semantic enrichment is necessary. At this level, the plethora of findings, the variety of research pipelines, and sometimes the discrepancies between conclusions can create a bewildering terrain. Therefore, a semantic integration through ontologies can provide common ground for direct comparison of findings and methods. Thus ontologies, for example, for biobanks (Pang et al., 2015), gene function (Gene Ontology Consortium, 2015), genetic variation (Byrne et al., 2012), and phenotypes (Robinson and Mundlos, 2010), can be extremely helpful. Of course, ontologies are not the panacea to all these problems since they have their own issues to be considered (Malone et al., 2016).

2.5.3 Support Virtualization

The computing requirements of a bioinformatics workflow are often unknown to the workflow author or user but include the processing power, memory consumption, and time required. In addition, the underlying computing environment has its own requirements (e.g., the operating system, installed libraries, and preinstalled packages). Since a workflow environment has its own requirements and dependencies, it is cumbersome for even skilled and well IT-trained

---

[11] See http://eev.ee/blog/2015/10/15/dont-use-pickle-use-camel/.

bioinformaticians to set it up and configure. Consequently, a considerable amount of valuable research time is spent in configuring and setting up a pipeline or a workflow environment. In addition, lack of documentation, IT inexperience, and time pressure lead to misconfigured environments that in turn leads to waste of resources and may even produce erroneous results.[12] A solution for this can be virtualization. Virtualization is the "bundling" of all required software, operating system, libraries, and packages into a unique object (usually a file) that is called an "image" or "container." This image can be executed in almost all known operating systems with the help of specific software, making this technique a very nice approach for the "be embeddable" feature discussed previously. Any workflow environment that can be virtualized is automatically easily embeddable in any other system by just including the created image.

Some nice examples include the I2B2 (Informatics for Integrating Biology and the Bedside) consortium,[13] which offers the complete framework in a VMare image (Uzuner et al., 2011). Another is the transMART software that is offered in a VWware or VirtualBox container (Athey et al., 2013). Docker is also an open-source project that offers virtualization, as well as an open repository where users can browse, download, and execute a vast collection of community generated containers. Docker borrows concepts from the GIT version control system. Namely, users can download a container, apply changes, and "commit" the changes to a new container. The potential value of Docker in science has already been discussed (Boettiger, 2015; Di Tommaso et al., 2015); for example, BioSha-Dock is a Docker repository of tools that can be seamlessly executed in Galaxy (Moreews et al., 2015). Other bioinformatics initiatives that are based on Docker are the Critical Assessment of Metagenomic Interpretation (CAMI[14]), nucleotid.es,[15] and bioboxes.org. All these initiatives offer a test-bed for comparison and evaluation of existing techniques mainly for NGS tasks.

2.5.4   Offer Easy Access to Commonly Used Datasets

Over the last years, an increasing number of large research projects have generated large volumes of qualitative data that are part of almost all bioinformatics analysis pipelines. So far although locating and collecting the data is straightforward, their volume and their dispersed descriptions make their inclusion a difficult task. Modern workflow environments should offer automatic methods to collect and use them and examples of relevant datasets are ExAC

---

[12] Paper retracted due to software incompatibility: http://www.nature.com/news/error-found-in-study-of-first-ancient-african-genome-1.19258.

[13] See https://www.i2b2.org/

[14] See http://cami-challenge.org/.

[15] Very interesting presentation on virtualization: http://nucleotid.es/blog/why-use-containers/.

(Exome Aggregation Consortium, 2016) with 60,000 exomes, 1000 Genomes Project (Abecasis et al., 2012), GoNL (The Genome of the Netherlands Consortium, 2014), ENCODE project (ENCODE Project Consortium, 2004), METABRIC dataset (Curtis et al., 2012), and data released from large consortia like GIANT.[16] Another impediment is that different consortia release data with different consent forms, for example, in the European Genome-phenome Archive (EGA) (Lappalainen et al., 2015), each provider has a different access policy. Filling and submitting these forms is another task that can be automated (with information provided by users). Another essential feature should be the automatic generation of artificial data whenever this is requested for testing and simulation purposes (Mu et al., 2015).

2.5.5 Support and Standardize Data Visualization

A feature that is partially supported by existing workflow environments is the inherent support for data visualization. Over the course of genetics research, certain visualization methods have become standard and are easily interpreted and widely understood by the community. Workflow environments should not only support the creation and inclusion of these visualization techniques, but also suggest them whenever a pipeline is being written.

For example in genome-wide analysis studies (GWAS), Manhattan plots for detecting significance, Q-Q plots for detecting inflation, and Principal Component Analysis plots for detecting population stratification or the inclusion of visualizations from tools that have become standard in certain analysis pipelines, like LocusView and Haploview plots for GWAS. The environment should also support visualization for research data like haplotypes (Jäger et al., 2014; Barrett et al., 2005), reads from sequencing (Hilker et al., 2014), NGS data (Shen, 2014; Thorvaldsdóttir et al., 2013), and biological networks (Mitra et al., 2013).

As an implementation note, in the last few years we have seen a huge improvement in the ability of Internet browsers to visualize large and complex data. This was possible mainly due to the wide adoption and standardization of JavaScript. Today JavaScript can be used as a stand-alone genome browser (e.g., JBrowse (Skinner et al., 2009) although UCSC also allows this functionality; Raney et al., 2014) and genome viewer (e.g., pileup.js; Vanderkam et al., 2016). JavaScript has allowed the inclusion of interactive plots like those from RShiny[17] and the creation of esthetically and informative novel forms of visualization with libraries like D3.[18] Today there are hundreds of very useful, context-specific minor online tools in the area of bioinformatics. The superior

---

[16] See https://www.broadinstitute.org/collaboration/giant/index.php/Main_Page.

[17] See http://shiny.rstudio.com/.

[18] See https://d3js.org/.

advantage of JavaScript is that it is already installed (and used daily) in almost all Internet browsers by default. We therefore strongly advocate workflow environments that support JavaScript data visualization.

2.5.6   Enable "Batteries Included" Workflow Environments

A workflow environment, no matter how advanced or rich-in features, is of no use to the bioinformatics community if it does not offer a basic collection of widely adopted and ready-to-use tools and workflows. The inclusion of these, even in early releases, will offer a valuable testing framework and will quickly attract users to become a contributing community. These tools and workflows can be split into the following categories:

Tools:

- Tools that are essential and can be used for a variety of purposes such as plink (Purcell et al., 2007) and GATK (DePristo et al., 2011).
- Tools for basic text file manipulation (e.g., column extraction), format validation, conversion, and quality control.
- Tools for basic analysis (e.g., for variant detection, sequence alignment, variant calling, Principal Component Analysis, eQTL analysis, phasing, imputation, association analysis, and meta-analysis).
- Wrappers for online tools. Interestingly, a large part of modern analysis demands the use of tools that require interaction through a web browser. These tools either do not offer a local offline version or operate on large datasets that it is inconvenient to store locally. A modern workflow should include these tools via special methods that emulate and automate browser interaction. Examples of these tools are: for annotation (e.g., Avia (Vuong et al., 2015), ANNOVAR (Wang et al., 2010), GEMINI (Paila et al., 2013)), function analysis of variation (e.g., Polyphen-2 (Adzhubei et al., 2010), SIFT, CONSURF (Celniker et al., 2013)) and gene-set analysis (e.g., WebGestalt (Wang et al., 2013)).

Workflows:

- Existing workflows, for example, for high throughput sequencing (Mimori et al., 2013), RNA-sequencing data (Deelen et al., 2015; Torres-Garcia et al., 2014), identification of de novo mutations (Samocha et al., 2014), and genotype imputation (Kanterakis et al., 2015).
- Protocols for downstream analysis, for example, for functional variation (Lappalainen et al., 2013), finding of novel regulatory elements (Samuels et al., 2013), investigation of somatic mutations for cancer research (Choudhary and Noronha, 2014), and delivering clinically significant findings (Zemojtel et al., 2014).
- Meta-workflows for comparison and benchmarking (Bahcall, 2015).

The environment should also provide basic indications of their performance and system requirements.

2.5.7 Facilitate Data Integration, Both for Import and Export

Today, despite years of developing of formats, ontologies, and standards in bioinformatics, there is still uncertainty over the correct or "right" standard when it comes to tool or service interoperation (Tenenbaum et al., 2014). To demonstrate this, we present a list of seven commonly used bioinformatics databases for various -omics datasets in Table 2.1. Each database is qualitative, built and maintained by professionals, offers modern user interfaces, and has been a standard source for information for numerous research workflows. Yet each database offers a different standard as a primary method for automated data exchange. Some are using raw data in a specific format, some have enabled API access, and some delegate the access to secondary tools (i.e., BioMart). As long as the genetics field generates data with varying rates and complexity, we should not expect this situation to change in the near future. For this reason, a workflow environment that needs to include these databases must offer methods for accessing these sources tailored to the specifics of each database. This underlines the need for "write once, use many" tools. A user who generates a tool that automates the access to a database and includes it in a workflow environment should also make it public for other users too, while the environment should allow and promote this policy after some basic quality checks of course. The conclusion is that an environment that promotes the inclusion of many diverse tools, workflows, and databases will also promote the creation of tools that automate data access.

Table 2.1 A List of Popular Databases for -Omics Data Along With Their Major Access Methods and Formats

-Omics	Title	Access	Reference
Mutations	Locus Specific Mutation Database	Atom, VarioML, JSON	Fokkema et al. (2011)
lncRNA	lncRNASNP: a database of SNPs in lncRNAs and their potential functions in human and mouse	CSV files	Gong et al. (2015)
Phenotype	The NCBI dbGaP database of genotypes and phenotypes	FTP	Mailman et al. (2007)
Diseases	Online Mendelian Inheritance in Man (OMIM)	XML, JSON, JSONP	Hamosh et al. (2005)
Variation	ClinVar: public archive of relationships among sequence variation and human phenotype	FTP, VCF	Landrum et al. (2014)
Annotation	GENCODE: the reference human genome annotation for the ENCODE Project	GTF, GFF3	Harrow et al. (2012)
Functional annotation	A promoter-level mammalian expression atlas (FANTOM)	BioMart	Forrest et al. (2014)

2.5.8   Offer Gateways for High Performance Computing Environments

The computational resources for a large part of modern workflows make their execution prohibitive in a local commodity computer. Fortunately, we have experienced an explosion of HPC options in the area of -omics data (Berger et al., 2013). These options can be split into two categories: the first refers to software abstractions that allow parallel execution in an HPC environment, and the second refers to existing HPC options.

Regarding the first category the options are:

- Map-reduce systems (Zou et al., 2014), for example, GATK (McKenna et al., 2010).
- Specific map-reduce implementations, for example, Hadoop for NGS (Niemenmaa et al., 2012) data.
- SPARK (Wiewiórka et al., 2014).
- Multithread or multiprocessing techniques, tailored for high performance workstations, for example, WiggleTools (Zerbino et al., 2014).

The second category includes the available HPC options:

- Computer clusters and computational grids, or else job-oriented approaches (Byelas et al., 2013).
- Cloud, for example, whole genome analysis (Zhao et al., 2013).
- Single high performance workstations.
- Workstations with optimized hardware for specific tasks, for example, GPUs.[19]

A workflow environment should be compatible with tools that belong to the first category and offer execution options that belong to the second. This integration is perhaps the most difficult challenge for a workflow environment because each HPC environment has a completely different architecture and access policies. Access policies include free-for-academic-use through nation-wide cluster infrastructure like Surf for the Netherlands and EDET for Greece, and paid for corporate providers like Amazon EC2 or other bioinformatics tailored cloud solutions like sense[20] and arvados.[21] Existing environments that offer execution for a plethora of HPC options are Yabi (Hunter et al., 2012) for analysis and BioPlanet (Highnam et al., 2015) for benchmarking and comparison.

---

[19] See http://richardcaseyhpc.com/parallel2/.

[20] See https://sense.io/.

[21] See https://arvados.org/.

Another major consideration is the data policy. Submitting data for analysis in an external environment often violate strict (and sometimes very old) institutional data management policies. For this reason some researchers suggest that for sensitive data we should "bring computation to the data" (Gaye et al., 2014) and not the opposite (Editorial, 2016).

2.5.9 Engage Users in Collaborative Experimentation and Scientific Authoring

Existing workflow environments focus on automating the analysis part of the scientific process. We anticipate that the new generation of environments will also include the second and maybe more vital part of this process: active and fruitful engagement in a creative discourse that will enable the collaborative coauthoring of complete scientific reports, reaching perhaps the level of published papers. For these purposes, workflow environments should enable users build to communities, invite other users, and easily share workflows and results. The system should include techniques to allow users to comment, discuss, and rate not only workflows, but results, ideas, and techniques. It should also let users include rich text formatting descriptions of all parts of the analysis and automatically generate publishable reports. Finally it should also give proper credits to each user in the respective parts of the analysis to which they have contributed (Bechhofer et al., 2013). By offering these features workflow environments will become the center of scientific exploration rather than just a useful analysis tool.

This emerging trend—qualitative authoring pooled effort of an engaged community—is called crowdsourcing. It has been used successfully to accomplish tasks that otherwise would have required enormous amounts of work by a small group of researchers for example the monitoring of infectious diseases (McIver and Brownstein, 2014), investigating novel drug-drug interactions (Pfundner et al., 2015), and creating valuable resources for personal genomics (e.g., OpenSNP (Greshake et al., 2014) and SNPedia (Cariaso and Lennon, 2012)). Of course crowdsourcing techniques can introduce a certain bias in the analysis (Hasty et al., 2014) and they should be used done with extreme caution especially in the area of personal genomics (Corpas et al., 2015). This means the workflow environment should be able to apply moderation and detect cases where bias arises from a polarized user's contribution.

The most prominent tools for online interactive analysis are Jupyter[22] and R Markdown[23] The first mainly targets the Python community and the second the R community. In these environments, the complete analysis can be easily

---

[22] See http://jupyter.org/.

[23] See http://rmarkdown.rstudio.com/.

shared and reproduced. In addition, collaborative online scientific authoring is possible with services like Overleaf[24] and the direct hosting of generated reports can take place in a GIT enabled repository for impartial analysis or in arXiv[25] for close-to-publication reports. Hence an ideal workflow environment could integrate these techniques and enrich them with user feedback via comments and ratings. Although this scenario sounds futuristic, it has long been envisioned by the open science movement.

2.6 BENEFITS FROM INTEGRATED WORKFLOW ANALYSIS IN BIOINFORMATICS

In this section we present the foreseeable short- and long-term benefits for the bioinformatics community by following the criteria presented for workflow environments.

2.6.1 Enable Meta-Studies, Combine Datasets, and Increase Statistical Power

One obvious benefit is to be able to perform integrated analysis and uncover previously unknown correlations or even causal mechanisms (MacArthur et al., 2014). For example, existing databases already include a huge amount if research data from GWAS that enable the meta-analysis and reassessment of their findings (Leslie et al., 2014; Johnson and O'Donnell, 2009) and index them according to diseases and traits (Bulik-Sullivan et al., 2015). The improved statistical power can locate correlations for very diverse and environmentally affected phenotypes such as body mass index (Speliotes et al., 2010) or even educational attainment (Rietveld et al., 2013) although the interpretation of socially sensitive findings should always be very cautious.

2.6.2 Include Methods and Data From Other Research Disciplines

Automated analysis methods can generate the necessary abstraction layer, which will attract specialists from distant research areas with a minimum knowledge of genetics. For example, bioinformatics research can benefit greatly from "knowledge import" from areas like machine learning, data mining, medical informatics, medical sensors, artificial intelligence, and supercomputing. The Encode project had a special "machine learning" track that generated very interesting findings (ENCODE Project Consortium, 2004). Existing studies can also be enriched with data from social networks (Coiera, 2013).

---

[24] See https://www.overleaf.com/.

[25] See http://arxiv.org/.

2.6.3 Fight the Reproducibility Crisis

The sparseness of data and analysis workflows from high-profile published studies is a contributing factor to what has been characterized as the reproducibility crisis in bioinformatics. Namely, researchers are failing to reproduce results from many prominent papers and some times-groundbreaking findings.[26] The consequences of this can be really harsh, since many retractions occur that stigmatize the community in general (Azoulay et al., 2012), which in turn affect the overall financial viability of the field. One report stated, "…mistakes such as failure to reproduce the data, were followed by a 5073% drop in NIH funding for related studies."[27]

Moreover, irreproducible science is of practically no use for the community nor for society and constitutes a waste of valuable funding (Chalmers and Glasziou, 2009). Many researchers are the blaming the publication process (which favors positive findings and original work rather that negative findings or confirmation of results) and are demanding either the adoption of open peer-review policies (Groves, 2010; Bohannon, 2013) or even the redesign of the complete publishing industry (Editorial, 2013). Yet, a more simple and applicable remedy for this crisis is the wide adoption (and maybe enforcement) of open workflow environments that contain the complete analyses of published papers.

2.6.4 Spread of Open Source Policies in Genetics and Privacy Protection

Today a significant part of NGS data analysis for research purposes is taking place with closed-source commercial or even academic software. The presence of this software in research workflows has been characterized as an impediment to science (Vihinen, 2015). There is also an increase in commercial services that offer genetic data storage (e.g., Google[28]) and analysis.[29] Other worrisome trends are gene patents (Marchant, 2007) and possible malevolent exploitation of easily deanonymized traits and genomes that reside in public databases (Angrist, 2014). These trends raise ethical concerns and they also contribute to the reproducibility crises. Workflows that enable user collaboration can also monitor and guide the development of novel software from an initial draft to a mature and tested phase. This development can only happen of course in the open source domain. In the same way that GIT promoted open source policies

---

[26] For a review of reports, see http://simplystatistics.org/2013/12/16/a-summary-of-the-evidence-that-most-published-research-is-false/.

[27] See http://blogs.nature.com/news/2012/11/retractions-stigmatize-scientific-fields-study-finds.html.

[28] See https://cloud.google.com/genomics/.

[29] For example, see https://insilicodb.com.

by "socializing" software development, we believe that collaborative workflow environments can promote similar open source policies in genetics.

2.6.5 Help Clinical Genetics Research

The clinical genetics field is tightly related to the availability of highly documented, well-tested, and directly executable workflows. Even today there are genetic analysis workflows that deliver clinically applicable results in a few days or even hours. For example, complete whole genome sequencing pipeline for neonatal diagnosis (Saunders et al., 2012; Chiang et al., 2015) or pathogen detection in intensive care units (Naccache et al., 2014), where speed is a of paramount importance. There are also clinical genetics workflows for the early diagnosis of Mendelian disorders (Yang et al., 2013), while the automation of genetic analysis benefits the early diagnosis of complex non-Mendelian rare (Boycott et al., 2013) and common diseases like Alzheimer (Doecke et al., 2012) and can simplify or replace complex procedures like amniocentesis for the prenatal diagnosis of Trisomy 21 (Chiu et al., 2011). Moreover, cancer diagnosis and treatment has been revolutionized with the introduction of NGS data (Meyerson et al., 2010) and guidelines for clinical applications are already emerging (Ding et al., 2014; Shyr and Liu, 2013).

Workflows in clinical genetics do not have to be "disease specific." Although still in its infancy, the idea of the universal inclusion of genetics in modern healthcare systems (called personalized or precision medicine) has been envisioned (Hayden, 2012). Probably the most prominent public figure in this area,[30] Eric Topol, has emphasized the importance of DNA and RNA level data on individualized medicine throughout the lifespan of an individual (Topol, 2014). This will allow the delivery of personalized drugs (Esplin et al., 2014) and therapies (Schork, 2015) and will remedy current trends that threaten the survival of modern healthcare systems.[31] The introduction of genetic screening as a routine test in healthcare will guide medical practitioners (MPs) to take more informed decisions. Also, MPs will have to adjust phenotyping in order to fully exploit the potentials of genetic screening (Hennekam and Biesecker, 2012). Therefore, we expect that clinical genetics have the potential to fundamentally change the discipline of medicine.

---

[30] Although there might be more prominent figures! http://www.nytimes.com/2015/01/25/us/obama-to-request-research-funding-for-treatments-tailored-to-patients-dna.html.

[31] "An astonishing 86% of all full-time employees in the United States are now either overweight or suffer from a chronic (but often preventable) disease," "the average cost of a family health insurance premium will surpass average household income by 2033": http://www.forbes.com/sites/forbesleadershipforum/2012/06/01/its-time-to-bet-on-genomics/.

These visionary and futuristic ideas of course entail many challenges and considerations (Khoury et al., 2012). So far, the most important are (Taylor et al., 2015) the conflicting models (Editorial, 2014), the lack of standards, and the uncertainty regarding the safety, effectiveness (Dewey et al., 2014) and cost (Douglas et al., 2015) of the clinical genetics guidelines generated. This is despite the initial promising results from GWAS (Manolio, 2013) and much effort to create guidelines and mechanisms for reporting clinically significantly findings (CFS) to cohort or healthcare participants (Brownstein et al., 2014; Kaye et al., 2014).

To tackle this, we suggest that modern workflow environments enabling multilevel data integration are the key to a solution. Adopting open data policies will allow large population-level studies and metastudies that will resolve existing discrepancies. A necessary prerequisite for this is also the inclusion of data from electronic Health Records (Scheuner et al., 2009) and BioBanks (van Ommen et al., 2014) (one successful initiative is the Dutch Lifelines cohort; Scholtens et al., 2014). As long as findings from genetic research remain institutionally isolated and disconnected from medical records; the realization of clinical genetics and personalized medicine will remain only as vision.

2.7 DISCUSSION

Today we have an abundance of high-quality data that describe nearly all stages of genetic regulation and transcription for various cell types, diseases, and populations. DNA sequences, RNA sequences and expression, protein sequence and structure, and small molecule metabolite structure are all routinely measured in unprecedented rates (Eisenstein, 2015). Moreover, the prospects in the area of data collection seem very bright as the throughput rates are increasing, the quality metrics are improving and the prices of obtaining such data are dropping. This leads us to the conclusion that the intermediate step that is missing between enormous data collections and mechanistic or causal models is the data analysis. The recommended way to proceed should be via data-driven approaches. This in turn leads to a new question: How can we bring together diverse analysis methods in order to qualitatively process heterogeneous big data in a constantly changing technological landscape? This challenge requires the collective effort of the bioinformatics community and the effort can only flourish if development is focused on building reusable components rather than isolated solutions.

References

Abecasis, G.R., Auton, A., Brooks, L.D., DePristo, M.A., Durbin, R.M., Handsaker, R.E., Kang, H.M., Marth, G.T., McVean, G.A., 2012. An integrated map of genetic variation from 1,092 human genomes. Nature 1476-4687. 491 (7422), 56–65. https://doi.org/10.1038/nature11632.

Adzhubei, I.A., Schmidt, S., Peshkin, L., Ramensky, V.E., Gerasimova, A., Bork, P., Kondrashov, A.S., Sunyaev, S.R., 2010. A method and server for predicting damaging missense mutations. Nat. Methods 1548-7105. 7 (4), 248–249. https://doi.org/10.1038/nmeth0410-248.

Angrist, M., 2014. Open window: when easily identifiable genomes and traits are in the public domain. PLoS ONE 1932-6203. 9 (3), e92060. https://doi.org/10.1371/journal.pone.0092060. http://journals.plos.org/plosone/article?id=10.1371/journal.pone.0092060.

Athey, B.D., Braxenthaler, M., Haas, M., Guo, Y., 2013. tranSMART: an open source and community-driven informatics and data sharing platform for clinical and translational research. AMIA Jt Summits Transl. Sci. Proc. 2153-4063. 2013, 6–8.

Axton, M., 2011. No second thoughts about data access. Nat. Genet. 43 (5), 389.

Azoulay, P., Furman, J.L., Krieger, J.L., Murray, F.E., 2012. Retractions. Working Paper 18499, National Bureau of Economic Research. http://www.nber.org/papers/w18499.

Bahcall, O.G., 2015. Genomics: benchmarking genome analysis pipelines. Nat. Rev. Genet. 1471-0056. 16 (4), 194. https://doi.org/10.1038/nrg3930.

Barnes, N., 2010. Publish your computer code: it is good enough. Nature 467 (7317), 753.

Barrett, J.C., Fry, B., Maller, J., Daly, M.J., 2005. Haploview: analysis and visualization of LD and haplotype maps. Bioinformatics (Oxford, England) 1367-4803. 21 (2), 263–265. https://doi.org/10.1093/bioinformatics/bth457.

Bechhofer, S., Buchan, I., De Roure, D., Missier, P., Ainsworth, J., Bhagat, J., Couch, P., Cruickshank, D., Delderfield, M., Dunlop, I., Gamble, M., Michaelides, D., Owen, S., Newman, D., Sufi, S., Goble, C., 2013. Why linked data is not enough for scientists. Futur. Gener. Comput. Syst. 0167739X. 29 (2), 599–611. https://doi.org/10.1016/j.future.2011.08.004. http://www.sciencedirect.com/science/article/pii/S0167739X11001439.

Berger, B., Peng, J., Singh, M., 2013. Computational solutions for omics data. Nat. Rev. Genet. 1471-0064. 14 (5), 333–346. https://doi.org/10.1038/nrg3433. http://www.pubmedcentral.nih.gov/articlerender.fcgi?artid=3966295tool=pmcentrezrendertype=abstract.

Boettiger, C., 2015. An introduction to Docker for reproducible research. ACM SIGOPS Oper. Syst. Rev. 01635980. 49 (1), 71–79. https://doi.org/10.1145/2723872.2723882.

Bohannon, J., 2013. Who's afraid of peer review? Science (New York, NY) 1095-9203. 342 (6154), 60–65. https://doi.org/10.1126/science.342.6154.60. http://science.sciencemag.org/content/342/6154/60.abstract.

Bonnal, R.J.P., Aerts, J., Githinji, G., Goto, N., et al., 2012. Biogem: an effective tool-based approach for scaling up open source software development in bioinformatics. Bioinformatics (Oxford, England) 1367-4811. 28 (7), 1035–1037. https://doi.org/10.1093/bioinformatics/bts080.

Boycott, K.M., Vanstone, M.R., Bulman, D.E., MacKenzie, A.E., 2013. Rare-disease genetics in the era of next-generation sequencing: discovery to translation. Nat. Rev. Genet. 1471-0064. 14 (10), 681–691. https://doi.org/10.1038/nrg3555.

Brazas, M.D., Yamada, J.T., Ouellette, B., 2010. Providing web servers and training in bioinformatics: 2010 update on the bioinformatics links directory. Nucleic Acids Res. 38 (suppl 2), W3–W6.

Brownstein, C.A., Beggs, A.H., Homer, N., et al., 2014. An international effort towards developing standards for best practices in analysis, interpretation and reporting of clinical genome sequencing results in the CLARITY challenge. Genome Biol. 1474-760X. 15 (3), R53. https://doi.org/10.1186/gb-2014-15-3-r53. http://www.pubmedcentral.nih.gov/articlerender.fcgi?artid=4073084tool=pmcentrezrendertype=abstract.

Bulik-Sullivan, B., Finucane, H.K., et al., 2015. An atlas of genetic correlations across human diseases and traits. Nat. Genet. 1546-1718. 47 (11), 1236–1241. https://doi.org/10.1038/ng.3406. http://www.ncbi.nlm.nih.gov/pubmed/26414676.

Byelas, H., Kanterakis, A., Swertz, M., 2011. Towards a MOLGENIS based computational framework. 2011 19th International Euromicro Conference on Parallel, Distributed and Network-Based Processing, IEEE, pp. 331–338.

Byelas, H.V., Dijkstra, M., Swertz, M.A., 2012. Introducing data provenance and error handling for NGS workflows within the Molgenis computational framework. Proceedings of the International Conference on Bioinformatics Models, Methods and Algorithms, SciTePress—Science and Technology Publications, pp. 42–50.

Byelas, H., Dijkstra, M., Neerincx, P.B., Van Dijk, F., Kanterakis, A., Deelen, P., Swertz, M.A., 2013. Scaling bio-analyses from computational clusters to grids. In: IWSG, vol. 993.

Byrne, M., Fokkema, I.F., Lancaster, O., Adamusiak, T., et al., 2012. VarioML framework for comprehensive variation data representation and exchange. BMC Bioinf. 1471-2105. 13 (1), 254. https://doi.org/10.1186/1471-2105-13-254.

Cariaso, M., Lennon, G., 2012. SNPedia: a wiki supporting personal genome annotation, interpretation and analysis. Nucleic Acids Res. 1362-4962. 40, D1308–D1312. https://doi.org/10.1093/nar/gkr798. http://nar.oxfordjournals.org/content/40/D1/D1308.short.

Celniker, G., Nimrod, G., Ashkenazy, H., Glaser, F., Martz, E., Mayrose, I., Pupko, T., Ben-Tal, N., 2013. ConSurf: using evolutionary data to raise testable hypotheses about protein function. Isr. J. Chem. 00212148. 53 (3–4), 199–206. https://doi.org/10.1002/ijch.201200096.

Chalmers, I., Glasziou, P., 2009. Avoidable waste in the production and reporting of research evidence. Lancet (Oxford, England) 1474-547X. 374 (9683), 86–89. https://doi.org/10.1016/S0140-6736(09)60329-9. http://www.thelancet.com/article/S0140673609603299/fulltext.

Chiang, C., Layer, R.M., Faust, G.G., Lindberg, M.R., Rose, D.B., Garrison, E.P., Marth, G.T., Quinlan, A.R., Hall, I.M., 2015. SpeedSeq: ultra-fast personal genome analysis and interpretation. Nat. Methods 1548-7091. 12 (10), 966–968. https://doi.org/10.1038/nmeth.3505. http://www.ncbi.nlm.nih.gov/pubmed/26258291.

Chiu, R.W.K., Akolekar, R., Zheng, Y.W.L., et al., 2011. Non-invasive prenatal assessment of trisomy 21 by multiplexed maternal plasma DNA sequencing: large scale validity study. BMJ (Clin. Res. ed.) 1756-1833. 342, c7401. https://doi.org/10.1136/bmj.c7401. http://www.bmj.com/content/342/bmj.c7401.long.

Choudhary, S.K., Noronha, S.B., 2014. Galdrive: pipeline for comparative identification of driver mutations using the galaxy framework. bioRxiv 010538.

Cingolani, P., Sladek, R., Blanchette, M., 2015. BigDataScript: a scripting language for data pipelines. Bioinformatics (Oxford, England) 1367-4811. 31 (1), 10–16. https://doi.org/10.1093/bioinformatics/btu595.

Cobb, M., 2015. NGSdb: An NGS Data Management and Analysis Platform for Comparative Genomics (Ph.D. thesis). University of Washington.

Coiera, E., 2013. Social networks, social media, and social diseases. BMJ (Clin. Res. ed.) 1756-1833. 346, f3007. https://doi.org/10.1136/bmj.f3007. http://www.bmj.com/content/346/bmj.f3007.abstract.

Corpas, M., Fatumo, S., Schneider, R., 2012. How not to be a bioinformatician. Source Code Biol. Med. 1751-0473. 7 (1), 3. https://doi.org/10.1186/1751-0473-7-3.

Corpas, M., Valdivia-Granda, W., Torres, N., et al., 2015. Crowdsourced direct-to-consumer genomic analysis of a family quartet. BMC Genomics 1471-2164. 16 (1), 910. https://doi.org/10.1186/s12864-015-1973-7. http://bmcgenomics.biomedcentral.com/articles/10.1186/s12864-015-1973-7.

Curtis, C., Shah, S.P., Chin, S.-F., et al., 2012. The genomic and transcriptomic architecture of 2,000 breast tumours reveals novel subgroups. Nature 1476-4687. 486 (7403), 346–352. https://doi.org/10.1038/nature10983.

Deelen, P., Zhernakova, D.V., de Haan, M., van der Sijde, M., Bonder, M.J., Karjalainen, J., van der Velde, K.J., Abbott, K.M., Fu, J., Wijmenga, C., Sinke, R.J., Swertz, M.A., Franke, L., 2015. Calling genotypes from public RNA-sequencing data enables identification of genetic variants that affect gene-expression levels. Genome Med. 1756-994X. 7 (1), 30. https://doi.org/10.1186/s13073-015-0152-4.

DePristo, M.A., Banks, E., Poplin, R., et al., 2011. A framework for variation discovery and genotyping using next-generation DNA sequencing data. Nat. Genet. 1546-1718. 43 (5), 491–498. https://doi.org/10.1038/ng.806.

Dewey, F.E., Grove, M.E., Pan, C., et al., 2014. Clinical interpretation and implications of whole-genome sequencing. JAMA 1538-3598. 311 (10), 1035–1045. https://doi.org/10.1001/jama.2014.1717. http://www.pubmedcentral.nih.gov/articlerender.fcgi?artid=4119063tool=pmcentrezrendertype=abstract.

Di Tommaso, P., Palumbo, E., Chatzou, M., Prieto, P., Heuer, M.L., Notredame, C., 2015. The impact of Docker containers on the performance of genomic pipelines. PeerJ 2167-8359. 3, e1273. https://doi.org/10.7717/peerj.1273.

Ding, L., Wendl, M.C., McMichael, J.F., Raphael, B.J., 2014. Expanding the computational toolbox for mining cancer genomes. Nat. Rev. Genet. 1471-0056. 15 (8), 556–570. https://doi.org/10.1038/nrg3767.

Doecke, J.D., Laws, S.M., Faux, N.G., et al., 2012. Blood-based protein biomarkers for diagnosis of Alzheimer disease. Arch. Neurol. 1538-3687. 69 (10), 1318–1325. https://doi.org/10.1001/archneurol.2012.1282. http://archneur.jamanetwork.com/article.aspx?articleid=1217314.

Douglas, M.P., Ladabaum, U., Pletcher, M.J., Marshall, D.A., Phillips, K.A., 2015. Economic evidence on identifying clinically actionable findings with whole-genome sequencing: a scoping review. Genet. Med. 1530-0366. 18 (2), 111–116. https://doi.org/10.1038/gim.2015.69.

Editorial, 2013. The future of publishing: a new page. Nature 1476-4687. 495 (7442), 425. https://doi.org/10.1038/495425a. http://www.nature.com/news/the-future-of-publishing-a-new-page-1.12665.

Editorial, 2014. Standards for clinical use of genetic variants. Nat. Genet. 1546-1718. 46 (2), 93. https://doi.org/10.1038/ng.2893.

Editorial, 2016. Peer review in the cloud. Nat. Genet. 1061-4036. 48 (3), 223. https://doi.org/10.1038/ng.3524. http://www.nature.com/articles/ng.3524

Eisenstein, M., 2015. Big data: the power of petabytes. Nature 0028-0836. 527 (7576), S2–S4. https://doi.org/10.1038/527S2a.

ENCODE Project Consortium, 2004. The ENCODE (ENCyclopedia Of DNA Elements) project. Science (New York, NY) 1095-9203. 306 (5696), 636–640. https://doi.org/10.1126/science.1105136.

Esplin, E.D., Oei, L., Snyder, M.P., 2014. Personalized sequencing and the future of medicine: discovery, diagnosis and defeat of disease. Pharmacogenomics 1744-8042. 15 (14), 1771–1790. https://doi.org/10.2217/pgs.14.117. http://www.pubmedcentral.nih.gov/articlerender.fcgi?artid=4336568tool=pmcentrezrendertype=abstract.

Exome Aggregation Consortium, 2016. Analysis of protein-coding genetic variation in 60,706 humans. Nature 0028-0836. 536 (7616), 285–291. https://doi.org/10.1038/nature19057.

Fisch, K.M., Meißner, T., Gioia, L., Ducom, J.-C., Carland, T.M., Loguercio, S., Su, A.I., 2015. Omics pipe: a community-based framework for reproducible multi-omics data analysis. Bioinformatics (Oxford, England) 1367-4811. 31 (11), 1724–1728. https://doi.org/10.1093/bioinformatics/btv061.

Fokkema, I.F.A.C., Taschner, P.E.M., Schaafsma, G.C.P., Celli, J., Laros, J.F.J., den Dunnen, J.T., 2011. LOVD v.2.0: the next generation in gene variant databases. Hum. Mutat. 1098-1004. 32 (5), 557–563. https://doi.org/10.1002/humu.21438.

Forrest, A.R.R., Kawaji, H., Rehli, M., et al., 2014. A promoter-level mammalian expression atlas. Nature 1476-4687. 507 (7493), 462–470. https://doi.org/10.1038/nature13182.

Galperin, M.Y., Fernández-Suárez, X.M., Rigden, D.J., 2017. The 24th annual nucleic acids research database issue: a look back and upcoming changes. Nucleic Acids Res. 45 (D1), D1. https://doi.org/10.1093/nar/gkw1188.

Gaye, A., Marcon, Y., Isaeva, J., LaFlamme, P., et al., 2014. DataSHIELD: taking the analysis to the data, not the data to the analysis. Int. J. Epidemiol. 1464-3685. 43 (6), 1929–1944. https://doi.org/10.1093/ije/dyu188. http://ije.oxfordjournals.org/content/early/2014/09/26/ije.dyu188.short?rss=1.

Gene Ontology Consortium, 2015. Gene ontology consortium: going forward. Nucleic Acids Res. 43 (D1), D1049–D1056. https://doi.org/10.1093/nar/gku1179.

Giardine, B., Riemer, C., Hardison, R.C., Burhans, R., et al., 2005. Galaxy: a platform for interactive large-scale genome analysis. Genome Res. 1088-9051. 15 (10), 1451–1455. https://doi.org/10.1101/gr.4086505.

Goble, C.A., Bhagat, J., Aleksejevs, S., Cruickshank, D., Michaelides, D., Newman, D., Borkum, M., Bechhofer, S., Roos, M., Li, P., et al., 2010. myexperiment: a repository and social network for the sharing of bioinformatics workflows. Nucleic Acids Res. 38 (suppl 2), W677–W682.

Gong, J., Liu, W., Zhang, J., Miao, X., Guo, A.-Y., 2015. lncRNASNP: a database of SNPs in lncRNAs and their potential functions in human and mouse. Nucleic Acids Res. 1362-4962. 43, D181–D186. https://doi.org/10.1093/nar/gku1000.

Goodman, A., Pepe, A., Blocker, A.W., Borgman, C.L., Cranmer, K., Crosas, M., Di Stefano, R., Gil, Y., Groth, P., Hedstrom, M., Hogg, D.W., Kashyap, V., Mahabal, A., Siemiginowska, A., Slavkovic, A., 2014. Ten simple rules for the care and feeding of scientific data. PLoS Comput. Biol. 10 (4), 1–5. https://doi.org/10.1371/journal.pcbi.1003542.

Greshake, B., Bayer, P.E., Rausch, H., Reda, J., 2014. openSNP—a crowdsourced web resource for personal genomics. PLoS ONE 1932-6203. 9 (3), e89204. https://doi.org/10.1371/journal.pone.0089204. http://journals.plos.org/plosone/article?id=10.1371/journal.pone.0089204.

Groves, T., 2010. Is open peer review the fairest system? Yes. BMJ (Clin. Res. ed.) 1756-1833. 341, c6424. https://doi.org/10.1136/bmj.c6424. http://www.bmj.com/content/341/bmj.c6424?sid=60af7fc1-eb55-40af-b372-3db2a60593cc.

Guimera, R.V., 2012. bcbio-nextgen: automated, distributed next-gen sequencing pipeline. EMBnet.journal 2226-6089. 17 (B), 30. https://doi.org/10.14806/cj.17.B.286.

Hamosh, A., Scott, A.F., Amberger, J.S., Bocchini, C.A., McKusick, V.A., 2005. Online Mendelian inheritance in man (OMIM), a knowledgebase of human genes and genetic disorders. Nucleic Acids Res. 1362-4962. 33, D514–D517. https://doi.org/10.1093/nar/gki033. http://www.ncbi.nlm.nih.gov/pubmed/15608251.

Harris, N.L., Cock, P.J.A., Lapp, H., Chapman, B., Davey, R., Fields, C., Hokamp, K., Munoz-Torres, M., 2016. The 2015 bioinformatics open source conference (BOSC 2015). PLoS Comput. Biol. 12 (2), 1–6. https://doi.org/10.1371/journal.pcbi.1004691.

Harrow, J., Frankish, A., Gonzalez, J.M., Tapanari, E., et al., 2012. GENCODE: the reference human genome annotation for the ENCODE Project. Genome Res. 1549-5469. 22 (9), 1760–1774. https://doi.org/10.1101/gr.135350.111. http://www.pubmedcentral.nih.gov/articlerender.fcgi?artid=3431492tool=pmcentrezrendertype=abstract.

Hasty, R.T., Garbalosa, R.C., Barbato, V.A., et al., 2014. Wikipedia vs peer-reviewed medical literature for information about the 10 most costly medical conditions. J. Am. Osteopath. Assoc. 1945-1997. 114 (5), 368–373. https://doi.org/10.7556/jaoa.2014.035. http://www.ncbi.nlm.nih.gov/pubmed/24778001.

Hayden, E.C., 2012. Sequencing set to alter clinical landscape. Nature 1476-4687. 482 (7385), 288. https://doi.org/10.1038/482288a. http://www.nature.com/news/sequencing-set-to-alter-clinical-landscape-1.10032.

Hennekam, R.C.M., Biesecker, L.G., 2012. Next-generation sequencing demands next-generation phenotyping. Hum. Mutat. 1098-1004. 33 (5), 884–886. https://doi.org/10.1002/humu.22048. http://www.pubmedcentral.nih.gov/articlerender.fcgi?artid=3327792 tool=pmcentrezrendertype=abstract.

Henry, V.J., Bandrowski, A.E., Pepin, A.-S., Gonzalez, B.J., Desfeux, A., 2014. Omictools: an informative directory for multi-omic data analysis. Database 2014, bau069.

Hettne, K., Soiland-Reyes, S., Klyne, G., Belhajjame, K., Gamble, M., Bechhofer, S., Roos, M., Corcho, O., 2012. Workflow forever: semantic web semantic models and tools for preserving and digitally publishing computational experiments. In: SWAT4LS '11. Proceedings of the 4th International Workshop on Semantic Web Applications and Tools for the Life Sciences, ACM, New York, NY, USA, pp. 36–37.

Highnam, G., Wang, J.J., Kusler, D., Zook, J., Vijayan, V., Leibovich, N., Mittelman, D., 2015. An analytical framework for optimizing variant discovery from personal genomes. Nat. Commun. 2041-1723. 6, 6275. https://doi.org/10.1038/ncomms7275. http://www.nature.com/ncomms/2015/150225/ncomms7275/full/ncomms7275.html.

Hilker, R., Stadermann, K.B., Doppmeier, D., Kalinowski, J., Stoye, J., Straube, J., Winnebald, J., Goesmann, A., 2014. ReadXplorer-visualization and analysis of mapped sequences. Bioinformatics (Oxford, England) 1367-4811. 30 (16), 2247–2254. https://doi.org/10.1093/bioinformatics/btu205.

Hunter, A.A., Macgregor, A.B., Szabo, T.O., Wellington, C.A., Bellgard, M.I., 2012. Yabi: an online research environment for grid, high performance and cloud computing. Source Code Biol. Med. 1751-0473. 7 (1), 1. https://doi.org/10.1186/1751-0473-7-1. http://www.pubmedcentral.nih.gov/articlerender.fcgi?artid=3298538tool=pmcentrezrendertype=abstract.

Ince, D.C., Hatton, L., Graham-Cumming, J., 2012. The case for open computer programs. Nature 1476-4687. 482 (7386), 485–488. https://doi.org/10.1038/nature10836.

Ison, J., Rapacki, K., Ménager, H., Kalaš, M., Rydza, E., Chmura, P., Anthon, C., Beard, N., Berka, K., Bolser, D., et al., 2016. Tools and data services registry: a community effort to document bioinformatics resources. Nucleic Acids Res. 44 (D1), D38–D47.

Jäger, G., Peltzer, A., Nieselt, K., 2014. inPHAP: interactive visualization of genotype and phased haplotype data. BMC Bioinf. 1471-2105. 15 (1), 200. https://doi.org/10.1186/1471-2105-15-200.

Johnson, A.D., O'Donnell, C.J., 2009. An open access database of genome-wide association results. BMC Med. Genet. 1471-2350. 10 (1), 6. https://doi.org/10.1186/1471-2350-10-6. http://bmcmedgenet.biomedcentral.com/articles/10.1186/1471-2350-10-6.

Kallio, M.A., Tuimala, J.T., Hupponen, T., Klemelä, P., Gentile, M., Scheinin, I., Koski, M., Käki, J., Korpelainen, E.I., 2011. Chipster: user-friendly analysis software for microarray and other high-throughput data. BMC Genomics 1471-2164. 12 (1), 507. https://doi.org/10.1186/1471-2164-12-507.

Kanterakis, A., Kuiper, J., Potamias, G., Swertz, M.A., 2015. PyPedia: using the wiki paradigm as crowd sourcing environment for bioinformatics protocols. Source Code Biol. Med. 1751-0473. 10 (1), 14. https://doi.org/10.1186/s13029-015-0042-6.

Karim, M., Michel, A., Zappa, A., Baranov, P., Sahay, R., Rebholz-Schuhmann, D., et al., 2017. Improving data workflow systems with cloud services and use of open data for bioinformatics research. Briefings Bioinf.

Kaye, J., Hurles, M., Griffin, H., Grewal, J., et al., 2014. Managing clinically significant findings in research: the UK10K example. Eur. J. Hum. Genet. 1476-5438. 22 (9), 1100–1104. https://doi.org/10.1038/ejhg.2013.290. http://www.pubmedcentral.nih.gov/articlerender.fcgi?artid=4026295tool=pmcentrezrendertype=abstract.

Khoury, M.J., Gwinn, M., Bowen, M.S., Dotson, W.D., 2012. Beyond base pairs to bedside: a population perspective on how genomics can improve health. Am. J. Public Health 1541-0048. 102 (1), 34–37. https://doi.org/10.2105/AJPH.2011.300299. http://www.pubmedcentral.nih.gov/articlerender.fcgi?artid=3490552tool=pmcentrezrendertype=abstract.

Krieger, M.T., Torreno, O., Trelles, O., Kranzlmüller, D., 2016. Building an open source cloud environment with auto-scaling resources for executing bioinformatics and biomedical workflows. Futur. Gener. Comput. Syst. 0167739X. https://doi.org/10.1016/j.future.2016.02.008.

Landrum, M.J., Lee, J.M., Riley, G.R., Jang, W., Rubinstein, W.S., Church, D.M., Maglott, D.R., 2014. ClinVar: public archive of relationships among sequence variation and human phenotype. Nucleic Acids Res. 1362-4962. 42, D980–D985. https://doi.org/10.1093/nar/gkt1113. http://nar.oxfordjournals.org/content/early/2013/11/14/nar.gkt1113.short.

Lappalainen, T., Sammeth, M., Friedländer, M.R., 't Hoen, P.A., Monlong, J., Rivas, M.A., Gonzàlez-Porta, M., Kurbatova, N., Griebel, T., Ferreira, P.G., et al., 2013. Transcriptome and genome sequencing uncovers functional variation in humans. Nature 501 (7468), 506–511.

Lappalainen, I., Almeida-King, J., Kumanduri, V., Senf, A., et al., 2015. The European genome-phenome archive of human data consented for biomedical research. Nat. Genet. 1061-4036. 47 (7), 692–695. https://doi.org/10.1038/ng.3312.

Leipzig, J., 2016. A review of bioinformatic pipeline frameworks. Brief. Bioinf. 1467-5463, bbw020. https://doi.org/10.1093/bib/bbw020.

Leprevost, F.d.V., Barbosa, V.C., Francisco, E.L., Perez-Riverol, Y., Carvalho, P.C., 2014. On best practices in the development of bioinformatics software. Front. Genet. 1664-8021. 5, 199. https://doi.org/10.3389/fgene.2014.00199.

Leslie, R., O'Donnell, C.J., Johnson, A.D., 2014. GRASP: analysis of genotype-phenotype results from 1390 genome-wide association studies and corresponding open access database. Bioinformatics (Oxford, England) 1367-4811. 30 (12), i185–i194. https://doi.org/10.1093/bioinformatics/btu273. http://bioinformatics.oxfordjournals.org/content/30/12/i185.short.

Li, P.-E., Lo, C.-C., Anderson, J.J., Davenport, K.W., Bishop-Lilly, K.A., Xu, Y., Ahmed, S., Feng, S., Mokashi, V.P., Chain, P.S.G., 2017. Enabling the democratization of the genomics revolution with a fully integrated web-based bioinformatics platform. Nucleic Acids Res. 0305-1048. 45 (1), 67–80. https://doi.org/10.1093/nar/gkw1027.

Liu, Z., Pounds, S., 2014. An R package that automatically collects and archives details for reproducible computing. BMC Bioinf. 1471-2105. 15 (1), 138. https://doi.org/10.1186/1471-2105-15-138.

Longo, D.L., Drazen, J.M., 2016. Data sharing. N. Engl. J. Med. 374 (3), 276–277. https://doi.org/10.1056/NEJMe1516564.

MacArthur, D.G., Manolio, T.A., Dimmock, D.P., Rehm, H.L., Shendure, J., et al., 2014. Guidelines for investigating causality of sequence variants in human disease. Nature 1476-4687. 508 (7497), 469–476. https://doi.org/10.1038/nature13127.

Mailman, M.D., Feolo, M., Jin, Y., Kimura, M., et al., 2007. The NCBI dbGaP database of genotypes and phenotypes. Nat. Genet. 1546-1718. 39 (10), 1181–1186. https://doi.org/10.1038/ng1007-1181. http://www.pubmedcentral.nih.gov/articlerender.fcgi?artid=2031016tool=pmcentrezrendertype=abstract.

Malone, J., Stevens, R., Jupp, S., Hancocks, T., Parkinson, H., Brooksbank, C., 2016. Ten simple rules for selecting a bio-ontology. PLoS Comput. Biol. 1553-7358. 12 (2), e1004743. https://doi.org/10.1371/journal.pcbi.1004743.

Manolio, T.A., 2013. Bringing genome-wide association findings into clinical use. Nat. Rev. Genet. 1471-0064. 14 (8), 549–558. https://doi.org/10.1038/nrg3523.

Manrai, A.K., Ioannidis, J.P.A., Kohane, I.S., 2016. Clinical genomics, from pathogenicity claims to quantitative risk estimates. JAMA. 0098-7484. https://doi.org/10.1001/jama.2016.1519.

Marchant, G., et al., 2007. Genomics, ethics, and intellectual property. In: Krattiger, A., Mahoney, R.T., Nelsen, L. (Eds.), Intellectual Property Management in Health and Agricultural Innovation: A Handbook of Best Practices. pp. 29–38.

McIver, D.J., Brownstein, J.S., 2014. Wikipedia usage estimates prevalence of influenza-like illness in the United States in near real-time. PLoS Comput. Biol. 1553-7358. 10 (4), e1003581. https://doi.org/10.1371/journal.pcbi.1003581. http://journals.plos.org/ploscompbiol/article?id=
10.1371/journal.pcbi.1003581.

McKenna, A., Hanna, M., Banks, E., Sivachenko, A., et al., 2010. The genome analysis toolkit: a MapReduce framework for analyzing next-generation DNA sequencing data. Genome Res. 1549-5469. 20 (9), 1297–1303. https://doi.org/10.1101/gr.107524.110. http://genome.cshlp.org/content/20/9/1297.

Merali, Z., 2010. Computational science: error. Nature 1476-4687. 467 (7317), 775–777. https://doi.org/10.1038/467775a.

Meyerson, M., Gabriel, S., Getz, G., 2010. Advances in understanding cancer genomes through second-generation sequencing. Nat. Rev. Genet. 1471-0064. 11 (10), 685–696. https://doi.org/10.1038/nrg2841.

Mimori, T., Nariai, N., Kojima, K., Takahashi, M., Ono, A., Sato, Y., Yamaguchi-Kabata, Y., Nagasaki, M., 2013. iSVP: an integrated structural variant calling pipeline from high-throughput sequencing data. BMC Syst. Biol. 1752-0509. 7 (Suppl 6), S8. https://doi.org/10.1186/1752-0509-7-S6-S8.

Mitra, K., Carvunis, A.-R., Ramesh, S.K., Ideker, T., 2013. Integrative approaches for finding modular structure in biological networks. Nat. Rev. Genet. 1471-0064. 14 (10), 719–732. https://doi.org/10.1038/nrg3552.

Moreews, F., Sallou, O., Bras, Y.L., Marie, G., Monjeaud, C., Darde, T.A., Collin, O., Blanchet, C., 2015. A curated Domain centric shared Docker registry linked to the Galaxy toolshed. Galaxy Community Conference 2015.

Mu, J.C., Mohiyuddin, M., Li, J., Bani Asadi, N., Gerstein, M.B., Abyzov, A., Wong, W.H., Lam, H.Y.K., 2015. VarSim: a high-fidelity simulation and validation framework for high-throughput genome sequencing with cancer applications. Bioinformatics (Oxford, England) 1367-4811. 31 (9), 1469–1471. https://doi.org/10.1093/bioinformatics/btu828.

Naccache, S.N., Federman, S., Veeraraghavan, N., et al., 2014. A cloud-compatible bioinformatics pipeline for ultrarapid pathogen identification from next-generation sequencing of clinical samples. Genome Res. 1549-5469. 24 (7), 1180–1192. https://doi.org/10.1101/gr.171934.113. http://genome.cshlp.org/content/early/2014/05/16/gr.171934.113.

Niemenmaa, M., Kallio, A., Schumacher, A., Klemelä, P., Korpelainen, E., Heljanko, K., 2012. Hadoop-BAM: directly manipulating next generation sequencing data in the cloud. Bioinformatics (Oxford, England) 1367-4811. 28 (6), 876–877. https://doi.org/10.1093/bioinformatics/bts054. http://bioinformatics.oxfordjournals.org/content/28/6/876.

Paila, U., Chapman, B.A., Kirchner, R., Quinlan, A.R., 2013. GEMINI: integrative exploration of genetic variation and genome annotations. PLoS Comput. Biol. 1553-7358. 9 (7), e1003153. https://doi.org/10.1371/journal.pcbi.1003153.

Pang, C., Hendriksen, D., Dijkstra, M., van der Velde, K.J., Kuiper, J., Hillege, H.L., Swertz, M.A., 2015. BiobankConnect: software to rapidly connect data elements for pooled analysis across biobanks using ontological and lexical indexing. J. Am. Med. Inform. Assoc. 1527-974X. 22 (1), 65–75. https://doi.org/10.1136/amiajnl-2013-002577.

Perkel, J.M., 2015. Programming: pick up python. Nature 1476-4687. 518 (7537), 125–126. https://doi.org/10.1038/518125a.

Perkel, J.M., 2017. How bioinformatics tools are bringing genetic analysis to the masses. Nat. News 543 (7643), 137.

Pfundner, A., Schönberg, T., Horn, J., Boyce, R.D., Samwald, M., 2015. Utilizing the wikidata system to improve the quality of medical content in Wikipedia in diverse languages: a pilot study. J. Med. Internet Res. 1438-8871. 17 (5), e110. https://doi.org/10.2196/jmir.4163. http://www.jmir.org/2015/5/e110/.

Purcell, S., Neale, B., Todd-Brown, K., Thomas, L., Ferreira, M.A.R., Bender, D., Maller, J., Sklar, P., de Bakker, P.I.W., Daly, M.J., Sham, P.C., 2007. PLINK: a tool set for whole-genome association and population-based linkage analyses. Am. J. Hum. Genet. 0002-9297. 81 (3), 559–575.

Ram, K., 2013. Git can facilitate greater reproducibility and increased transparency in science. Source Code Biol. Med. 1751-0473. 8 (1), 7. https://doi.org/10.1186/1751-0473-8-7.

Raney, B.J., Dreszer, T.R., Barber, G.P., Clawson, H., et al., 2014. Track data hubs enable visualization of user-defined genome-wide annotations on the UCSC Genome Browser. Bioinformatics (Oxford, England) 1367-4811. 30 (7), 1003–1005. https://doi.org/10.1093/bioinformatics/btt637.

Rietveld, C.A., Medland, S.E., Derringer, J., et al., 2013. GWAS of 126,559 individuals identifies genetic variants associated with educational attainment. Science (New York, NY) 1095-9203. 340 (6139), 1467–1471. https://doi.org/10.1126/science.1235488. http://www.pubmedcentral.nih.gov/articlerender.fcgi?artid=3751588tool=pmcentrezrendertype=abstract.

Robinson, P.N., Mundlos, S., 2010. The human phenotype ontology. Clin. Genet. 1399-0004. 77 (6), 525–534. https://doi.org/10.1111/j.1399-0004.2010.01436.x.

Sadedin, S.P., Pope, B., Oshlack, A., 2012. Bpipe: a tool for running and managing bioinformatics pipelines. Bioinformatics (Oxford, England) 1367-4811. 28 (11), 1525–1526. https://doi.org/10.1093/bioinformatics/bts167.

Samocha, K.E., Robinson, E.B., Sanders, S.J., Stevens, C., Sabo, A., et al., 2014. A framework for the interpretation of de novo mutation in human disease. Nat. Genet. 1546-1718. 46 (9), 944–950. https://doi.org/10.1038/ng.3050.

Samuels, D.C., Han, L., Li, J., Quanghu, S., Clark, T.A., Shyr, Y., Guo, Y., 2013. Finding the lost treasures in exome sequencing data. Trends Genet. 0168-9525. 29 (10), 593–599. https://doi.org/10.1016/j.tig.2013.07.006.

Saunders, C.J., Miller, N.A., Soden, S.E., Dinwiddie, D.L., et al., 2012. Rapid whole-genome sequencing for genetic disease diagnosis in neonatal intensive care units. Sci. Transl. Med. 1946-6234. 4 (154), 154ra135. https://doi.org/10.1126/scitranslmed.3004041. http://stm.sciencemag.org/content/4/154/154ra135.long.

Scheuner, M.T., de Vries, H., Kim, B., Meili, R.C., Olmstead, S.H., Teleki, S., 2009. Are electronic health records ready for genomic medicine? Genet. Med 1530-0366. 11 (7), 510–517. https://doi.org/10.1097/GIM.0b013e3181a53331.

Scholtens, S., Smidt, N., Swertz, M.A., Bakker, S.J., Dotinga, A., Vonk, J.M., van Dijk, F., van Zon, S.K., Wijmenga, C., Wolffenbuttel, B.H., Stolk, R.P., 2014. Cohort profile: LifeLines, a three-generation cohort study and biobank. Int. J. Epidemiol. 0300-5771. https://doi.org/10.1093/ije/dyu229http://ije.oxfordjournals.org/content/early/2014/12/13/ije.dyu229.abstract?sid=4d7f00a7-ce80-4c6d-9b3e-3232bc06419f.

Schork, N.J., 2015. Personalized medicine: time for one-person trials. Nature 0028-0836. 520 (7549), 609–611. https://doi.org/10.1038/520609a. http://www.nature.com/news/personalized-medicine-time-for-one-person-trials-1.17411.

Seemann, T., 2013. Ten recommendations for creating usable bioinformatics command line software. GigaScience 2047-217X. 2 (1), 15. https://doi.org/10.1186/2047-217X-2-15.

Shen, H., 2014. Interactive notebooks: sharing the code. Nature 1476-4687. 515 (7525), 151–152. https://doi.org/10.1038/515151a.

Shyr, D., Liu, Q., 2013. Next-Generation Sequencing in cancer research and clinical application. Biol. Proced. Online 1480-9222. 15 (1), 4. https://doi.org/10.1186/1480-9222-15-4. http://biologicalproceduresonline.biomedcentral.com/articles/10.1186/1480-9222-15-4.

Skinner, M.E., Uzilov, A.V., Stein, L.D., Mungall, C.J., Holmes, I.H., 2009. JBrowse: a next-generation genome browser. Genome Res. 1549-5469. 19 (9), 1630–1638. https://doi.org/10.1101/gr.094607.109.

Speliotes, E.K., Willer, C.J., Berndt, S.I., et al., 2010. Association analyses of 249,796 individuals reveal 18 new loci associated with body mass index. Nat. Genet. 1061-4036. 42 (11), 937–948. https://doi.org/10.1038/ng.686. http://www.pubmedcentral.nih.gov/articlerender.fcgi?artid=3014648tool=pmcentrezrendertype=abstract.

Spjuth, O., Krestyaninova, M., Hastings, J., Shen, H.-Y., et al., 2015. Harmonising and linking biomedical and clinical data across disparate data archives to enable integrative cross-biobank research. Eur. J. Hum. Genet. 1476-5438. 24 (4), 521–528. https://doi.org/10.1038/ejhg.2015.165.

Swertz, M.A., Dijkstra, M., Adamusiak, T., van der Velde, J.K., Kanterakis, A., et al., 2010. The MOLGENIS toolkit: rapid prototyping of biosoftware at the push of a button. BMC Bioinform. 1471-2105. 11 (suppl 12), S12. https://doi.org/10.1186/1471-2105-11-S12-S12.

Taylor, J.C., Martin, H.C., Lise, S., et al., 2015. Factors influencing success of clinical genome sequencing across a broad spectrum of disorders. Nat. Genet. 1546-1718. 47 (7), 717–726. https://doi.org/10.1038/ng.3304.

Tenenbaum, J.D., Sansone, S.-A., Haendel, M., 2014. A sea of standards for omics data: sink or swim? J. Am. Med. Inform. Assoc. 1527-974X. 21 (2), 200–203. https://doi.org/10.1136/amiajnl-2013-002066.

Tenopir, C., Allard, S., Douglass, K., Aydinoglu, A.U., Wu, L., Read, E., Manoff, M., Frame, M., 2011. Data sharing by scientists: practices and perceptions. PLoS ONE 1932-6203. 6 (6), e21101. https://doi.org/10.1371/journal.pone.0021101.

The Genome of the Netherlands Consortium, 2014. Whole-genome sequence variation, population structure and demographic history of the Dutch population. Nat. Genet. 1546-1718. 46 (8), 818–825. https://doi.org/10.1038/ng.3021.

Thorvaldsdóttir, H., Robinson, J.T., Mesirov, J.P., 2013. Integrative Genomics Viewer (IGV): high-performance genomics data visualization and exploration. Brief. Bioinform. 1477-4054. 14 (2), 178–192. https://doi.org/10.1093/bib/bbs017.

Topol, E.J., 2014. Individualized medicine from prewomb to tomb. Cell 1097-4172. 157 (1), 241–253. https://doi.org/10.1016/j.cell.2014.02.012. http://www.sciencedirect.com/science/article/pii/S0092867414002049.

Torres-Garcia, W., Zheng, S., Sivachenko, A., Vegesna, R., Wang, Q., Yao, R., Berger, M.F., Weinstein, J.N., Getz, G., Verhaak, R.G.W., 2014. PRADA: pipeline for RNA sequencing data analysis. Bioinformatics 1367-4803. 30 (15), 2224–2226. https://doi.org/10.1093/bioinformatics/btu169.

Uzuner, Ö., South, B.R., Shen, S., DuVall, S.L., 2011. 2010 i2b2/VA challenge on concepts, assertions, and relations in clinical text. J. Am. Med. Inform. Assoc. 1527-974X. 18 (5), 552–556. https://doi.org/10.1136/amiajnl-2011-000203.

van der Aalst, W.M.P., ter Hofstede, A.H.M., 2005. YAWL: yet another workflow language. Inform. Syst. 03064379. 30 (4), 245–275. https://doi.org/10.1016/j.is.2004.02.002.

van Ommen, G.-J.B., Törnwall, O., Bréchot, C., et al., 2014. BBMRI-ERIC as a resource for pharmaceutical and life science industries: the development of biobank-based expert centres. Eur. J. Hum. Genet. 1476-5438. 23 (7), 893–900. https://doi.org/10.1038/ejhg.2014.235.

Vanderkam, D., Aksoy, B.A., Hodes, I., Perrone, J., Hammerbacher, J., 2016. pileup.js: a JavaScript library for interactive and in-browser visualization of genomic data. Bioinformatics 1367-4803. 32 (15), 2378–2379. https://doi.org/10.1093/bioinformatics/btw167.

Vihinen, M., 2015. No more hidden solutions in bioinformatics. Nature 0028-0836. 521 (7552), 261. https://doi.org/10.1038/521261a. http://www.nature.com/news/no-more-hidden-solutions-in-bioinformatics-1.17587.

Vuong, H., Che, A., Ravichandran, S., Luke, B.T., Collins, J.R., Mudunuri, U.S., 2015. AVIA v2.0: annotation, visualization and impact analysis of genomic variants and genes. Bioinformatics (Oxford, England) 1367-4811. 31 (16), 2748–2750. https://doi.org/10.1093/bioinformatics/btv200.

Wang, K., Li, M., Hakonarson, H., 2010. ANNOVAR: functional annotation of genetic variants from high-throughput sequencing data. Nucleic Acids Res. 1362-4962. 38 (16), e164. https://doi.org/10.1093/nar/gkq603.

Wang, J., Duncan, D., Shi, Z., Zhang, B., 2013. WEB-based GEne SeT AnaLysis Toolkit (WebGestalt): update 2013. Nucleic Acids Res. 1362-4962. 41 (Web Server issue), W77–W83. https://doi.org/10.1093/nar/gkt439.

Wiewiórka, M.S., Messina, A., Pacholewska, A., Maffioletti, S., Gawrysiak, P., Okoniewski, M.J., 2014. SparkSeq: fast, scalable and cloud-ready tool for the interactive genomic data analysis with nucleotide precision. Bioinformatics (Oxford, England) 1367-4811. 30 (18), 2652–2653. https://doi.org/10.1093/bioinformatics/btu343. http://bioinformatics.oxfordjournals.org/content/early/2014/05/19/bioinformatics.btu343.

Wildeman, M., van Ophuizen, E., den Dunnen, J.T., Taschner, P.E.M., 2008. Improving sequence variant descriptions in mutation databases and literature using the Mutalyzer sequence variation nomenclature checker. Hum. Mutat. 1098-1004. 29 (1), 6–13. https://doi.org/10.1002/humu.20654.

Wilson, G., Aruliah, D., Brown, C.T., Hong, N.P.C., Davis, M., Guy, R.T., Haddock, S.H., Huff, K.D., Mitchell, I.M., Plumbley, M.D., et al., 2014. Best practices for scientific computing. PLoS Biol. 12 (1), e1001745.

Wolstencroft, K., Haines, R., Fellows, D., et al., 2013. The Taverna workflow suite: designing and executing workflows of web services on the desktop, web or in the cloud. Nucleic Acids Res. 1362-4962. 41 (Web Server issue), W557–W561. https://doi.org/10.1093/nar/gkt328.

Yang, Y., Muzny, D.M., Reid, J.G., Bainbridge, M.N., Willis, A., Ward, P.A., Braxton, A., Beuten, J., Xia, F., Niu, Z., et al., 2013. Clinical whole-exome sequencing for the diagnosis of Mendelian disorders. N. Engl. J. Med. 369 (16), 1502–1511.

Zemojtel, T., Köhler, S., Mackenroth, L., et al., 2014. Effective diagnosis of genetic disease by computational phenotype analysis of the disease-associated genome. Sci. Transl. Med. 1946-6242. 6 (252), 252ra123. https://doi.org/10.1126/scitranslmed.3009262.

Zerbino, D.R., Johnson, N., Juettemann, T., Wilder, S.P., Flicek, P., 2014. WiggleTools: parallel processing of large collections of genome-wide datasets for visualization and statistical analysis. Bioinformatics (Oxford, England) 1367-4811. 30 (7), 1008–1009. https://doi.org/10.1093/bioinformatics/btt737. http://www.pubmedcentral.nih.gov/articlerender.fcgi?artid=3967112tool=pmcentrezrendertype=abstract.

Zhao, S., Prenger, K., Smith, L., Messina, T., Fan, H., Jaeger, E., Stephens, S., 2013. Rainbow: a tool for large-scale whole-genome sequencing data analysis using cloud computing. BMC Genomics 1471-2164. 14, 425. https://doi.org/10.1186/1471-2164-14-425. http://www.pubmedcentral.nih.gov/articlerender.fcgi?artid=3698007tool=pmcentrezrendertype=abstract.

Zou, Q., Li, X.-B., Jiang, W.-R., Lin, Z.-Y., Li, G.-L., Chen, K., 2014. Survey of MapReduce frame operation in bioinformatics. Brief. Bioinform. 1477-4054. 15 (4), 637–647. https://doi.org/10.1093/bib/bbs088. http://bib.oxfordjournals.org/content/15/4/637.short?rss=1.

CHAPTER 3

How Cytogenetics Paradigms Shape Decision Making in Translational Genomics

Christophe G. Lambert

Center for Global Health, Division of Translational Informatics, Department of Internal Medicine, University of New Mexico Health Sciences Center, Albuquerque, NM, United States

3.1 INTRODUCTION

Advances in miniaturization and computation catalyzed by Moore's law (Moore, 2006), when wedded with genomic sequencing technologies, have led to an exponential increase in sequencing capacity and an exponential decrease in sequencing cost (NHGRI, 2016). In a few short years, it has become cost-effective to sequence entire human genomes and transcriptomes, and the informatics approaches to modeling genomic processes have evolved in parallel with the technologies.

Several sequencing applications, such as cytogenetics and detection of germline polymorphisms in clinical settings, cancer tumor-normal comparisons, and gene expression-based diagnostics, started with and continue to use lower-resolution technologies. However, these low-throughput technologies are steadily approaching the day of being replaced by the use of whole genome or whole transcriptome sequencing. Right now, a mixture of multiple generations of technologies is used within and across testing laboratories.

It is useful to examine the trajectories of these early applications to observe not only the changes in how genomic processes have been modeled, but how the value sets and paradigms of earlier phases of the technologies continue to shape translational applications of the higher-throughput technologies both now and in the future.

3.2 CLINICAL CYTOGENETIC TESTING

3.2.1 Karyotyping

Genomic rearrangements and copy number changes are a category of genetic variation that plays a key role in conditions such as Down syndrome, autism spectrum disorder, and many others. The study of such abnormalities is

45

Human Genome Informatics. https://doi.org/10.1016/B978-0-12-809414-3.00003-6

covered by the field of *cytogenetics*. Normally, the human genome has two copies of each chromosome, one inherited from each parent. In Down syndrome, a person has three copies of some or all of chromosome 21. Many causes of autism, physical dysmorphology, and developmental delay have been linked to deletions, duplications, and rearrangements of portions of chromosomes.

These abnormalities were first detected with microscopy in the 1950s using a staining process to create a banded karyotype, pictured in Fig. 3.1. The different regions of darker and lighter color, called cytobands, are comprised of millions of base pairs. Large deletions can be detected by observing bands that are shorter than normal, and duplications can be seen as extra or lengthened bands in the karyogram. Other techniques, such as fluorescence in situ hybridization (FISH), bind fluorescent probes to known sequences to verify duplications and translocations.

Detection of chromosomal abnormalities by microscopy with the human eye is subject to error (and high variability) and can only resolve abnormalities that span millions of base pairs. The combination of karyotyping and FISH assays was exclusively used for many years, and yet, resulted in a rather dismal clinical yield of approximately 3%–5%. That is, of 100 patients with obvious birth defects, 95–97 were not provided with a genetic reason for their condition. Despite this low yield, these tests remained a technology standard for years and still have utility in confirming certain structural genomic abnormalities.

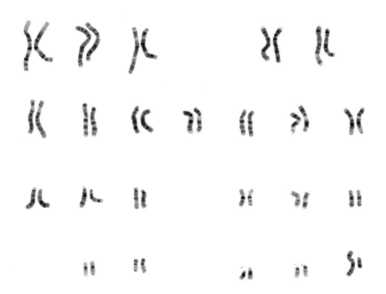

FIG. 3.1

Chromosome staining. *Courtesy: National Human Genome Research Institute—Extracted image from http://www.genome.gov/glossary/resources/karyotype.pdf., Public Domain, https://commons. wikimedia.org/w/index.php?curid=583512.*

Note, however, that if one could detect a major structural abnormality of millions of base pairs or the duplication or deletion of an entire chromosome, a medical geneticist could pronounce with near certainty that it was the cause of a patient's ailment. Conversely, if nothing was detected, a karyogram was termed "normal." This binary mental model of normal and abnormal karyograms still persists today, despite the introduction of higher-resolution detection methods, with some unfortunate consequences, described later.

3.2.2 Chromosomal Microarrays

The field of cytogenetics was next revolutionized by the introduction of genome-wide bacterial artificial chromosome arrays followed by even higher-resolution DNA microarrays, which are the dominant technology in use at the time of writing. Fig. 3.2 shows a microarray graph of two samples with large-scale cytogenetic abnormalities.

The multimegabase changes in Fig. 3.2 are readily detectable even by karyotyping. Chromosomal microarray (CMA) tests typically are used to detect copy number changes 100 kilobases (kb) in size and larger, but in principle can detect gains and losses of 10's of kb or smaller by increasing probe density in targeted regions of interest. As a result of these increases in resolution, the clinical yield for individuals with developmental disabilities or congenital anomalies is approximately 15%, with a range of 9%–25% reported over the years by various organizations (Mc Cormack et al., 2016). In 2010, a consensus statement by the International Standard Cytogenomic Array (ISCA) Consortium recommended CMAs as a first-tier clinical diagnostic test for individuals with developmental disabilities or congenital anomalies, largely due to its enhanced clinical yield over a G-banded karyotype (Miller et al., 2010).

Interestingly, despite the higher yield, in 2010 only about 20% of the market share for cytogenetic testing was held by facilities using CMA testing. The rest of the market was held by those doing traditional testing with the 2%–5% clinical yield. One would have thought that very rapid adoption would have occurred, but as Lambert described at the time (Lambert, 2010), the field had an internal conflict slowing adoption, which still persists to this day, even though CMA testing now has the dominant market share. In fact, the same conflict, plus challenges in insurance reimbursement, may delay the displacement of CMA testing with whole-genome sequencing.

When there was just one technology, there was no conflict. But as the new technology came into being, the following practices were observed in the field that were symptomatic of an underlying conflict:

- Different cytogenetic laboratories had dramatically different size thresholds for computationally detecting and reporting on deletions and duplications.

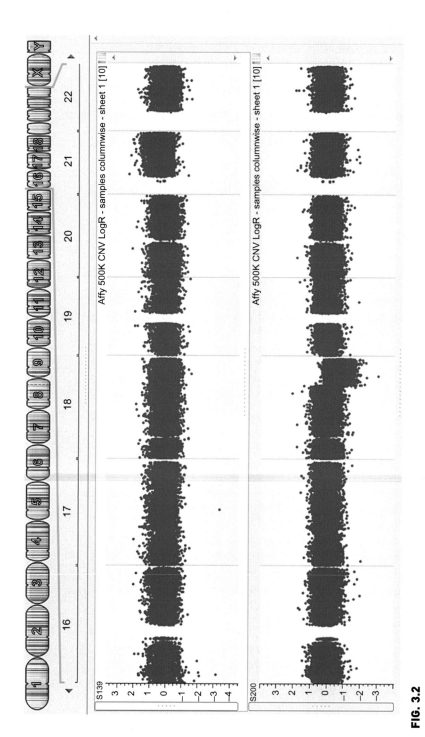

FIG. 3.2

Cytogenetic testing with microarrays. Log_2 ratios are visualized for two patient DNA samples for chromosomes 16–22. Samples that have a normal copy number of two will have log_2 ratios of DNA binding intensity of zero, relative to a neutral reference sample. The upper DNA sample with elevated intensity in chromosome 21 is consistent with trisomy 21 or Down syndrome. The lower sample appears to have a heterozygous deletion in the q arm of chromosome 18, known as chromosome 18q deletion syndrome.

- Some physicians wanted to know about every variant in the patient's genome, others only wanted to know of abnormal variants that were well-characterized and known.
- Some laboratories reported only "normal" versus "abnormal," others reported multiple levels of uncertainty.
- Some high-resolution cytogenetic providers were moving to ultra-high-density arrays, others were limiting resolution and/or masking regions to limit uncertainty.
- Imperfect penetrance and interaction effects were both acknowledged and ignored.
- When a known pathogenic variant was identified, there was a tendency to ignore other variants of less clear significance.
- Several laboratories provided dramatically different answers when tests were ordered on the same difficult-to-diagnose samples—with most of the reports not reflecting the inherent uncertainty or thinking process that went into what was, in essence, a judgment call.
- Cytogeneticists were and still are conflicted over stating a variant is possibly or probably pathogenic when doctors are then likely to consider the case closed, and parents were likely to cling to these judgment calls that may later be determined erroneous.
- Doctors are often reluctant to prescribe high-density microarray testing for prenatal testing because pregnancy termination decisions are made on this data. Some labs use more stringent criteria for prenatal calls and even use different arrays.
- Cytogenetic service providers' marketing pieces often promoted "clarity" or "certainty" (perhaps in response to objections of those who are starting from the "near-certain" traditional methods). Yet, this clarity comes with a sacrifice in that only about 15% of patients are diagnosed.
- Most laboratory policies required that humans evaluate all copy number variants (CNVs) identified by the computer; computers are not allowed to make these decisions. However, choosing lower-resolution arrays and less sensitive algorithms that reduce false discovery at the expense of sensitivity are already, in effect, automating these decisions.
- The majority of undiagnosed patients' genetic tests were surprisingly still called "normal," though the patients were almost certainly suffering from a genetic abnormality that was simply not detected or understood. This use of language appears to be a holdover from the technological limitations of karyotyping, where the dismal 95%–97% failure rate in diagnosis and the psychological burden of being unhelpful to parents and child are perhaps repressed under the label "normal," rather than using a more apt description like "failed to diagnose."

All of these symptoms could be reduced to one primary conflict, depicted in Fig. 3.3.

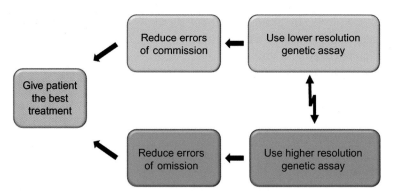

FIG. 3.3

Genetic testing conflict. Everyone can agree on a common goal of giving the patient the best advice/treatment. To do so, one clearly needs to both reduce errors of commission and errors of omission. However, these two legitimate needs have led to conflicting behaviors. On one hand, in order to reduce errors of commission, there is pressure to use a lower-resolution genetic assay and limit genetic information to only that which is already known and well-characterized. On the other hand, to ensure that errors of omission are reduced, there is pressure to take the opposite approach and use a higher-resolution assay to derive as much information as possible to leave no stone unturned in being informed to give the patient the best advice and treatment.

This conflict stemmed from the extra information, both knowledge and uncertainty, derived from using higher-resolution microarrays instead of the blunt (but certain) instrument of karyotyping. Some people in the field were on one side of the conflict, others on the other side, but most were oscillating in a compromise position within the dichotomy.

The needs expressed in this conflict are universal, not only to CMA genetic testing versus karyotyping, but to CMA testing versus whole-genome sequencing, and cannot be ignored when producing informatics solutions. At the end of the day, decisions must be made under uncertainty, and the choice of thresholds of type I and type II error has different implications under different situations. That is, what is valued will determine the choice of control mechanism to manage the high *variety* in decision making (see Chapter 1 for a discussion of *requisite variety*).

The number of differences in thresholds among genetic testing laboratories is startling. For instance, one laboratory will decide to only report on deletions larger than 250 kb, and another will choose 100 kb. The same laboratory might decide to report copy number deletions of 250 kb or larger, but only copy number gains of 500 kb or larger (apparently an historical artifact of array performance where gains were harder to detect than deletions on some platforms). Mosaicism, the result of different cells having different copy numbers for certain regions, is also notoriously inconsistently reported. In part, this is due to variation in microarray platform sensitivity and the fact that the larger the region of mosaicism and the greater the departure from the reference samples (usually copy 2), the easier it is to detect. In most laboratories, there is a whole set of threshold cutoffs along respective receiver operator characteristic curves which create apparent certainty, but really is a choice of which information to show. Why not disclose all of the data? The full CMA data is large, but not that large—a few megabytes per patient. The issue is that the CMA testing companies' job is to distill that data into information—to reduce the variety to a small set of relevant signals that can inform a physician on choice of care. Yet, which signals are relevant is uncertain and evolves over time with research and growth of the clinical knowledge base.

In order to improve patient care and alleviate the conflict of Fig. 3.3, the underlying assumptions behind the arrows must first be surfaced and challenged. Specifically, would having more genetic information truly help with understanding and diagnosing disease better and making fewer errors of omission? Or is it more likely to insert noise into the decision making with questionable information and result in making more errors of commission?

In 2010, many traditional cytogeneticists doubted that more than a 3%–5% yield was possible. Perhaps, some of this doubt was not so much about whether the microarrays could find small variants, but rather if one could have certainty about the relevance of those findings—still a valid concern in microarray-based cytogenetic testing.

Let us set aside for the moment how to deal with the uncertainty of going to higher-resolution arrays and focus instead on the question of whether there is actually more signal in the data that could account for a large portion of the 80% undiagnosed. Following are compelling reasons to believe there is more to be found:

- Unless mosaicism (chromosomal abnormalities in only a subset of a person's cellular DNA) spans very large regions, current practices often leave them undetected or ignored. Mosaicism accounts for several percent of the current abnormal findings. One would expect higher clinical yield by detection of smaller regions and smaller mosaic fractions than current levels.
- Loss of heterozygosity (LOH) accounts for 2% or more of abnormal karyotypes. Some companies detect these, others don't, but most focus on only large-scale LOH. Similar to mosaicism, smaller-resolution LOH detection is likely to improve clinical yields by additional percentage points.
- Current cytogenetic practices mostly ignore multicopy variants, considering only gain, neutral, or loss, plus some large mosaicism. Further, areas containing benign common copy variants are avoided in

some custom arrays in an effort to unburden the analyst. However, while having one, two, or even three copies of a given segment may be benign, having zero or more than four might be pathogenic (the literature contains several examples of different threshold effects for CNVs).

- Interaction effects (epistasis) are acknowledged, but current cytogenetic practices do not provide a way to delineate them with confidence and thereby add to the clinical yield.
- Ultimately, copy number variation exists at the scale of single base pair indels. There are already disorders associated with single base pair deletions—spurring the need for whole-genome sequencing.

Whether the field will be called "cytogenetics" in the future or be supplanted by a new one, it is clear that diagnosis of genetic defects must one day embrace higher-resolution techniques, including and up to next-generation sequencing.

3.2.2.1 Resolving the Conflict

If one accepts that greater genomic content can lead to improved clinical yield, the conflict diagrammed in Fig. 3.3 must be resolved before that content can be effectively utilized; a way must be found to utilize the information in higher density genomic testing to identify more variants, and diagnose more patients, but without sacrificing clarity, creating confusion, or eroding the confidence of the cytogenetic laboratory's customers; in other words, increasing clinical yield while avoiding the incorrect attribution of the clinical relevance of copy number calls (either benign or pathogenic).

Thus, the heart of the conflict is the concern over finding and revealing CNVs of unknown significance, resulting in a vicious cycle: CNVs cannot be shared until their clinical significance is known, yet their clinical significance cannot be determined until they are collected and analyzed. Further, clinical significance may not be binary, but rather need to be quantified along a continuum of risk.

Thus, algorithms and methods of lesser sensitivity become desirable in an effort to avoid dealing with uncertainty—if a CNV of unclear significance isn't detected, it doesn't create uncertainty. However, by not finding and characterizing new, unknown CNVs, the field is locked in a period of stagnation, relying on external research organizations to find and characterize new syndromes associated with genetic anomalies.

Several ideas are presented in the direction of a solution on how to turn these technology improvements into increased clinical yield and thus benefit more patients. Some ideas represent policy changes to break out of the limitations inherited from outworn traditional cytogenetics paradigms, others involve analytics research and development.

3.2.2.2 *Capitalizing on the Conflict*

Cytogenetic laboratories should implement highly sensitive algorithms, capable of accurately finding CNVs of smaller size and of lesser relative intensity. In addition, this new information could be masked or hidden from both the cytogenetic directors and their customers for a period of time. Alternatively, one might envision that this data be placed in some kind of research track that is not normally (or initially) made part of the decision process.

During an average month, larger cytogenetic laboratories might process approximately 1000 or more samples, of which 850 samples might go undiagnosed. Over the course of 6 months, 5100 undiagnosed samples are processed, resulting in over 10,000 over the course of a year. With this ever-increasing number of samples processed at the proposed finer detail, patterns in the data will begin to emerge as CNVs of previously unknown clinical significance converge, as do their associated phenotypes. These could be uncovered in an automated way through statistical association methods, particularly if laboratories pool their findings and phenotypes. At the point where patterns emerge and these CNVs have become increasingly and repetitively associated with a region and phenotype, the filters or mechanisms that hid them can be selectively removed, allowing the users to see these correlations and enabling them to associate these new variants with a repository of similar cases, not too dissimilar from how many laboratories work today. Capturing quality phenotypes will be essential for this and is discussed later.

The result, from the referring physician's perspective, is that they are told not only of their patient's known variants, but, benefiting from a given laboratory's experience and public databases, are also told of additional CNVs that may not be of known published syndromes, but that may be related to the patient's condition. While the results themselves may actually be somewhat clinically ambiguous, the laboratory is able to present them in a fashion which resolves the ambiguity. Uncertainty is further reduced as data sharing takes place across laboratories, incorporating multiple public repositories and supported by vendors in tools such as Golden Helix VarSeq (http://goldenhelix.com/products/VarSeq/).

Building on this capability or, perhaps, in place of this model, the following ideas are also presented:

- The cytogenetic service provider can create a subscription plan or premium service that includes longitudinal follow-up. For the life of a patient, if new information changes the patient diagnosis, the new findings and their impact will be communicated to the referring doctor, letting the physician know there is an ongoing commitment to find the cause of their patients' conditions. Note: with every patient as a customer for life, additional follow-on offerings will also be easier to introduce. On the other hand, this represents a significant shift from the current laboratory testing paradigm wherein one orders a test, gets an answer, and pays once, with no further follow-up being required. Insurance reimbursement may also need modification to support the costs of such efforts.

- Dedicate R&D efforts to increasing clinical yield. Focus on detecting smaller variants of all kinds including mosaics, LOH, and epistatic effects.
- Segment the market by tolerance for uncertainty—some customers prefer information that is certain, others prefer to understand the ambiguity and uncertainty of their cases. Since it takes more time to look at higher-resolution calls, the cytogenetic laboratory can charge accordingly for this extra service. A test might start with standard resolution and move to higher-resolution analysis, conditional on finding or not finding something at low resolution—the physician can make the choice to go to higher resolution if nothing is found.
- Quantify the clinical uncertainty with statistical methods. Arm the medical geneticist with a true estimate of uncertainty based on global statistical analysis of thousands of samples. Educate physicians on the aforementioned conflict and how the laboratory helps them break it.
- Address the large problem of ambiguity of phenotypes. Currently, phenotypic information provided by the referring physician is highly variable, ranging from very specific to extremely general, making statistical inferences to characterize new abnormal regions problematic. Given the volume of samples, payback on this extra information is relatively quick.
  - Deploy a methodology to capture phenotypes consistently, simply, and efficiently. For example, through a drop-down taxonomy that allows quick exclusion of entire categories of symptoms via a hierarchical list, building on tools such as PhenoDB (Hamosh et al., 2013) for data capture and the human phenotype ontology, which provides a common vocabulary for representing abnormalities (Köhler et al., 2014, 2017), see Fig. 3.4. Hamosh reported that head-to-toe assessment of phenotypic abnormalities could be accomplished in about 3 min per patient (personal communication).
  - Help doctors to resolve the conflict of needing to submit one phenotype for insurance purposes and another to help diagnose their patient properly.
  - Incentivize full participation by the physicians, perhaps offering a better price for those that take the time to submit detailed phenotypic information.

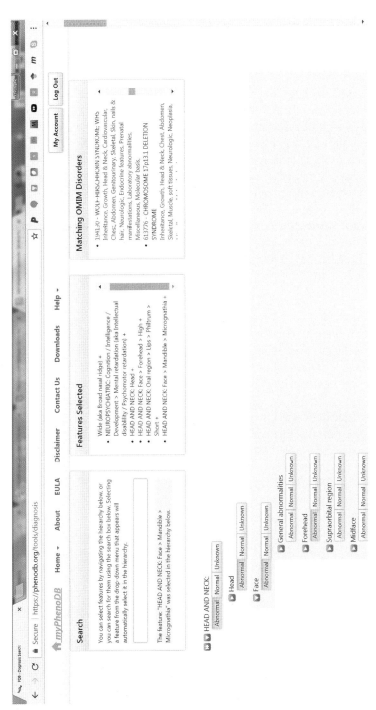

FIG. 3.4

PhenoDB hierarchical phenotype capture. A drop-down taxonomy enables a rapid head-to-toe characterization of the abnormal features of an individual—in this case various attributes of an individual with Wolf-Hirschhorn syndrome are captured.

3.3   FROM CYTOGENETICS TO CYTOGENOMICS IN THE ERA OF NEXT-GENERATION SEQUENCING

The author recently had a conversation with a genetic testing laboratory director who reported that their organization could perform whole-genome sequencing at 0.25X coverage for $187/sample and detect all of the abnormalities that a $300 CMA test could, plus find additional variants and identify various chromosomal rearrangements that are not detectable with CMA testing. Note that this low coverage is insufficient for genotype calling (usually 30X or more is required), but is superior for detecting copy variants. As a result, the laboratory was discontinuing CMA testing, which is now more expensive and, as a recent study has shown, provides less power to detect chromosomal abnormalities (Zhou et al., 2018). Besides the drop in costs, this step forward has largely been a function of the development of better bioinformatics pipelines for CNV calling from next-generation sequencing (Ghoneim et al., 2014; Tattini et al., 2015; Zhao et al., 2013).

The transition from microarrays to whole-genome sequencing represents a bridge where most of the decision-making processes of CMA can be translated to next-generation sequencing. However, cytogenetic testing firms may run into an identity crisis as the costs for, say 30-fold genome-wide coverage, start to achieve parity with CMA testing. Then the challenge becomes not only making sense of CNVs, but genetic polymorphisms. A separate field devoted to the clinical interpretation of whole-genome and whole-exome data has been developing in parallel, with standards on reporting spearheaded by the Human Genome Variation Society (HGVS). In 2013, discussions were underway to harmonize the efforts of the International System for Human Cytogenetic Nomenclature (ISCN) and the HGVS (Ellard et al., 2013), culminating in combined recommendations in 2016 and renaming ISCN to the International System for Human *Cytogenomic* Nomenclature (den Dunnen et al., 2016; International Standing Committee on Human Cytogenomic Nomenclature, 2016). The algorithms for detecting copy number variation with whole-genome sequencing are now so mature that they may completely supplant both karyotyping and CMA testing. In addition, sequencing can detect single nucleotide polymorphisms (SNPs), tiny insertions and deletions (indels), and even methylation.

The worldview of cytogeneticists as detectors of large-scale modifications of chromosomes is ultimately giving way to an integrated view of genetic testing as a whole. However, the analysis approaches to how CNVs are detected in single patients, perhaps looking at a few relatives as opposed to the association testing to find SNPs significantly associated with disease, creates a conflict between two worldviews. The former view holds a mindset that there is certainty and single causes can be identified, while the latter recognizes that complex diseases are driven by many genetic modifiers of modest effect in complex interactions driving outcomes.

Much speculation has been made about where the so-called "missing heritability" of genetic diseases is to be found. The consensus appears to be that there are many sources: epistasis, many variants of modest effect, undetected small insertions/deletions, variation that is hard to measure in high-repeat centromere regions, mosaicism, as well as epigenetic effects passed down through DNA methylation or the microbiome (Eichler et al., 2010).

Despite this almost overwhelming complexity and our increasing ability to measure these phenomena, in the clinical realm, the search for single causes continues unabated, and the pipelines for next-generation sequencing seem to be following the aforementioned paradigm set early on by cytogenetics: find the variants not seen in "healthy" people and filter them by various processes that determine which is the most deleterious variant and which fall in genes or promoters with known association to the condition and then try to find a single variant that appears to be the "smoking gun" that is causative of the disease. While this will bring benefit to some, it also seems clear that the majority of disease cases will not be explained by this approach. Nevertheless, there is hope that as thousands of scientific minds augmented with ever more powerful computation and more sensitive high-throughput instrumentation in a massively parallel community, effort to model this complexity will gradually win the day. Also, just as important as determining the cause for a genetic disease will be the identification of interventions to treat or cure them, whether through medications that target mechanisms identified with genomics or through directly editing the genome with advances such as CRISPR/Cas-9 (Cong et al., 2013).

3.4 CONCLUSIONS

While there will likely be a long-term place for low-throughput tests with high sensitivity, it appears that it is only a matter of time before whole-genome sequencing largely supplants earlier technologies for broad-based testing. However, the single-variant cytogenetics mindset will have to be changed in order to capture the full opportunity afforded by whole-genome sequencing and to more comprehensively map out the complexity of the architecture of the complex diseases that plague mankind. Further, the transactional laboratory test model will not work in the future for genetic testing: after patients have had their whole genome sequenced, they will need to be regularly updated with actionable information that is created with continuous learning from the collective scientific effort to model complex genetic diseases.

References

Cong, L., Ran, F.A., Cox, D., Lin, S., Barretto, R., Habib, N., Hsu, P.D., Wu, X., Jiang, W., Marraffini, L.A., Zhang, F., 2013. Multiplex genome engineering using CRISPR/Cas systems. Science 339, 819–823.

den Dunnen, J.T., Dalgleish, R., Maglott, D.R., Hart, R.K., Greenblatt, M.S., McGowan-Jordan, J., Roux, A.-F., Smith, T., Antonarakis, S.E., Taschner, P.E.M., 2016. HGVS recommendations for the description of sequence variants: 2016 update. Hum. Mutat. 37, 564–569.

Eichler, E.E., Flint, J., Gibson, G., Kong, A., Leal, S.M., Moore, J.H., Nadeau, J.H., 2010. Missing heritability and strategies for finding the underlying causes of complex disease. Nat. Rev. Genet. 11, 446–450.

Ellard, S., Patrinos, G.P., Oetting, W.S., 2013. Clinical applications of next-generation sequencing: the 2013 human genome variation society scientific meeting. Hum. Mutat. 34, 1583–1587.

Ghoneim, D.H., Myers, J.R., Tuttle, E., Paciorkowski, A.R., 2014. Comparison of insertion/deletion calling algorithms on human next-generation sequencing data. BMC Res. Notes 7, 864.

Hamosh, A., Sobreira, N., Hoover-Fong, J., Sutton, V.R., Boehm, C., Schiettecatte, F., Valle, D., 2013. PhenoDB: a new web-based tool for the collection, storage, and analysis of phenotypic features. Hum. Mutat. 34, 566–571.

International Standing Committee on Human Cytogenomic Nomenclature, 2016. ISCN: an international system for human cytogenomic nomenclature (2016). Karger, Basel, New York.

Köhler, S., Doelken, S.C., Mungall, C.J., Bauer, S., Firth, H.V., Bailleul-Forestier, I., Black, G.C.M., Brown, D.L., Brudno, M., Campbell, J., FitzPatrick, D.R., Eppig, J.T., Jackson, A.P., Freson, K., Girdea, M., Helbig, I., Hurst, J.A., Jähn, J., Jackson, L.G., Kelly, A.M., Ledbetter, D.H., Mansour, S., Martin, C.L., Moss, C., Mumford, A., Ouwehand, W.H., Park, S.-M., Riggs, E.R., Scott, R.H., Sisodiya, S., Van Vooren, S., Wapner, R.J., Wilkie, A.O.M., Wright, C.F., Vulto-van Silfhout, A.T., de Leeuw, N., de Vries, B.B.A., Washingthon, N.L., Smith, C.L., Westerfield, M., Schofield, P., Ruef, B.J., Gkoutos, G.V., Haendel, M., Smedley, D., Lewis, S.E., Robinson, P.N., 2014. The human phenotype ontology project: linking molecular biology and disease through phenotype data. Nucleic Acids Res. 42, D966–D974.

Köhler, S., Vasilevsky, N.A., Engelstad, M., Foster, E., McMurry, J., Aymé, S., Baynam, G., Bello, S.M., Boerkoel, C.F., Boycott, K.M., Brudno, M., Buske, O.J., Chinnery, P.F., Cipriani, V., Connell, L.E., Dawkins, H.J.S., DeMare, L.E., Devereau, A.D., de Vries, B.B.A., Firth, H.V., Freson, K., Greene, D., Hamosh, A., Helbig, I., Hum, C., Jähn, J.A., James, R., Krause, R., F Laulederkind, S.J., Lochmüller, H., Lyon, G.J., Ogishima, S., Olry, A., Ouwehand, W.H., Pontikos, N., Rath, A., Schaefer, F., Scott, R.H., Segal, M., Sergouniotis, P.I., Sever, R., Smith, C.L., Straub, V., Thompson, R., Turner, C., Turro, E., Veltman, M.W.M., Vulliamy, T., Yu, J., von Ziegenweidt, J., Zankl, A., Züchner, S., Zemojtel, T., Jacobsen, J.O.B., Groza, T., Smedley, D., Mungall, C.J., Haendel, M., Robinson, P.N., 2017. The human phenotype ontology in 2017. Nucleic Acids Res. 45, D865–D876.

Lambert, C.G., 2010. Rising Above Uncertainty; Increasing Clinical Yield in Array-Based Cytogenetics. [WWW Document]. Golden Helix Inc., Bozeman, Montana, USA. http://blog.goldenhelix.com/clambert/sustaining-competitive-advantage-in-array-based-cytogenetics/. [(Accessed 11 February 2018)].

Mc Cormack, A., Claxton, K., Ashton, F., Asquith, P., Atack, E., Mazzaschi, R., Moverley, P., O'Connor, R., Qorri, M., Sheath, K., Love, D.R., George, A.M., 2016. Microarray testing in clinical diagnosis: an analysis of 5,300 New Zealand patients. Mol. Cytogenet. 9, 29.

Miller, D.T., Adam, M.P., Aradhya, S., Biesecker, L.G., Brothman, A.R., Carter, N.P., Church, D.M., Crolla, J.A., Eichler, E.E., Epstein, C.J., Faucett, W.A., Feuk, L., Friedman, J.M., Hamosh, A., Jackson, L., Kaminsky, E.B., Kok, K., Krantz, I.D., Kuhn, R.M., Lee, C., Ostell, J.M., Rosenberg, C., Scherer, S.W., Spinner, N.B., Stavropoulos, D.J., Tepperberg, J.H.,

Thorland, E.C., Vermeesch, J.R., Waggoner, D.J., Watson, M.S., Martin, C.L., Ledbetter, D.H., 2010. Consensus statement: chromosomal microarray is a first-tier clinical diagnostic test for individuals with developmental disabilities or congenital anomalies. Am. J. Hum. Genet. 86, 749–764.

Moore, G.E., 2006. Cramming more components onto integrated circuits, reprinted from electronics, volume 38, number 8, April 19, 1965, pp.114 ff. IEEE Solid-State Circuits Soc. Newslett. 11, 33–35.

NHGRI, 2016. The Cost of Sequencing a Human Genome4. [WWW Document]. National Human Genome Research Institute (NHGRI), Bethesda, Maryland, USA. https://www.genome.gov/sequencingcosts/. [(Accessed 22 January 2017)].

Tattini, L., D'Aurizio, R., Magi, A., 2015. Detection of genomic structural variants from next-generation sequencing data. Front. Bioeng. Biotechnol. 3, 92.

Zhao, M., Wang, Q., Wang, Q., Jia, P., Zhao, Z., 2013. Computational tools for copy number variation (CNV) detection using next-generation sequencing data: features and perspectives. BMC Bioinf. 14 (Suppl 11), S1.

Zhou, B., Ho, S.S., Zhang, X., Pattni, R., Haraksingh, R.R., Urban, A.E., 2018. Whole-genome sequencing analysis of copy number variation (CNV) using low-coverage and paired-end strategies is highly efficient and outperforms array-based CNV analysis. [Preprint]. April 28, 2018. Available from: https://doi.org/10.1101/192310.

CHAPTER 4

An Introduction to Tools, Databases, and Practical Guidelines for NGS Data Analysis

Alexandros Kanterakis*,†, George Potamias*, George P. Patrinos†

*Institute of Computer Science, Foundation for Research and Technology Hellas (FORTH), Heraklion, Greece, †Department of Pharmacy, School of Health Sciences, University of Patras, Patras, Greece

4.1 INTRODUCTION

With the term next generation sequencing (NGS), we commonly refer to a family of technologies that allows the rapid profiling of DNA sequences in the order of hundreds of thousands reads per run or more. A throughput of sequencing data of this range is enough to assess the complete genome of a human in a time period that spans no more than a day. What is perhaps more remarkable about NGS is that we are currently experiencing an active race of groundbreaking technologies that are constantly competing on driving the cost of sequencing as low as possible. Today, the milestone of $1000 cost for the reliable Whole Genome Sequencing (WGS) of a human genome has already been reached (Hayden, 2014). Cost is diminished into a secondary concern in large sequencing studies, whereas data interpretation and data management are considered a more difficult and costly concern (Stephens et al., 2015). This sharp cost reduction has allowed sequencing pipelines to enter into all areas of life sciences, getting gradually closer to everyday clinical practice, and is steadily embraced by modern healthcare systems. More importantly, the automation and rapid streamlining of analysis pipelines have turned sequencing into a "Swiss army knife" that can be applied from rudimental to very sophisticated experiments (Goodwin et al., 2016). Namely, although sequencing was initially designed to assess the order of the nucleotides in DNA, the same technology can be applied to quantify all stages of genetic regulation (Soon et al., 2013). For example, with ChIP-seq, we can identify which and how proteins interact with DNA and enact the transcription process, then use RNA-seq to inspect the end result of transcription and quantify gene expression, and finally use Ribo-seq to analyze the regions that have been converted to proteins after translation. All these "-seq" technologies apply NGS at some part of the analysis. As an effect, acquiring the genome is no longer the final objective of most studies. Instead, sequencing is rather an

61

Human Genome Informatics. https://doi.org/10.1016/B978-0-12-809414-3.00004-8

intermediate step in complex research protocols that investigate the functional role of the genome and not the genome per se. All these technologies are gradually integrating towards a unified toolbox with the purpose of unraveling the complex genetic factors that affect common and complex diseases (Hasin et al., 2017).

This review targets newcomers in the area of NGS and attempts to provide an introduction to the most common technical considerations and analysis tools of NGS pipelines. Its starting point is the presentation of the semantics and the most common formats for storing raw sequencing data. It proceeds to present widely used online data repositories for sequencing data. These repositories are excellent resources for newcomers for two reasons. The first is that it contains data that researchers can experiment with and the second is that it gives a sense of the scale, structure, and good data management practices of this domain. We continue by presenting Read Alignment and Variant Calling, which are the two basic building blocks of NGS pipelines along with common analysis steps for annotation and complementary analysis. In this review, we attempt to portray NGS as a highly modular technique with limitless applications. We proceed to describe two of them: RNA-seq and ChIP-seq. RNA-seq is an analysis protocol that combines NGS techniques with special analysis tools in order to measure the gene expression and identify alternative splicing events. With ChIP-seq, we can identify and quantify the regions of the DNA that are targeted by proteins.

This review does not discuss the many issues that arise prior to raw data generation from a sequencing device. These are sample preparation (van Dijk et al., 2014), standardization (Aziz et al., 2015), and quality management (Endrullat et al., 2016). Instead, it focuses on available software tools, parameter tuning, and options for results interpretation in regard to data analysis in NGS pipelines. For this reason, readers are encouraged to get accustomed with tool installation and configuration methods in common operating systems used in data analysis such as Linux. Many of the tools presented here require knowledge of the R programming environment (R Core Team, 2013) and, specifically, the Bioconductor set of analysis tools (Gentleman et al., 2004).

4.2 DATA FORMATS

In general, the raw data that are produced from NGS experiments are short reads of sequences. Typically, these reads do not exceed 200 bases, although new promising techniques that produce longer reads of high quality are in the rise. A NGS experiment can produce billion of sequences that result in many gigabytes of data. The most common format for storing sequencing reads is FASTQ. Every entry in a FASTQ file contains both the sequence and a quality score of each nucleotide. The quality score that is typically used is Phred (Ewing et al., 1998). Phred gives an indication of the probability of correct calling of the given base. If P is the probability of correct base call, then $Q = -10 \log_{10}(P)$. For example, a Phred score of 30 indicates a 99.9% base call accuracy, or else a 1 in 1000 probability of incorrect calling. Due to the specifics of the sequencing technology, the closer a base to the beginning of a read, the higher the expected Phred score. Visualizing the distribution of this score is the first Quality Check (QC) step in a sequencing experiment. For this purpose, the most well-known tool is fastQC (Andrews, 2014). Due to the enormous size of FASTQ files, it is common to apply compression before making them available in a public repository, usually with the GNU zip tool algorithm. Therefore, most FASTQ files have the extension "fastq.gz."

Sequences that represent assembled genomes are stored in the FASTA format. For example, a FASTA file may contain the complete genome of an organism, a chromosome, or single gene. FASTA files contain lines starting with ">" that contain information for the genome (name of gene, or the version of the genome assembly). These lines are followed with usually very long lines of genome sequences that contain the letter "A", "C", "G", and "T". Special letters are used to indicate base uncertainty (e.g., the letter "M" means "C" or "A"). The letter "N" or "X" is used to indicate an unknown sequence. FASTA format is also used to store peptide sequences. In that case, the sequence letters consist of amino acids codes. This sequence encoding is maintained by the International Union of Pure and Applied Chemistry (IUPAC) (IUPAC code table, 2017). The most notable tool for FASTA file exploration is Integrative Genomic Viewer (IGV) (Thorvaldsdóttir et al., 2013).

After obtaining a set of raw DNA sequences in FASTQ format, we can align them in a reference genome (encoded in FASTA format). This process is called alignment and is the first computational resourceful step of most NGS pipelines. The purpose of this step is to assign a best-matching genomic position to most of the reads in the FASTQ files. The most used data format for aligned reads is SAM/BAM (Li et al., 2009). SAM stands for Sequence Alignment Map. SAM files contain alignment information for each read in a collection of FASTQ file. In contrast to FASTQ and FASTA files, SAM files have a wide variety of fields that contain very useful metainformation regarding the performed experiment. For example, it contains the sequencing technology, sample identifiers, and flow-cell barcodes. It also contains the version of the software and even the command lines that generated this file. This information is very useful for tracking possible errors and augments the reproducibility of results. BAM is a binary version of SAM. It contains exactly the same information, but it has smaller size relieving, thus, the computational cost of loading and parsing the contained data. SAMtools (Li et al., 2009) is typically used for managing SAM/BAM files and IGV (Thorvaldsdóttir et al., 2013) for exploration.

Special formats exist for the annotation of genomic sequences. Entries on these files consist of a declaration of a genomic region accompanied with the annotation of this region. For example, these files provide information on the ranges that consist of various genes, exons, and transcripts of a reference genome. Annotation files are a valuable accompaniment to NGS pipelines for the purpose of prioritizing specific regions or giving insights on the biological significance of a specific locus. These formats are GFF (GFF2—GMOD, 2017), GTF (GFF/GTF, 2017), and BED (BED, 2017). GFF version 2 is identical to GFF. For example, Ensembl contains a list of annotation in all these three formats for 200 organisms (Ensembl, 2017).

The final analysis steps of most NGS pipelines generate files that contain genomic variants. These are sequences that differ in a significant level from the reference genome. The most common format for describing these variants is VCF (Variant Call Format) (Danecek et al., 2011). It is a simple text, tab delimited format where each line contains the position, the alternative sequence, the reference sequence, various quality metrics and, most importantly, the genotype of all samples in our experiment on these locations. For example, the 1000 Genomes Project (Abecasis et al., 2012) has made available all the variants that it detected in VCF format (VCF, 2017). The VCF files are usually the end result of a NGS pipeline since it contains detected variants for all sequenced samples. This information can be coupled with files that contain phenotype information for these samples in order to locate statistically significant associations between genomic variants and measured phenotypes.

4.3   DATA SOURCES

Traditionally, there have been three major institutions, worldwide, that maintain large databases for raw sequencing data. These are the DNA Data Bank in Japan, GenBank in United States, and the European Nucleotide Archive (ENA) in United Kingdom. These three organisms joined efforts and formed the International Nucleotide Sequence Database Collaboration (INSDC) (Karsch-Mizrachi et al., 2012) in 2011, which was a milestone event for genetics. The main part of the agreement was to synchronize and integrate the sequencing data that are individually maintained by each of these institutions. The database platform that was chosen for this purpose was the Sequence Read Archive (SRA) (Leinonen et al., 2011a). Each institution maintains a separate instance of SRA and all instances are synchronized periodically.

SRA started as a repository for short reads, mainly for RNA-seq and ChIP-seq studies, and it has been expanded to cover all ranges of NGS experiment types. Except from raw sequencing data, SRA also lists library preparation protocols and sample description, used sequencing machines, and bundles everything in an intuitive online user interface. SRA has made available a collection of tools (SRA toolkit) that preprocess, download, and convert data to various formats. Researchers have also the option to use SRAdb (Zhu et al., 2013) to query and use SRA data in the R environment. The most known repository for functional genomics is GEO (Edgar et al., 2002), which stores a large variety of data containing gene expression. Apart from gene expression which contains 90% of its content, GEO contains data on genome methylation, protein profiling, and genome variation (Barrett et al., 2012). It supports microarrays, genotyping arrays, RT-PCR, and NGS data, although GEO uses SRA to store and manage raw NGS data. This GEO-SRA synergy is only a small part of a bigger picture that is composed of a collection of very large and tightly connected databases from NCBI (National Center for Biotechnology Information). These databases cover the complete spectrum of genomic information, from raw sequencing data to findings of medical significance (NCBI Resource Coordinators, 2017).

Similarly to NCBI, the European Bioinformatics Institute (EBI) also offers a rich collection of highly connected tools and databases that cover the complete spectrum of genomic information. The equivalent of NCBI's GEO-SRA synergy for EBI is ArrayExpress-ENA. ArrayExpress (Kolesnikov et al., 2014) is a database for functional genomics data and the European Nucleotide Archive (ENA) (Leinonen et al., 2011b) operates an instance of the SRA database. On Table 4.1, we present a small collection of databases for the main types of genetic data from both NCBI and EMBL. Similar to the SRA integration

Table 4.1 EMBL and NCBI Have a Wide Range of Databases for the Complete Spectrum of Genetics Research

DB Types/Institutes	EMBL/EBI	NIH/NCBI
Sequences	ENA	SRA
	ENA operates an instance of SRA. These two databases along with DDBJ (Japan) are regularly synchronized under the INSDC initiative	
Gene expression	ArrayExpress	GEO
Genotype/phenotype	EGA	dbGaP
Variants	EVA	dbSNP
Publications	EuropePMC	PubMed
Genome browser	ENSEMBL	dbVAR
	UCSC is more advanced and well-integrated with both ENSEMBL and NCBI	
API	ENSEMBL REST, dbfetch	Entrez
Link	https://www.ebi.ac.uk/services	https://www.ncbi.nlm.nih.gov/guide/all/

This table presents the major databases offered by both institutes for the main types of genetic data presented in this report. Also, databases from different institutes for the same type of data differ significantly on the semantics and integration level with analysis tools. Researchers are encouraged to consider databases from both institutes and also to review the complete range by following the links in the last row of the table.

between Europe, United States, and Japan, we should expect in the near future a tightened cooperation between tools and databases that provide similar functionality.

Another comprehensive set of sequencing data is from the 1000 Genomes Project (1KGP). 1KGP investigated the human genetic variation across many populations by employing mainly sequencing techniques. At its final stage, it sequenced 2.504 individuals from 26 different populations covering a significant part of the world (Sudmant et al., 2015). What is more important is that 1KGP has made available all data from raw sequencing reads in FASTQ format to final variant calls in VCF format.

4.4 VARIANT DATA

For validation, annotation, and downstream analysis, researchers might be interested in acquiring large sets of variant data, which are usually the end results of NGS pipelines. Three of the most known sources for variant data are the 1000 Genomes Project, the ExAC consortium, and the TCGA project.

The ExAC (Exome Aggregation Consortium) is a collaboration that enabled the harmonization of variants from 60,706 unrelated individuals from 14 large sequencing studies (Lek et al., 2016). ExAC does not provide genotypes for every individual; instead, it provides aggregates statistics, namely, allele frequencies, protein alterations, and functional consequences. Among the 14 sequencing studies, ExAC includes samples from the 1kGP and the TCGA project.

The Cancer Genome Atlas (TCGA) (Weinstein et al., 2013) is an ongoing project to characterize the genetic and molecular profile for a wide variety of cancer types. As of 2017, it has collected more than 11,500 tumor samples, paired with normal samples. These samples belong to 33 difference cancer types, ten of which are considered rare. TCGA has applied a multitude of analysis techniques on these samples, which include gene expression profiling, DNA methylation, SNP genotyping, and exon sequencing and on some samples (10%) of genome-wide sequencing. Data that are produced during TCGA's activities belong to various "data levels" which characterize, among others, their availability status. What is more important is that most of the data from the final stage of analysis belong to the level 3 status which means that they are publicly available, although data in the initial phase can be made available through a specific request. TCGA offers both VCF files and files that describe somatic mutations in individual samples in a specialized format called Mutation Annotation Format (Wan and Pihl, 2017).

4.5 NGS PIPELINES

In the following sections, we will present the standard analysis practices and most prominent tools suited for data from NGS experiments.

4.5.1 Read Alignment

Aligning a DNA sequence to a reference sequence has been one of the central problems of genetics which has attracted a lot of interest from computer science researchers. Starting with the Needleman-Wunsch algorithm (Needleman and Wunsch, 1970), which was introduced in 1970, computers could efficiently perform exact sequence alignment (or else global alignment). This was followed by the introduction of the Smith-Waterman algorithm (Smith and Waterman, 1981), introduced in 1981, which is a variation of Needleman-Wunsch that allowed sequence mismatches (local alignment). The Smith-Waterman algorithm also allowed the computational detection of sequence variants, mainly insertions and deletions. It is interesting that the demanding time requirements of these algorithms gave rise to the prominent FASTA (Lipman and Pearson, 1985) and BLAST (Altschul et al., 1990) algorithms (introduced in 1985 and 1990, respectively) that are approximately 50 times faster, although they produce suboptimal results. Today, these algorithms are available either as online web services or programs implemented in low-level programming languages, which helps in the optimal utilization of the underlying computer architecture. These algorithms and their variations are the foundational components of modern NGS techniques.

Although, theoretically, read alignment has many computational efficient solutions, a modern experimental sequencing setup might require the alignment of reads that can easily reach the order of billions. Therefore, any optimization in software or hardware level can speed up this process considerably. Today, there are more than 90 tools that perform read alignment, the vast majority (~80%) of which have been developed over the last 10 years (Otto et al., 2014). Choosing the optimal tool requires the careful consideration of many criteria regarding our experimental setup (Fonseca et al., 2012). Some of these criteria are:

- Type of sequencing experiment. This type can range from fragmented DNA to spliced DNA (RNA-seq).
- Type of sequencing technology. Each technology produces reads of different length and different quality. Reads vary from 10,000 bps for Single-molecule real-time sequencing, 700 bps for Pyrosequencing to the most commonly used 50–300 bp sequencing in Illumina Miniseq and Hiseq platforms. Fortunately, most platforms are accompanied with "best practices" guides that include which alignment tool is best suited for the kind of data that it produces.

- Type of preparation protocol. Some protocols might result in the enrichment of particular genomic regions, which in turn produces an uneven distribution of the abundance of the reads. DNA fragmentation and treatment methods might also alter read quantity for certain regions.
- Quality of genome assembly. Species like human, mouse, fruit fly, and zebra fish are considered model organisms and have top quality reference assemblies. Species with complex genome or low scientific interest might have a low quality or even nonexistent reference genome. Also, lately we have experienced an increase of tools that are able to perform alignment in multiple reference genomes like DynMap (Flouri et al., 2011) and GenomeMapper (Schneeberger et al., 2009).
- Type of downstream analysis. Depending on the type of research and the scientific question, different alignment tools can be more suitable. Additionally, some tools for downstream analysis (i.e., variant calling) are better coupled with specific aligners. This includes supporting the data formats on which the aligned data are described.
- Type of available computational infrastructure. This is perhaps one of the most important criteria. Some aligners require the availability of a high performance computation environment like a cluster, whereas some tools can run in a typical commodity computer. Of course, the time required and the quality of the results differ a lot. For example, in a recent benchmark, the tool Segemehl (Otto et al., 2014) had the least False Positive hits in mRNA-seq data compared to other popular aligners. Yet, it required 5 times more computational time and 10 times more memory. Therefore, weighting the pros and cons for each tool before making a choice is very important.
- Availability and familiarity. Over the course of time, certain tools have matured and offer very detailed documentation and online tutorials. Additionally, they have formed very knowledgeable user groups that act as community ready to help inexperienced users or tackle advanced technical issues. Finally, the software license that accompanies the tool might range from free use for even nonacademic environments to requiring substantial fees that might constitute the tool prohibitive in a restrained budgeted laboratory.

The development of novel read alignment tools continues to be an active research field. A prominent method to validate the efficacy of these tools is to use artificially generated NGS read data. The researchers in this case can generate read data "on-demand" with predefined characteristics like length, depth, and quality. Tools for this purpose are ART (Huang et al., 2011) and FluxSimulator (Griebel et al., 2012). Researchers can also submit their alignment protocols and have them assessed through open challenges such as RGASP (Engström et al., 2013).

4.5.2 Variant Calling

The alignment of reads in a specific genomic reference results in a SAM/BAM file. Assuming that all quality control steps for eliminating biases in these files have been applied, the next analysis step is variant calling. This step has been the subject of thorough scrutiny from researchers since many options are available, each implementing a different analytic approach. On one of the most known benchmarks, Hwang et al. (2015) used a golden standard to compare the efficacy of existing variant calling methods. This golden standard was the set of variants detected on a sample from the 1000 Genomes Project after applying five different NGS technologies, seven read mapping, and three variant calling methods. The authors then applied a plethora of established variant calling methods on the aligned reads of the same sample and compared the results with the golden standard. Results showed that the task of SNP calling was performed better with the combination of BWA-MEM (Li, 2013) for alignment and Samtools (Li et al., 2009) for variant call. Another software for variant calling, Freebayes (Marth et al., 1999), had similar performance when combined with any alignment tool. For the task of Indel calling, the combination of BWA-MEM and GATK (Genome Analysis Toolkit) (McKenna et al., 2010) tools had better performance. It is important to note that this comparison was made with tool versions older than the current. Given that all of the presented tools are constantly improved, readers should always consult comparisons that include the latest versions before making decisions. Another crucial factor is that tool simplicity differs from large and complicated software frameworks to tools that are simple to install and use. For example, GATK has hundreds of possible options and requires above the average skills in software installation and configuration in contrast to much simpler tools like Freebayes and bam2mpg (Teer et al., 2010). Yet, the versatility of GATK parameters allows for some complex tasks like low-depth variant calling.

There are two types of possible computational pipelines for variant calling. The first is for germline mutation and the second is for somatic mutations. Somatic mutations happen in any cell of a somatic tissue and, since these mutations are not present in germ line, they are not passed to the progenies of the organism. Somatic mutations are of interest mainly in cancer studies where we are looking for mutating events that disrupt the apoptosis pathway and lead to uncontrolled cell proliferation. This line of work entails additional challenges due to high cell differentiation in cancer tissues and the low quality of acquired DNA sample (Meldrum et al., 2011). Nevertheless, existing computation pipeline sufficiently detects somatic mutations and corrects for possible DNA sampling artifacts (Ding et al., 2010).

The primary use of NGS pipelines is to acquire a partial (e.g., Whole Exome Sequencing, WES) or a complete (Whole Genome Sequencing, WGS) image of an individual's DNA. In both cases, the last step of the pipeline (variant calling) can give insights on already known or unknown mutations. NGS can be also applied to perform a typical genotyping experiment, where the final result is the assessment of alleles in a given set of genotypes. This technique is called genotyping-by-sequencing and involves the use of Restriction Enzymes in order to fragment the DNA in regions of interest and then sequencing these regions (Baird et al., 2008). Therefore, by compromising the ability to discover novel variants, we can reduce significantly the cost of the experiment and increase the throughput compared to WGS for the variants of interest. Tools for this purpose are 2b-RAD (Wang et al., 2012) and tGBS (Ott et al., 2017).

Another crucial question is the type of variation that can be captured by existing sequencing techniques and variant calling tools. Typical NGS pipelines contain tools that focus on SNPs and Copy Number Variations (CNV) of a relative small size (usually up to 1 kbp). Although WGS performs well for identification of SNPs and CNVs, the nature of WES experiments limits significantly the ability of many tools to identify CNVs (Tan et al., 2014). Special tools that attempt to tackle these limitations are ERDS-exome (Tan et al., 2018), CODEX (Jiang et al., 2015), CNVnator (Abyzov et al., 2011), and PennCNV-seq (de Araújo Lima and Wang, 2017).

Another consideration is that there is a big class of variants that contain more complex or large variations that are referred to as structural variants. For example, large insertions, deletions, duplications, inversions, and translocations comprise a considerable proportion of the human genome and have been associated with important phenotypes (Sudmant et al., 2015; Weischenfeldt et al., 2013). Structural variants can exhibit extreme complexity and require specialized calling software. Variants in this category are interspersed (nontandem) duplications, inverted duplications, and inversions that contain deletions (Arthur et al., 2017). Tools specialized for structural variation detection are ARC-SV (Arthur et al., 2017), Pindel (Ye et al., 2009), and LUMPY (Layer et al., 2014).

Larger structural variations like chromosomal aberrations are even more difficult to detect from short reads. Locating this kind of variation in somatic cells is detrimental for tumor profiling in cancer studies. Also, large SVs in germline cells can give insights on various mutagenesis mechanisms. Sequencing technologies that produce longer reads are more suitable to detect large SVs. An instance of long-read sequencing technique is mate pair sequencing, which generates reads in the order of 3–10 kbps. Tools for detecting long SVs from long-read mapping data are BreakDancer (Chen et al., 2009), SVDetect (Zeitouni et al., 2010), and SVachra (Hampton et al., 2017).

Another class of variants of special interest is Short Tandem Repeats (STRs). STRs exhibit a very low conservation rate and, consequently, they have large variability even in homogeneous populations. This variability constitutes STRs ideal for forensic uses and paternal testing. Tools that specialize in STR calling from NGS data are lobSTR (Gymrek et al., 2012), STR-FM (Fungtammasan et al., 2015), and STRScan (Tang and Nzabarushimana, 2017).

A very crucial aspect that affects directly the quality of sequencing results is the sequencing depth and the achieved coverage. As a rule of thumb, the higher the depth, the better the quality of inferred variants. Nevertheless, an increased depth does not guarantee that all regions in the genome have been sequenced adequately since some regions are less easily targeted than others. Examples are highly repeated elements, or elements that belong to the telomere or centromere regions of the chromosomes. A common approach to increase coverage (or else "mappability") is to increase read length or to use sequence with long inserts in pair-end sequencing. A very good review of these considerations along with best practices is in (Sims et al., 2014).

4.5.3 Downstream Analysis

Downstream analysis refers to possible computation pipelines that can be applied after variant calling. Usually, the first step that follows variant calling is annotation. With annotation, we enrich the set of identified variants with information from existing genomic databases. This information usually includes Allele frequency distribution of the variants across various populations, known clinical implications, known functional consequences, and predicted functional consequences. One useful source of information for population distribution of detected variants is from the 1000 Genomes Project (Auton et al., 2015), with tools like GGV (Wolfe et al., 2013), which also allows the geographical visualization of this information. Ongoing and future large population-specific sequencing experiments are also expected to enrich our knowledge on the geographic distribution of common and rare alleles, improving thus our knowledge of the phenotypic implication of these variants. A project of this scale is the UK10K (Walter et al., 2015).

Annotation can be performed not only on mutations identified after variant calling, but also on the sequences themselves. For example, large sequences that represent genes can be annotated with information like conservation scores. In addition, for large sequences of unknown function, we can apply known gene models in order to detect the number and location of included introns/exons as well as possible alternative spliced transcripts. This is called Genomic Range annotation and a review of tools for this process is available in (Lawrence et al., 2013).

Population structure analysis is a method to infer the genetic similarity of a set of genotyped samples. This information can identify the presence of multiple populations in an experiment as well as the presence of population admixed samples. Two well-known tools for this task are ADMIXTURE (Alexander et al., 2009) and Principal Component Analysis (PCA) (Conomos et al., 2015).

An immediate step that follows variant calling mainly from large or from population-based studies is kinship analysis. Kinship analysis reveals all possible pedigree information among the participating samples. This information is useful for family-based genetic analysis or as a quality control since cryptic family relations can significantly obscure the statistics of association testing. Two of the most known tools to infer family relations from genotype data are KING (Manichaikul et al., 2010) and GCTA (Yang et al., 2010). One important prerequisite for these tools is that all samples should belong to the same population. Alternatively, tools like REAP (Thornton et al., 2012) and PC-Relate (Conomos et al., 2016) can perform the same task for admixed population. One disadvantage of these methods is that they do not account for genotyping errors, which might be considerable especially in sequencing studies of low depth. In these cases, an option is SEEKIN (Dou et al., 2017), which can perform kinship analysis directly from aligned reads and is tailored for low-depth (~0.15X) sequencing data.

After variants and/or genes have been prioritized through a genotype-phenotype association test, we can apply tools that locate common functional roles in groups (or sets) of genes. A valuable source of information for gene function is gene ontology (Consortium, 2015) (GO). GO maintains a hierarchical ontology of terms which describes the function of a gene along with a database of genes and gene products from various organisms that are pre-annotated. Tools like PANTHER (Mi et al., 2013) accepts as input a list of genes and generates clusters with common functional annotation. This type of analysis is called gene set analysis (or gene set enrichment). In the effort to locate functional similarities between large sets of genes, other valuable sources of information are Gene Regulatory Networks, protein interaction networks, pathways, and molecular interaction data. Fortunately, tools like GeneMania (Warde-Farley et al., 2010) offer a holistic gateway to these sources and provide a comprehensive functional annotation.

4.5.4   RNA-seq

RNA-seq is a powerful technique to measure the abundance of mRNA in a sample (Trapnell et al., 2012; Garber et al., 2011; Conesa et al., 2016). This technique is used for two primary purposes. The first is to generate a very detailed map of all splice isoforms that exist in the entirety of transcripts in the sample. The second is to measure the gene or transcript expression based on the counts of reads that were mapped in various areas in the genome. Both types give valuable insights when applied in a differential analysis. Differences from normal samples in splice isoforms can reveal insertions, deletions, and mismatches. On the other side, differential gene expression patterns can provide insights on more complex biologic mechanisms that affect physiology. Before moving to computational approaches for RNA-seq data analysis, we should note that sample preparation and sequencing is a complex procedure on which stringent quality steps should be applied in order to eliminate biases that could, in the worst case, render the data unusable (Li et al., 2015). Also, RNA-seq experiments contain hundreds of millions of short reads (<150 bp) per sample that require a storage capacity in the order of hundreds of gigabytes. Therefore, although there is extremely optimized software for this purpose, analysis might require days in a High Performance Computing Environment.

RNA-seq pipelines usually start with the alignment of sequencing reads in a reference assembly. Computational methods for alignment belong in two main categories: unspliced and spliced. Unspliced aligners do not allow for any large gap in the mapped sequence, or else, the whole read must map to a single region. With splice aligners, a part of a read can map to a region and another part in another. Given the pros and cons of each category, a hybrid approach is usually applied. One example is the Bowtie-TopHat tool combination. Bowtie (Langmead and Salzberg, 2012) implements a memory-efficient (loads the complete reference in the memory) alignment which is suitable for short reads. Nevertheless, reads from RNA transcripts can belong to multiple exoms. Given that in the reference genome these exoms may contain large intronic regions in between, Bowtie fails to map these reads. TopHat (Trapnell et al., 2009) resolves this issue by splitting the reads in many short segments and uses Bowtie to map them in the reference genome. Another tool that belongs to the unspliced category is BWA (Li and Durbin, 2009), whereas MapSplice (Wang et al., 2010) and SpliceMap (Au et al., 2010) are suitable for spliced alignment. The end result, which is the aligned transcriptome of a sample, may reveal interesting information such as new splice isoforms or even new genes.

After alignment, the next step is transcript assembly. The main predicament in this process is to determine for each read the transcript where it originates. This is challenging for two main reasons: The first is that the reads might originate from mRNA containing only exons or from un-spliced RNA that contain intronic regions. The second is that short reads may have many candidate mapping transcripts. Scripture (Guttman et al., 2010) and Cufflinks (Trapnell et al., 2010) are notable tools that are able to locate the different splicing variants of a gene from the aligned reads of a single sample. The main difference between these tools is sensitivity. Scripture exports all possible isoforms, whereas Cufflinks reports a minimal set. Scripture is more suitable for novel gene and transcript discovery. Nevertheless, both generate an assembly of all isoforms that contain reads for every sample. Cufflinks is accompanied with a tool, Cuffmerge, to merge the transcript assemblies of all samples and the final result is a high-quality unique transcriptome assembly. Optionally, this step might include annotation from a reference genome (Roberts et al., 2011). Another tool that performs this task is MISO (Li et al., 2008), whereas other tools such as ERANGE (Mortazavi et al., 2008) and NEUMA (Lee et al., 2010) focus only in genes (not transcripts).

In order to locate differences of gene/transcript expression between samples that belong to different phenotypic categories, we can use a third tool from the same family, Cuffdiff. This tool compares the number of reads that mapped to the transcriptome assembly that was generated in the previous step. For every identified transcript, it reports *P*-values of the statistical test under the null-hypothesis that all samples generated the same mRNA of this transcript regardless of which phenotypic category they belonged to. Cuffdiff applies methods to eliminate biases originating from library preparation and varying sequencing depth among samples that are known to affect RNA-seq results. Finally, the output files of Cuffdiff can be fed to downstream analysis tools such as CummeRbund for visualization and quality control. Apart from Cuffdiff, another popular tool for the same task is DESeq (Anders and Huber, 2010) and its successor DESeq2 (Love et al., 2014). A complete review of computational approaches and challenges is in (Costa-Silva et al., 2017).

It is important to note that transcript assembly is possible without aligning first the reads in a reference genome. This de novo assembly is suitable when a reference genome is not available or is of poor quality. The most prominent tool for this analysis is TransAbyss (Robertson et al., 2010).

One fundamental concept in RNA-seq experiments is read count normalization. This is because the differential expression of a gene between different phenotypic categories can be measured by comparing the read counts that mapped on that gene between the samples belonging to these categories. One source of an issue here is that different samples might be sequenced in different depths in the same RNA-seq experiment. Moreover, the number of reads that map to each gene is proportional to the size of this gene, making longer genes to receive more reads. Therefore, these two factors introduce significant bias that needs to be corrected. The correction is based on normalization on both gene length and sequencing depth.

To illustrate this process, we can represent the results of a RNA-seq experiment as a two-dimensional matrix. The first dimension contains genes and the second samples. Each element in the matrix is the number of reads that mapped to a specific gene for a specific sample. For each sequenced sample, we measure the total number of reads that mapped to this sample. Then we divide each sum with 1,000,000. This generates a single scaling factor per sample. Subsequently, we divide each element in the matrix with the scaling factor that corresponds to the sample of this element. This is called the RPM (Reads Per Million) measurement. Then, we divide each RPM value in the matrix with the length of the corresponding gene, measured in kilobases. This is called the RPKM (Reads Per Kilobase Million) (Lee et al., 2010). RPKM is suitable for single-end RNA-seq experiments. In paired-end RNA-seq experiments, a DNA fragment can be captured by either two reads (at both ends of the fragment) or by one. Using the RPKM measurement in pair-end RNA-seq experiments is not suitable since the number of reads does not correspond to the number of mapped fragments. For this reason, the FPKM (Fragments per Kilobase Million) is the appropriate measurement since it adjusts for differences of reads per fragment (Trapnell et al., 2010). In single-end experiments, both metrics produce the same results. Recently, a new measurement is in the rise, namely the TPM (Transcripts Per Million). The basic difference of TPM is that the order of normalization is reversed. Initially, we divide each read count in the matrix to the length (again in kilobases) of the gene that it corresponds. The resulted matrix is called RPK (Reads Per Kilobase). The next step is to add up all RPK measurements for each sample and divide the result with 1,000,000. This is the "per sample scaling factor." Then we divide every RPK value in the matrix with the scaling factor of the corresponding sample. Although TPM and RPKM produce similar results, there is a profound qualitative difference that makes TPM more favorable. The sum of TPM values for each sample is always the same. This is because TPM depicts the proportion of a sample's reads that were mapped in a gene. Since the essence of RNA-seq experiments is to measure the relative proportions of mapped reads in a specific gene for different samples, this measure is more reliable and thus trustworthy.

These metrics rely heavily on averaging the expression profile from millions cells. Given the vast heterogeneity of cell types and the complexity of cell-to-cell interactions, the granularity of this approach on measuring the expression profile for specific cell types is limited. Therefore, another line of work is to limit analysis in not only specific cell types, but on individual cells. This analysis is called Single-cell RNA sequencing (or else, scRNA-seq) (Kolodziejczyk et al., 2015) and focuses on identifying rare cell types with low-expression levels that play a significant role in transcription regulation and cell differentiation. As with typical RNA-seq data, analysis starts by normalizing read counts (Vallejos et al., 2017). Single-Cell RNA-seq experiments have two main technical difficulties. The first is that the read count in single cells for specific transcripts might be very low or even zero (Bacher and Kendziorski, 2016). The second is that thousands of genes are simultaneously expressed in a single cell; therefore, sc-RNA seq data suffer from the "curse of dimensionality" which affects efforts to visualize and build classification models. To alleviate this problem, researchers apply typical dimensional reduction algorithms like PCA and t-SNE (van der Maaten and Hinton, 2008). Over the last year, special dimensional reduction methods for sc-RNA-seq have been crafted that also accommodate for the low read count problem. Examples are ZIFA (Pierson and Yau, 2015), SIMLR (Wang et al., 2017), SAIC (Yang et al., 2017), and VASC (Wang and Gu, 2017). Another approach is to apply clustering algorithms that group same or similar cells into clusters of common expression profile (Menon, 2017). Commonly used tools for this purpose are SC3 (Kiselev et al., 2017), RCA (Li et al., 2017), CIDR (Lin et al., 2017), and RaceID2 (Grün et al., 2016), although a metaanalysis revealed that for some sequencing platforms the consensus between these tools is low (Freytag et al., 2017). Discrepancies of this kind indicate that these methods are still in their infancy. Perhaps the only database for curation and annotation of human sc-RNA-seq data is scRNA-SeqDB (Cao et al., 2017).

Instead of locating specific cell types, a similar research goal is to identify genes that are expressed in specific tissues or diseases. An example of a tool for this task is SEGtool (Zhang et al., 2017). Particularly in cancer studies, this technique can be used primarily to distinguish tumor cells from normal cells, and subsequently to extract their expression profile. Additionally, we can generate the tumor mutation tree (or else tumor phylogenies), which in turn can be used to locate the original cell types where the starting tumor mutation appeared (Zafar et al., 2017). As an effect, we can extract a detailed cancer molecular signature and diagnose rare cancer types, which is an important step for improved prognosis. For a review of computational approaches in scRNA-seq, see (Bacher and Kendziorski, 2016).

By having in our disposal the expression profile of different kinds of cell types under different conditions (e.g., drug exposure), we can perform various types of differential analysis (Sá et al., 2017). Analysis includes building co-expression networks of different transcripts (van Dam et al., 2017). This can help to locate the transcripts that are in central positions in these networks and prioritizing on them for further investigation (van Dam et al., 2017). Another type of analysis measures the differential expression of transcripts that belong to the same genomic region, but carry different alleles. This is called allele-specific expression (Pastinen, 2010) and elucidates the functional role of variants that are usually identified in Genome Wide Association Studies (GWAS). The refinement of the regulatory role of a transcript in the development of a certain phenotype is a complex task. Usually, this role is investigated under the hypothesis that these elements act as transcription factors that target specific transcription binding sites and, as an effect, activate or deactivate the transcription process. This process can be direct by targeting sites of the same gene (cis-effect) (Majewski and Pastinen, 2011), or indirect by targeting intermediate factors which in turn affect the expression of a gene that is located in a distant genomic position (trans-effect) (Morley et al., 2004). Another large group of RNA-seq analysis involves time course data where we search for correlation between time and transcript expression (Spies et al., 2017).

Similarly to GWAS, expression measured from RNA-seq experiments can be used to identify differentially expressed transcripts between two or more types of phenotypes. This used to be the main type of analysis of expression-related studies that preceded the modern spread and diffusion of RNA-seq experiments. Before RNA-seq, Gene Expression Microarrays was the prevailing method for this purpose and Limma (Smyth, 2005) was the dominant tool for differential analysis. Today, although Limma has been extended for RNA-seq data (Ritchie et al., 2015), other tools like PLDA (Witten, 2011), NBLDA (Dong et al., 2016), and voomDDA (Zararsiz et al., 2017) offer more fine-grained analysis methods tailored for differential analysis of RNA-seq data.

4.5.5 ChIP-Seq

The third technology that somehow "completes" the puzzle of the main application of NGS technologies is ChIP-sequencing (or ChIP-seq). A ChIP-seq analysis protocol aims to locate the DNA binding site of any protein and, therefore, investigates epigenetic factors that affect gene expression. This can help to uncover the interaction pattern of elements like transcription factors, structural proteins, or any chromatin-associated protein and give insights on the epigenetic mechanisms that guide cell differentiation and physiology. A review of ChIP-Seq analysis tools is available in (Steinhauser et al., 2016), a pipeline in R/Bioconductor is available in (de Santiago and Carroll, 2018), and some practical guidelines are discussed in (Bailey et al., 2013). We also strongly recommend the careful study of the guidelines published from the ENCODE and modENCODE consortia (Landt et al., 2012). These consortia performed genome-wide chromatin profiling in human (Thurman et al., 2012), mouse, *Drosophila melanogaster*, and *Caenorhabditis elegans* by following multiple research protocols in a diverse set of equipment. The quality of reporting, the reproducibility of the analysis, and open data access policies of these projects are examples of future projects of this scale. There are also published "ready-to-run" computational pipelines that act as a front-end for the numerous tasks of quality control, analysis, and reporting of ChIP-Seq-based experiments. Two of the most prominent integrated pipelines are HiChIP (Yan et al., 2014) and ChiLin (Qin et al., 2016).

The first steps of ChIP-seq analysis pipeline resemble that of RNA-seq. Initially, we assess the quality of raw sequencing reads with tools like FastQC (Andrews, 2014). The second part is the alignment of these reads in a reference genome. Common alignments tools like BWA (Li and Durbin, 2009), Bowtie (Langmead and Salzberg, 2012), and STAR (Dobin et al., 2013) are well suited for this purpose. The proprietary alignment tool NovoAlign (2018) has shown good alignment results for short reads, but is considerably slower. The third and most sensitive part of the analysis in terms of parameter tuning is peak calling. With peak calling, we want to measure two inherently different properties. The first is the DNA regions that exhibit high read density and second is to identify regions that belong to a diffused binding site. The second part can be seen as the clustering of the regions identified from the first based on common epigenetic properties. The most common tool for read density peak calling is MACS (Zhang et al., 2008), whereas diffusion site identification can be performed with SICER (Zang et al., 2009) and BayesPeak (Cairns et al., 2011). After peak calling, the researcher has to assess two very important quality metrics: sensitivity and specificity. Sensitivity measures the ratio of the detected peaks that were correctly identified and specificity measures the ratio of the discarded peaks that were indeed experimental artifacts or other false positives. Various filtering choices like removing reads that mapped to multiple locations or removing different reads that mapped to the same location can have significant effects on sensitivity and specificity. Therefore, researchers should always find optimal parameters that favor both these metrics.

After peak calling, the results are open for visualization and interpretation, although peak calling, visualization, and annotation are tasks that are collectively offered from tools like ChIPseeker (Yu et al., 2015). The HiChIP pipeline generates results that can be directly visualized with IGV (Thorvaldsdóttir et al., 2013). Other tools for peak-call visualization are seqMiner (Ye et al., 2011) and Sole-Search (Blahnik et al., 2010). For annotation, researchers can query existing database of regulatory regions for various organisms that have been generated from collective processing of thousands of ChIP-seq samples. One of the most comprehensive databases of this kind for human is ReMap (Chèneby et al., 2018). Moreover, the MGA repository (Dréos et al., 2018) functions as an annotation server for ChIP-seq data; it offers wide range of well-integrated analysis tools and also contains 6500 human samples available in BED format. Another line of analysis is motif analysis for which commonly used tools are RSAT (Thomas-Chollier et al., 2012) and rGADEM (Mercier et al., 2011). Finally, a well-prepared and regularly updated set of technical notes, tools, and list of resources is available at (Tang, 2018).

4.6 DISCUSSION

DNA sequencing is one of the most revolutionary technologies of our age. In terms of efficiency and cost reduction, the advances of this technology have surpassed by far that of semiconductor, telecommunications, and automobile industry. DNA sequencing started as an experimental, costly, and error-prone procedure just 40 years ago, whereas today Genome Sequencing is gradually becoming a routinely screening method for healthcare and research

(Shendure et al., 2017). The prospects of this technology are beyond our grasp for reaching population-wide sequencing, portable sequencers, and even in vivo DNA editing. Tools and practices presented here are just a temporary snapshot on a constantly evolving terrain. Therefore, researchers should frequently take care to get accustomed with recent advances and new tools.

Here, we have reviewed three of the most known NGS applications that have helped tremendously in the discovery of genetic factors that affect disease and physiology. These technologies are NGS experiments for variant discovery RNA-seq for expression profiling and ChIP-seq for protein-DNA interaction analysis. These techniques are complementary since they shed light in three fundamentally different aspects of cell regulation. Fig. 4.1 presents a generic schematic of the computational pipelines for these tasks where we show their main components and the interdependencies. Also, these techniques are a small subset of the complete NGS "ecosystem" with which we can rapidly

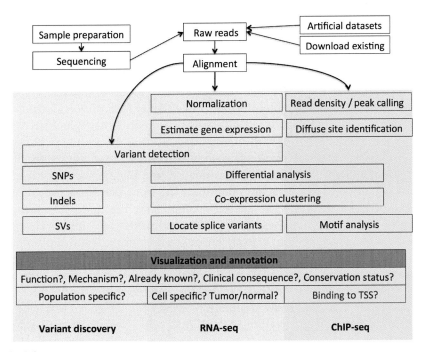

FIG. 4.1

A generic representation of the three families of NGS pipelines presented in this report. Raw reads can be obtained from sequencing, online, or from software that generates artificial datasets. Alignment is common to all NGS pipelines. Also, some analysis steps are also shared. Visualization and annotation are also steps that are present in all pipelines. This representation does not include quality control, which should be applied after every analysis step. Careful inspection of results, filtering of data with dubious quality, and findings duplication are essential steps that should be never omitted.

screen many more stages of genetic regulation. For example, another sequencing technique for epigenetic analysis is Bisulfite sequencing (BS-seq). With this technique, we can assess the methylation status of nearly any region in the genome (Wilhelm-Benartzi et al., 2013). A review of computational analysis methods for DNA methylation data targeted for users with basic knowledge of bioinformatics is in (Bock, 2012) and bicycle (Graña et al., 2017) is an integrated, ready-to-run pipeline. Researchers should keep in mind that there is still a lot of space for improvement in these technologies. For example, although theoretically variant discovery through RNA-seq should be equally efficient with NGS pipelines, there is still a considerable discrepancy. Approximately, 10% of genotypes inferred from RNA-seq data are different from the genotypes inferred from DNA sequencing data (Guo et al., 2017).

Recent advances in bioinformatics and, particularly, in workflow management systems (WMSs) allow the rapid automation and pipelining of these procedures. As an effect, most NGS pipelines are available as ready-to-run computational scripts that can be combined at will in sophisticated analysis protocols. For example, we can apply DNA sequencing in order to identify mutation in areas of interest and then perform RNA-seq in order to reveal the functional consequences of these mutations (Singh et al., 2017). Therefore, the presented techniques can be perceived as nothing more than a set of tools in the inventory of the researcher that can be combined at will for more sophisticated types of analysis. Modern WMS in bioinformatics like Galaxy (Giardine et al., 2005) act as execution front-ends and repositories of commonly used pipelines. These pipelines can be combined with simple "drag-n-drop" actions in a user-friendly environment. Overall, there are more than 100 WMSs today tailored for different types of analysis, computer infrastructure, and IT-competence of the user (Di Tommaso, 2018). This automation, apart from the rapid streamlining, maximizes the Findability, Accessibility, Interoperability, and Reproducibility aspects of the analysis. These four guiding principles are collectively called "FAIR" and they have been actively encouraged, endorsed, and promoted by many funding and policy-making institutions (Wilkinson et al., 2016). Modern research is beyond simple tool composition and experimentation. Instead, it requires a social mindset, collaborative skills, and the ability to construct analysis methods that are directly verifiable, reusable, and beneficial to a wide and diverse community. Confronting to the "FAIR" principles is an easy first step to fulfill these needs.

References

Abecasis, G.R., Auton, A., Brooks, L.D., DePristo, M.A., Durbin, R.M., Handsaker, R.E., et al., 2012. An integrated map of genetic variation from 1092 human genomes. Nature 491, 56–65. https://doi.org/10.1038/nature11632.

Abyzov, A., Urban, A.E., Snyder, M., Gerstein, M., 2011. CNVnator: an approach to discover, genotype, and characterize typical and atypical CNVs from family and population genome sequencing. Genome Res. 21, 974–984. Available from: http://www.ncbi.nlm.nih.gov/pubmed/21324876.

Alexander, D.H., Novembre, J., Lange, K., 2009. Fast model-based estimation of ancestry in unrelated individuals. Genome Res. 19, 1655–1664. Available from: http://www.ncbi.nlm.nih.gov/pubmed/19648217.

Altschul, S.F., Gish, W., Miller, W., Myers, E.W., Lipman, D.J., 1990. Basic local alignment search tool. J. Mol. Biol. 215, 403–410.

Anders, S., Huber, W., 2010. Differential expression analysis for sequence count data. Genome Biol. 11, R106.

Andrews S. FastQC: A Quality Control Tool for High Throughput Sequence Data. Version 0.11. 2. 2014, Babraham Institute, Cambridge http://www.bioinformatics.babraham.ac.uk/projects/fastqc.

Arthur, J.G., Chen, X., Zhou, B., Urban, A.E., Wong, W.H., 2017. Detection of complex structural variation from paired-end sequencing data. bioRxiv 200170.

Au, K.F., Jiang, H., Lin, L., Xing, Y., Wong, W.H., 2010. Detection of splice junctions from paired-end RNA-seq data by SpliceMap. Nucleic Acids Res. 38, 4570–4578.

Auton, A., Abecasis, G.R., Altshuler, D.M., Durbin, R.M., Abecasis, G.R., Bentley, D.R., et al., 2015. A global reference for human genetic variation. Nature 526, 68–74. Available from: http://www.ncbi.nlm.nih.gov/pubmed/26432245.

Aziz, N., Zhao, Q., Bry, L., Driscoll, D.K., Funke, B., Gibson, J.S., et al., 2015. College of American Pathologists' laboratory standards for next-generation sequencing clinical tests. Arch. Pathol. Lab. Med. 139, 481–493. Available from: http://www.ncbi.nlm.nih.gov/pubmed/25152313.

Bacher, R., Kendziorski, C., 2016. Design and computational analysis of single-cell RNA-sequencing experiments. Genome Biol. 17, 63. Available from: http://genomebiology.biomedcentral.com/articles/10.1186/s13059-016-0927-y.

Bailey, T., Krajewski, P., Ladunga, I., Lefebvre, C., Li, Q., Liu, T., et al., 2013. Practical guidelines for the comprehensive analysis of ChIP-seq data. Lewitter F, PLoS Comput. Biol. 9, e1003326. Available from: http://dx.plos.org/10.1371/journal.pcbi.1003326.

Baird, N.A., Etter, P.D., Atwood, T.S., Currey, M.C., Shiver, A.L., Lewis, Z.A., et al., 2008. Rapid SNP discovery and genetic mapping using sequenced RAD markers. PLoS One e3376, 3.

Barrett, T., Wilhite, S.E., Ledoux, P., Evangelista, C., Kim, I.F., Tomashevsky, M., et al., 2012. NCBI GEO: archive for functional genomics data sets—update. Nucleic Acids Res. 41, D991–D995.

BED, 2017. BED Format. Available from: http://genome.ucsc.edu/FAQ/FAQformat.html#format1.

Blahnik, K.R., Dou, L., O'Geen, H., McPhillips, T., Xu, X., Cao, A.R., et al., 2010. Sole-search: an integrated analysis program for peak detection and functional annotation using ChIP-seq data. Nucleic Acids Res. 38, e13. Available from: http://www.ncbi.nlm.nih.gov/pubmed/19906703.

Bock, C., 2012. Analysing and interpreting DNA methylation data. Nat. Rev. Genet. 13, 705–719. Available from: http://www.nature.com/doifinder/10.1038/nrg3273.

Cairns, J., Spyrou, C., Stark, R., Smith, M.L., Lynch, A.G., Tavaré, S., 2011. BayesPeak—an R package for analysing ChIP-seq data. Bioinformatics 27, 713–714. Available from: http://www.ncbi.nlm.nih.gov/pubmed/21245054.

Cao, Y., Zhu, J., Jia, P., Zhao, Z., 2017. scRNASeqDB: a database for RNA-Seq based gene expression profiles in human single cells. Genes (Basel) 8, 368. Available from: http://www.ncbi.nlm.nih.gov/pubmed/29206167.

Chen, K., Wallis, J.W., McLellan, M.D., Larson, D.E., Kalicki, J.M., Pohl, C.S., et al., 2009. BreakDancer: an algorithm for high-resolution mapping of genomic structural variation. Nat. Methods 6, 677–681. Available from: http://www.ncbi.nlm.nih.gov/pubmed/19668202.

Chèneby, J., Gheorghe, M., Artufel, M., Mathelier, A., Ballester, B., 2018. ReMap 2018: an updated atlas of regulatory regions from an integrative analysis of DNA-binding ChIP-seq experiments.

Nucleic Acids Res. 46, D267–D275. Available from: http://www.ncbi.nlm.nih.gov/pubmed/29126285.

Conesa, A., Madrigal, P., Tarazona, S., Gomez-Cabrero, D., Cervera, A., McPherson, A., et al., 2016. A survey of best practices for RNA-seq data analysis. Genome Biol. 17, 13. Available from: http://genomebiology.com/2016/17/1/13.

Conomos, M.P., Miller, M.B., Thornton, T.A., 2015. Robust inference of population structure for ancestry prediction and correction of stratification in the presence of relatedness. Genet. Epidemiol. 39, 276–293. Available from: http://www.ncbi.nlm.nih.gov/pubmed/25810074.

Conomos, M.P., Reiner, A.P., Weir, B.S., Thornton, T.A., 2016. Model-free estimation of recent genetic relatedness. Am. J. Hum. Genet. 98, 127–148. Available from: http://www.ncbi.nlm.nih.gov/pubmed/26748516.

Consortium, T.G.O., 2015. Gene ontology consortium: going forward. Nucleic Acids Res. 43, D1049–D1056. Available from: http://nar.oxfordjournals.org/content/43/D1/D1049.abstract.

Costa-Silva, J., Domingues, D., Lopes, F.M., 2017. RNA-Seq differential expression analysis: an extended review and a software tool. Wei Z, PLoS One. 12, e0190152. Available from: http://www.ncbi.nlm.nih.gov/pubmed/29267363.

Danecek, P., Auton, A., Abecasis, G., Albers, C.A., Banks, E., DePristo, M.A., et al., 2011. The variant call format and VCFtools. Bioinformatics 27, 2156–2158. Available from: http://bioinformatics.oxfordjournals.org/content/27/15/2156.

de Araújo Lima, L., Wang, K., 2017. PennCNV in whole-genome sequencing data. BMC Bioinf. 18, 383. Available from: http://www.ncbi.nlm.nih.gov/pubmed/28984186.

de Santiago, I., Carroll, T., 2018. Analysis of ChIP-seq Data in R/Bioconductor. Humana Press, New York, NY, pp. 195–226. Available from: http://link.springer.com/10.1007/978-1-4939-7380-4_17.

Di Tommaso, P., 2018. A curated list of awesome pipeline toolkits inspired by Awesome Sysadmin. Available from: https://github.com/pditommaso/awesome-pipeline.

Ding, L., Wendl, M.C., Koboldt, D.C., Mardis, E.R., 2010. Analysis of next-generation genomic data in cancer: accomplishments and challenges. Hum. Mol. Genet. 19, R188–R196. Available from: https://academic.oup.com/hmg/article-lookup/doi/10.1093/hmg/ddq391.

Dobin, A., Davis, C.A., Schlesinger, F., Drenkow, J., Zaleski, C., Jha, S., et al., 2013. STAR: ultrafast universal RNA-seq aligner. Bioinformatics 29, 15–21. Available from: http://www.ncbi.nlm.nih.gov/pubmed/23104886.

Dong, K., Zhao, H., Tong, T., Wan, X., 2016. NBLDA: negative binomial linear discriminant analysis for RNA-Seq data. BMC Bioinf. 17, 369. Available from: http://www.ncbi.nlm.nih.gov/pubmed/27623864.

Dou, J., Sun, B., Sim, X., Hughes, J.D., Reilly, D.F., Tai, E.S., et al., 2017. Estimation of kinship coefficient in structured and admixed populations using sparse sequencing data. PLoS Genet.. 13, e1007021. Available from: http://www.ncbi.nlm.nih.gov/pubmed/28961250.

Dréos, R., Ambrosini, G., Groux, R., Périer, R.C., Bucher, P., 2018. MGA repository: a curated data resource for ChIP-seq and other genome annotated data. Nucleic Acids Res. 46, D175–D180. Available from: http://www.ncbi.nlm.nih.gov/pubmed/29069466.

Edgar, R., Domrachev, M., Lash, A.E., 2002. Gene expression omnibus: NCBI gene expression and hybridization array data repository. Nucleic Acids Res. 30, 207–210.

Endrullat, C., Glökler, J., Franke, P., Frohme, M., 2016. Standardization and quality management in next-generation sequencing. Appl. Transl. Genom. 10, 2–9. Available from: http://www.ncbi.nlm.nih.gov/pubmed/27668169.

Engström, P.G., Steijger, T., Sipos, B., Grant, G.R., Kahles, A., Rätsch, G., et al., 2013. Systematic evaluation of spliced alignment programs for RNA-seq data. Nat. Methods 10, 1185–1191.

Ensembl, 2017. Ensembl Genome Assemblies and Annotations. Available from: https://www.ensembl.org/info/data/ftp/index.html.

Ewing, B., Hillier, L., Wendl, M.C., Green, P., 1998. Base-calling of automated sequencer traces usingPhred. I. Accuracy assessment. Genome Res. 8, 175–185.

Flouri, T., Iliopoulos, C.S., Pissis, S.P., 2011. DynMap: mapping short reads to multiple related genomes. Proceedings of the 2nd ACM Conf. Bioinformatics, Comput. Biol. Biomed. ACM, New York, NY, pp. 330–334.

Fonseca, N.A., Rung, J., Brazma, A., Marioni, J.C., 2012. Tools for mapping high-throughput sequencing data. Bioinformatics 28, 3169–3177.

Freytag, S., Lonnstedt, I., Ng, M., Bahlo, M., 2017. Cluster headache: comparing clustering tools for 10X single cell sequencing data. bioRxiv, 203752. Available from: https://www.biorxiv.org/content/early/2017/10/16/203752.

Fungtammasan, A., Ananda, G., Hile, S.E., Su, M.S.-W., Sun, C., Harris, R., et al., 2015. Accurate typing of short tandem repeats from genome-wide sequencing data and its applications. Genome Res. 25, 736–749. Available from: http://www.ncbi.nlm.nih.gov/pubmed/25823460.

Garber, M., Grabherr, M.G., Guttman, M., Trapnell, C., 2011. Computational methods for transcriptome annotation and quantification using RNA-seq. Nat. Methods 8, 469–477.

Gentleman, R.C., Carey, V.J., Bates, D.M., Bolstad, B., Dettling, M., Dudoit, S., et al., 2004. Bioconductor: open software development for computational biology and bioinformatics. Genome Biol. 5, R80. Available from: http://www.ncbi.nlm.nih.gov/pubmed/15461798.

GFF/GTF, 2017. GFF/GTF File Format. Available from: http://www.ensembl.org/info/website/upload/gff.html.

GFF2—GMOD. 2017. Available from: http://gmod.org/wiki/GFF2

Giardine, B., Riemer, C., Hardison, R.C., Burhans, R., Elnitski, L., Shah, P., et al., 2005. Galaxy: a platform for interactive large-scale genome analysis. Genome Res. 15, 1451–1455. Available from: http://www.pubmedcentral.nih.gov/articlerender.fcgi?artid=1240089&tool=pmcentrez&rendertype=abstract.

Goodwin, S., McPherson, J.D., WR, M.C., 2016. Coming of age: ten years of next-generation sequencing technologies. Nat. Rev. Genet. 17, 333–351. Available from: http://www.ncbi.nlm.nih.gov/pubmed/27184599.

Graña, O., López-Fernández, H., Fdez-Riverola, F., González Pisano, D., Glez-Peña, D., 2017. Bicycle: a bioinformatics pipeline to analyze bisulfite sequencing data. Bioinformatics. Available from: http://www.ncbi.nlm.nih.gov/pubmed/29211825.

Griebel, T., Zacher, B., Ribeca, P., Raineri, E., Lacroix, V., Guigó, R., et al., 2012. Modelling and simulating generic RNA-Seq experiments with the flux simulator. Nucleic Acids Res. 40, 10073–10083.

Grün, D., Muraro, M.J., Boisset, J.-C., Wiebrands, K., Lyubimova, A., Dharmadhikari, G., et al., 2016. De novo prediction of stem cell identity using single-cell transcriptome data. Cell Stem Cell 19, 266–277. Available from: http://www.ncbi.nlm.nih.gov/pubmed/27345837.

Guo, Y., Zhao, S., Sheng, Q., Samuels, D.C., Shyr, Y., 2017. The discrepancy among single nucleotide variants detected by DNA and RNA high throughput sequencing data. BMC Genomics 18, 690. Available from: http://www.ncbi.nlm.nih.gov/pubmed/28984205.

Guttman, M., Garber, M., Levin, J.Z., Donaghey, J., Robinson, J., Adiconis, X., et al., 2010. Ab initio reconstruction of cell type-specific transcriptomes in mouse reveals the conserved multi-exonic structure of lincRNAs. Nat. Biotechnol. 28, 503–510.

Gymrek, M., Golan, D., Rosset, S., Erlich, Y., 2012. lobSTR: a short tandem repeat profiler for personal genomes. Genome Res. 22, 1154–1162. Available from: http://www.ncbi.nlm.nih.gov/pubmed/22522390.

Hampton, O.A., English, A.C., Wang, M., Salerno, W.J., Liu, Y., Muzny, D.M., et al., 2017. SVachra: a tool to identify genomic structural variation in mate pair sequencing data containing inward and outward facing reads. BMC Genomics 18, 691. Available from: http://www.ncbi.nlm. nih.gov/pubmed/28984202.

Hasin, Y., Seldin, M., Lusis, A., 2017. Multi-omics approaches to disease. Genome Biol. 18, 83. Available from: http://genomebiology.biomedcentral.com/articles/10.1186/s13059-017-1215-1.

Hayden, E.C., 2014. The $1,000 genome. Nature 507, 294.

Huang, W., Li, L., Myers, J.R., Marth, G.T., 2011. ART: a next-generation sequencing read simulator. Bioinformatics 28, 593–594.

Hwang, S., Kim, E., Lee, I., Marcotte, E.M., 2015. Systematic comparison of variant calling pipelines using gold standard personal exome variants. Sci. Rep. 5 17875.

IUPAC code table, 2017. NIAS DNA Bank. Available from: https://www.bioinformatics.org/sms/ iupac.html.

Jiang, Y., Oldridge, D.A., Diskin, S.J., Zhang, N.R., 2015. CODEX: a normalization and copy number variation detection method for whole exome sequencing. Nucleic Acids Res. 43, e39. Available from: http://www.ncbi.nlm.nih.gov/pubmed/25618849.

Karsch-Mizrachi, I., Nakamura, Y., Cochrane, G., International Nucleotide Sequence Database Collaboration, 2012. The international nucleotide sequence database collaboration. Nucleic Acids Res. 40, D33–D37. Available from: http://www.ncbi.nlm.nih.gov/pubmed/22080546.

Kiselev, V.Y., Kirschner, K., Schaub, M.T., Andrews, T., Yiu, A., Chandra, T., et al., 2017. SC3: consensus clustering of single-cell RNA-seq data. Nat. Methods 14, 483–486. Available from: http:// www.ncbi.nlm.nih.gov/pubmed/28346451.

Kolesnikov, N., Hastings, E., Keays, M., Melnichuk, O., Tang, Y.A., Williams, E., et al., 2014. ArrayExpress update—simplifying data submissions. Nucleic Acids Res. 43, D1113–D1116.

Kolodziejczyk, A.A., Kim, J.K., Svensson, V., Marioni, J.C., Teichmann, S.A., 2015. The technology and biology of single-cell RNA sequencing. Mol. Cell. 58, 610–620.

Landt, S.G., Marinov, G.K., Kundaje, A., Kheradpour, P., Pauli, F., Batzoglou, S., et al., 2012. ChIP-seq guidelines and practices of the ENCODE and modENCODE consortia. Genome Res. 22, 1813–1831. Available from: http://www.ncbi.nlm.nih.gov/pubmed/22955991.

Langmead, B., Salzberg, S.L., 2012. Fast gapped-read alignment with Bowtie 2. Nat. Methods 9, 357–359.

Lawrence, M., Huber, W., Pagès, H., Aboyoun, P., Carlson, M., Gentleman, R., et al., 2013. Software for computing and annotating genomic ranges. Prlic A, PLoS Comput. Biol. 9, e1003118. Available from: http://dx.plos.org/10.1371/journal.pcbi.1003118.

Layer, R.M., Chiang, C., Quinlan, A.R., Hall, I.M., 2014. LUMPY: a probabilistic framework for structural variant discovery. Genome Biol. 15, R84.

Lee, S., Seo, C.H., Lim, B., Yang, J.O., Oh, J., Kim, M., et al., 2010. Accurate quantification of transcriptome from RNA-Seq data by effective length normalization. Nucleic Acids Res. 39, e9.

Leinonen, R., Sugawara, H., Shumway, M., International nucleotide sequence database collaboration, 2011a. The sequence read archive. Nucleic Acids Res. 39, D19–D21. Available from: http://www.ncbi.nlm.nih.gov/pubmed/21062823.

Leinonen, R., Akhtar, R., Birney, E., Bower, L., Cerdeno-Tárraga, A., Cheng, Y., et al., 2011b. The European nucleotide archive. Nucleic Acids Res. 39, D28–D31. Available from: http://www. ncbi.nlm.nih.gov/pubmed/20972220.

Lek, M., Karczewski, K.J., Minikel, E.V., Samocha, K.E., Banks, E., Fennell, T., et al., 2016. Analysis of protein-coding genetic variation in 60,706 humans. Nature 536, 285–291. Available from: http://www.nature.com/articles/nature19057.

Li, H., 2013. Aligning sequence reads, clone sequences and assembly contigs with BWA-MEM. arXiv Prepr arXiv1303.3997.

Li, H., Durbin, R., 2009. Fast and accurate short read alignment with Burrows—Wheeler transform. Bioinformatics 25, 1754–1760.

Li, H., Ruan, J., Durbin, R., 2008. Mapping short DNA sequencing reads and calling variants using mapping quality scores. Genome Res. 18, 1851–1858.

Li, H., Handsaker, B., Wysoker, A., Fennell, T., Ruan, J., Homer, N., et al., 2009. The sequence alignment/map format and SAMtools. Bioinformatics 25, 2078–2079.

Li, X., Nair, A., Wang, S., Wang, L., 2015. Quality control of RNA-Seq experiments. RNA Bioinformatics. Springer, pp. 137–146.

Li, H., Courtois, E.T., Sengupta, D., Tan, Y., Chen, K.H., Goh, J.J.L., et al., 2017. Reference component analysis of single-cell transcriptomes elucidates cellular heterogeneity in human colorectal tumors. Nat. Genet. 49, 708–718. Available from: http://www.nature.com/doifinder/10.1038/ng.3818.

Lin, P., Troup, M., JWK, H., 2017. CIDR: ultrafast and accurate clustering through imputation for single-cell RNA-seq data. Genome Biol. 18, 59. Available from: http://genomebiology.biomedcentral.com/articles/10.1186/s13059-017-1188-0.

Lipman, D.J., Pearson, W.R., 1985. Rapid and sensitive protein similarity searches. Science 227, 1435–1441. (80-).

Love, M.I., Huber, W., Anders, S., 2014. Moderated estimation of fold change and dispersion for RNA-seq data with DESeq2. Genome Biol. 15, 550. Available from: http://www.ncbi.nlm.nih.gov/pubmed/25516281.

Majewski, J., Pastinen, T., 2011. The study of eQTL variations by RNA-seq: from SNPs to phenotypes. Trends Genet. 27, 72–79. Available from: http://www.ncbi.nlm.nih.gov/pubmed/21122937.

Manichaikul, A., Mychaleckyj, J.C., Rich, S.S., Daly, K., Sale, M., Chen, W.-M., 2010. Robust relationship inference in genome-wide association studies. Bioinformatics 26, 2867–2873. Available from: http://www.ncbi.nlm.nih.gov/pubmed/20926424.

Marth, G.T., Korf, I., Yandell, M.D., Yeh, R.T., Gu, Z., Zakeri, H., et al., 1999. A general approach to single-nucleotide polymorphism discovery. Nat. Genet. 23, 452–456.

McKenna, A., Hanna, M., Banks, E., Sivachenko, A., Cibulskis, K., Kernytsky, A., et al., 2010. The genome analysis Toolkit: a MapReduce framework for analyzing next-generation DNA sequencing data. Genome Res. 20, 1297–1303. Available from: http://genome.cshlp.org/content/20/9/1297.

Meldrum, C., Doyle, M.A., Tothill, R.W., 2011. Next-generation sequencing for cancer diagnostics: a practical perspective. Clin. Biochem. Rev. 32, 177–195. Available from: http://www.ncbi.nlm.nih.gov/pubmed/22147957.

Menon, V., 2017. Clustering single cells: a review of approaches on high-and low-depth single-cell RNA-seq data. Brief. Funct. Genomics. Available from: http://www.ncbi.nlm.nih.gov/pubmed/29236955.

Mercier, E., Droit, A., Li, L., Robertson, G., Zhang, X., Gottardo, R., 2011. An integrated pipeline for the genome-wide analysis of transcription factor binding sites from ChIP-Seq. Xu Y, PLoS One. 6, e16432. Available from: http://dx.plos.org/10.1371/journal.pone.0016432.

Mi, H., Muruganujan, A., Casagrande, J.T., Thomas, P.D., 2013. Large-scale gene function analysis with the PANTHER classification system. Nat. Protoc. 8, 1551–1566. https://doi.org/10.1038/nprot.2013.092.

Morley, M., Molony, C.M., Weber, T.M., Devlin, J.L., Ewens, K.G., Spielman, R.S., et al., 2004. Genetic analysis of genome-wide variation in human gene expression. Nature 430, 743–747. Available from: http://www.ncbi.nlm.nih.gov/pubmed/15269782.

Mortazavi, A., Williams, B.A., McCue, K., Schaeffer, L., Wold, B., 2008. Mapping and quantifying mammalian transcriptomes by RNA-Seq. Nat. Methods 5, 621–628.

NCBI Resource Coordinators, 2017. Database resources of the National Center for Biotechnology Information. Nucleic Acids Res. 45, D12–D17. Available from: http://www.ncbi.nlm.nih.gov/pubmed/27899561.

Needleman, S.B., Wunsch, C.D., 1970. A general method applicable to the search for similarities in the amino acid sequence of two proteins. J. Mol. Biol. 48, 443–453.

NovoAlign, 2018. NovoAlign|Novocraft. Available from: http://www.novocraft.com/products/novoalign/.

Ott, A., Liu, S., Schnable, J.C., Yeh, C.-T., Wang, K.-S., Schnable, P.S., 2017. tGBS® genotyping-by-sequencing enables reliable genotyping of heterozygous loci. Nucleic Acids Res. 45, e178. Available from: http://academic.oup.com/nar/article/45/21/e178/4210942.

Otto, C., Stadler, P.F., Hoffmann, S., 2014. Lacking alignments? The next-generation sequencing mapper segemehl revisited. Bioinformatics 30, 1837–1843.

Pastinen, T., 2010. Genome-wide allele-specific analysis: insights into regulatory variation. Nat. Rev. Genet. 11, 533–538. Available from: http://www.ncbi.nlm.nih.gov/pubmed/20567245.

Pierson, E., Yau, C., 2015. ZIFA: dimensionality reduction for zero-inflated single-cell gene expression analysis. Genome Biol. 16, 241. Available from: http://www.ncbi.nlm.nih.gov/pubmed/26527291.

Qin, Q., Mei, S., Wu, Q., Sun, H., Li, L., Taing, L., et al., 2016. ChiLin: a comprehensive ChIP-seq and DNase-seq quality control and analysis pipeline. BMC Bioinf. 17, 404. Available from: http://www.ncbi.nlm.nih.gov/pubmed/27716038.

R Core Team, 2013. R: A Language and Environment for Statistical Computing. R Core Team, Vienna. Available from: http://www.r-project.org/.

Ritchie, M.E., Phipson, B., Wu, D., Hu, Y., Law, C.W., Shi, W., et al., 2015. Limma powers differential expression analyses for RNA-sequencing and microarray studies. Nucleic Acids Res. 43, e47. Available from: http://www.ncbi.nlm.nih.gov/pubmed/25605792.

Roberts, A., Pimentel, H., Trapnell, C., Pachter, L., 2011. Identification of novel transcripts in annotated genomes using RNA-Seq. Bioinformatics 27, 2325–2329.

Robertson, G., Schein, J., Chiu, R., Corbett, R., Field, M., Jackman, S.D., et al., 2010. De novo assembly and analysis of RNA-seq data. Nat. Methods 7, 909–912.

Sá, A.C.C., Sadee, W., Johnson, J.A., 2017. Whole transcriptome profiling: an RNA-Seq primer and implications for pharmacogenomics research. Clin. Transl. Sci. Available from: http://www.ncbi.nlm.nih.gov/pubmed/28945944.

Schneeberger, K., Hagmann, J., Ossowski, S., Warthmann, N., Gesing, S., Kohlbacher, O., et al., 2009. Simultaneous alignment of short reads against multiple genomes. Genome Biol. 10, R98.

Shendure, J., Balasubramanian, S., Church, G.M., Gilbert, W., Rogers, J., Schloss, J.A., et al., 2017. DNA sequencing at 40: past, present and future. Nature 550, 345–353. Available from: http://www.nature.com/doifinder/10.1038/nature24286.

Sims, D., Sudbery, I., Ilott, N.E., Heger, A., Ponting, C.P., 2014. Sequencing depth and coverage: key considerations in genomic analyses. Nat. Rev. Genet. 15, 121–132. Available from: http://www.nature.com/doifinder/10.1038/nrg3642.

Singh, B., Alonso, J.L.L.T., Tatlow, P., Piccolo, S.R., Eyras, E., 2017. Combined analysis of genome sequencing and RNA-motifs reveals novel damaging non-coding mutations in human tumors. bioRxiv. 200188. Available from: https://www.biorxiv.org/content/early/2017/10/09/200188?rss=1&utm_source=dlvr.it&utm_medium=twitter.

Smith, T.F., Waterman, M.S., 1981. Identification of common molecular subsequences. J. Mol. Biol. 147, 195–197.

Smyth, G.K., 2005. Limma: linear models for microarray data. In: Bioinformatics and Computational Biology Solutions Using R Bioconductor. Springer, New York, pp. 397–420.

Soon, W.W., Hariharan, M., Snyder, M.P., 2013. High-throughput sequencing for biology and medicine. Mol. Syst. Biol. 9, 640. Available from: http://www.ncbi.nlm.nih.gov/pubmed/23340846.

Spies, D., Renz, P.F., Beyer, T.A., Ciaudo, C., 2017. Comparative analysis of differential gene expression tools for RNA sequencing time course data. Brief. Bioinform. Available from: http://academic.oup.com/bib/article/doi/10.1093/bib/bbx115/4364840/Comparative-analysis-of-differential-gene.

Steinhauser, S., Kurzawa, N., Eils, R., Herrmann, C., 2016. A comprehensive comparison of tools for differential ChIP-seq analysis. Brief. Bioinform. 17, bbv110. Available from: https://academic.oup.com/bib/article-lookup/doi/10.1093/bib/bbv110.

Stephens, Z.D., Lee, S.Y., Faghri, F., Campbell, R.H., Zhai, C., Efron, M.J., et al., 2015. Big data: astronomical or genomical? PLoS Biol. 13, e1002195. Available from: http://journals.plos.org/plosbiology/article?id=10.1371/journal.pbio.1002195.

Sudmant, P.H., Rausch, T., Gardner, E.J., Handsaker, R.E., Abyzov, A., Huddleston, J., et al., 2015. An integrated map of structural variation in 2504 human genomes. Nature 526, 75–81.

Tan, R., Wang, Y., Kleinstein, S.E., Liu, Y., Zhu, X., Guo, H., et al., 2014. An evaluation of copy number variation detection tools from whole-exome sequencing data. Hum. Mutat. 35, 899–907. Available from: http://doi.wiley.com/10.1002/humu.22537.

Tan, R., Wang, J., Wu, X., Juan, L., Zheng, L., Ma, R., et al., 2018. ERDS-exome: a hybrid approach for copy number variant detection from whole-exome sequencing data. IEEE/ACM Trans. Comput. Biol. Bioinform. 1–1. Available from: http://www.ncbi.nlm.nih.gov/pubmed/28981421.

Tang, M., 2018. ChIP-seq Analysis Notes From Tommy Tang. Available from: https://github.com/crazyhottommy/ChIP-seq-analysis.

Tang, H., Nzabarushimana, E., 2017. STRScan: targeted profiling of short tandem repeats in whole-genome sequencing data. BMC Bioinf. 18, 398. Available from: http://www.ncbi.nlm.nih.gov/pubmed/28984185.

Teer, J.K., Bonnycastle, L.L., Chines, P.S., Hansen, N.F., Aoyama, N., Swift, A.J., et al., 2010. Systematic comparison of three genomic enrichment methods for massively parallel DNA sequencing. Genome Res. 20, 1420–1431.

Thomas-Chollier, M., Herrmann, C., Defrance, M., Sand, O., Thieffry, D., van Helden, J., 2012. RSAT peak-motifs: motif analysis in full-size ChIP-seq datasets. Nucleic Acids Res. 40, e31. Available from: http://www.ncbi.nlm.nih.gov/pubmed/22156162.

Thornton, T., Tang, H., Hoffmann, T.J., Ochs-Balcom, H.M., Caan, B.J., Risch, N., 2012. Estimating kinship in admixed populations. Am. J. Hum. Genet. 91, 122–138. Available from: http://www.ncbi.nlm.nih.gov/pubmed/22748210.

Thorvaldsdóttir, H., Robinson, J.T., Mesirov, J.P., 2013. Integrative genomics viewer (IGV): high-performance genomics data visualization and exploration. Brief. Bioinform. 14, 178–192. Available from: http://bib.oxfordjournals.org/content/14/2/178.full?keytype=ref&%2520ijkey=qTgjFwbRBAzRZWC.

Thurman, R.E., Rynes, E., Humbert, R., Vierstra, J., Maurano, M.T., Haugen, E., et al., 2012. The accessible chromatin landscape of the human genome. Nature 489, 75–82. Available from: http://www.ncbi.nlm.nih.gov/pubmed/22955617.

Trapnell, C., Pachter, L., Salzberg, S.L., 2009. TopHat: discovering splice junctions with RNA-Seq. Bioinformatics 25, 1105–1111.

Trapnell, C., Williams, B.A., Pertea, G., Mortazavi, A., Kwan, G., van Baren, M.J., et al., 2010. Transcript assembly and quantification by RNA-Seq reveals unannotated transcripts and isoform switching during cell differentiation. Nat. Biotechnol. 28, 511–515. Available from: http://www.ncbi.nlm.nih.gov/pubmed/20436464.

Trapnell, C., Roberts, A., Goff, L., Pertea, G., Kim, D., Kelley, D.R., et al., 2012. Differential gene and transcript expression analysis of RNA-seq experiments with TopHat and Cufflinks. Nat. Protoc. 7, 562–578. Available from: https://doi.org/10.1038/nprot.2012.016.

Vallejos, C.A., Risso, D., Scialdone, A., Dudoit, S., Marioni, J.C., 2017. Normalizing single-cell RNA sequencing data: challenges and opportunities. Nat. Methods 14, 565–571. Available from: http://www.ncbi.nlm.nih.gov/pubmed/28504683.

van Dam, S., Võsa, U., van der Graaf, A., Franke, L., de Magalhães, J.P., 2017. Gene co-expression analysis for functional classification and gene–disease predictions. Brief. Bioinform. bbw139, https://doi.org/10.1093/bib/bbw139.

van der Maaten, L., Hinton, G., 2008. Visualizing data using t-SNE. J. Mach. Learn. Res. 9, 2579–2605.

van Dijk, E.L., Jaszczyszyn, Y., Thermes, C., 2014. Library preparation methods for next-generation sequencing: tone down the bias. Exp. Cell Res. 322, 12–20. Available from: http://www.ncbi.nlm.nih.gov/pubmed/24440557.

VCF, 2017. VCF | 1000 Genomes. Available from: http://www.internationalgenome.org/category/vcf/.

Walter, K., Min, J.L., Huang, J., Crooks, L., Memari, Y., McCarthy, S., et al., 2015. The UK10K project identifies rare variants in health and disease. Nature 526, 82–90. Available from: http://www.ncbi.nlm.nih.gov/pubmed/26367797.

Wan, Y., Pihl, T., 2017. Mutation annotation format (MAF) Specification—TCGA—National Cancer Institute—Confluence Wiki. Available from: https://wiki.nci.nih.gov/display/TCGA/Mutation+Annotation+Format+%28MAF%29+Specification.

Wang, D., Gu, J., 2017. VASC: dimension reduction and visualization of single cell RNA sequencing data by deep variational autoencoder. bioRxiv. 199315. Available from: https://www.biorxiv.org/content/early/2017/10/06/199315.

Wang, K., Singh, D., Zeng, Z., Coleman, S.J., Huang, Y., Savich, G.L., et al., 2010. MapSplice: accurate mapping of RNA-seq reads for splice junction discovery. Nucleic Acids Res. 38, e178.

Wang, S., Meyer, E., McKay, J.K., Matz, M.V., 2012. 2b-RAD: a simple and flexible method for genome-wide genotyping. Nat. Methods 9, 808–810.

Wang B, Ramazzotti D, De Sano L, Zhu J, Pierson E, Batzoglou S. SIMLR: a tool for large-scale single-cell analysis by multi-kernel learning. 2017; Available from: http://arxiv.org/abs/1703.07844

Warde-Farley, D., Donaldson, S.L., Comes, O., Zuberi, K., Badrawi, R., Chao, P., et al., 2010. The GeneMANIA prediction server: biological network integration for gene prioritization and predicting gene function. Nucleic Acids Res. 38, W214–W220. Available from: http://www.ncbi.nlm.nih.gov/pubmed/20576703.

Weinstein, J.N., Collisson, E.A., Mills, G.B., Shaw, K.R.M., Ozenberger, B.A., Ellrott, K., et al., 2013. The cancer genome atlas pan-cancer analysis project. Nat. Genet. 45, 1113–1120.

Weischenfeldt, J., Symmons, O., Spitz, F., Korbel, J.O., 2013. Phenotypic impact of genomic structural variation: insights from and for human disease. Nat. Rev. Genet. 14, 125–138. Available from: http://www.ncbi.nlm.nih.gov/pubmed/23329113.

Wilhelm-Benartzi, C.S., Koestler, D.C., Karagas, M.R., Flanagan, J.M., Christensen, B.C., Kelsey, K.T., et al., 2013. Review of processing and analysis methods for DNA methylation array data. Br. J. Cancer 109, 1394–1402. Available from: http://www.nature.com/doifinder/10.1038/bjc.2013.496.

Wilkinson, M.D., Dumontier, M., Aalbersberg, I.J., Appleton, G., Axton, M., Baak, A., et al., 2016. The FAIR guiding principles for scientific data management and stewardship. Sci. Data. 3, 160018. Available from: http://www.nature.com/articles/sdata201618.

Witten, D.M., 2011. Classification and clustering of sequencing data using a Poisson model. Ann. Appl. Stat. 5, 2493–2518.

Wolfe, D., Dudek, S., Ritchie, M.D., Pendergrass, S.A., 2013. Visualizing genomic information across chromosomes with PhenoGram. BioData Min. 6, 18. Available from: http://www. biodatamining.org/content/6/1/18.

Yan, H., Evans, J., Kalmbach, M., Moore, R., Middha, S., Luban, S., et al., 2014. HiChIP: a high-throughput pipeline for integrative analysis of ChIP-Seq data. BMC Bioinf. 15, 280. Available from: http://bmcbioinformatics.biomedcentral.com/articles/10.1186/1471-2105-15-280.

Yang, J., Benyamin, B., McEvoy, B.P., Gordon, S., Henders, A.K., Nyholt, D.R., et al., 2010. Common SNPs explain a large proportion of the heritability for human height. Nat. Genet. 42, 565–569. Available from: http://www.ncbi.nlm.nih.gov/pubmed/20562875.

Yang, L., Liu, J., Lu, Q., Riggs, A.D., Wu, X., 2017. SAIC: an iterative clustering approach for analysis of single cell RNA-seq data. BMC Genomics 18, 689. Available from: http://www.ncbi.nlm.nih. gov/pubmed/28984204.

Ye, K., Schulz, M.H., Long, Q., Apweiler, R., Ning, Z., 2009. Pindel: a pattern growth approach to detect break points of large deletions and medium sized insertions from paired-end short reads. Bioinformatics 25, 2865–2871.

Ye, T., Krebs, A.R., Choukrallah, M.-A., Keime, C., Plewniak, F., Davidson, I., et al., 2011. seqMINER: an integrated ChIP-seq data interpretation platform. Nucleic Acids Res. 39, e35. Available from: http://www.ncbi.nlm.nih.gov/pubmed/21177645.

Yu, G., Wang, L.-G., He, Q.-Y., 2015. ChIPseeker: an R/Bioconductor package for ChIP peak annotation, comparison and visualization. Bioinformatics 31, 2382–2383. Available from: http:// www.ncbi.nlm.nih.gov/pubmed/25765347.

Zafar, H., Tzen, A., Navin, N., Chen, K., Nakhleh, L., 2017. SiFit: inferring tumor trees from single-cell sequencing data under finite-sites models. Genome Biol. 18, 178. Available from: http:// www.ncbi.nlm.nih.gov/pubmed/28927434.

Zang, C., Schones, D.E., Zeng, C., Cui, K., Zhao, K., Peng, W., 2009. A clustering approach for identification of enriched domains from histone modification ChIP-Seq data. Bioinformatics 25, 1952–1958. Available from: http://www.ncbi.nlm.nih.gov/pubmed/19505939.

Zararsiz, G., Goksuluk, D., Klaus, B., Korkmaz, S., Eldem, V., Karabulut, E., et al., 2017. voomDDA: discovery of diagnostic biomarkers and classification of RNA-seq data. PeerJ. 5, e3890. Available from: http://www.ncbi.nlm.nih.gov/pubmed/29018623.

Zeitouni, B., Boeva, V., Janoueix-Lerosey, I., Loeillet, S., Legoix-né, P., Nicolas, A., et al., 2010. SVDetect: a tool to identify genomic structural variations from paired-end and mate-pair sequencing data. Bioinformatics 26, 1895–1896. Available from: http://www.ncbi.nlm.nih.gov/pubmed/ 20639544.

Zhang, Y., Liu, T., Meyer, C.A., Eeckhoute, J., Johnson, D.S., Bernstein, B.E., et al., 2008. Model-based analysis of ChIP-Seq (MACS). Genome Biol. 9, R137. Available from: http://www. ncbi.nlm.nih.gov/pubmed/18798982.

Zhang, Q., Liu, W., Liu, C., Lin, S.-Y., Guo, A.-Y., 2017. SEGtool: a specifically expressed gene detection tool and applications in human tissue and single-cell sequencing data. Brief. Bioinform.. . Available from: https://academic.oup.com/bib/article-lookup/doi/10.1093/bib/bbx074.

Zhu, Y., Stephens, R.M., Meltzer, P.S., Davis, S.R., 2013. SRAdb: query and use public next-generation sequencing data from within R. BMC Bioinf. 14, 19. https://doi.org/ 10.1186/1471-2105-14-19.

CHAPTER 5

Proteomics and Metabolomics Data Analysis for Translational Medicine

Theodora Katsila*, George P. Patrinos*,†,‡

*Department of Pharmacy, School of Health Sciences, University of Patras, Patras, Greece,
†Department of Pathology—Bioinformatics Unit, Faculty of Medicine and Health Sciences,
Erasmus University Medical Center, Rotterdam, The Netherlands, ‡Department of Pathology,
College of medicine and Health Sciences, United Arab Emirates University, Al-Ain,
United Arab Emirates

5.1 INTRODUCTION

Tremendous advances in our understanding of the molecular basis of diseases, computational power, and data analytics hold promise towards tailored-made theranostics. Yet, there are still substantial gaps, when the development, validation, and application of cost-effective health strategies are considered.

To meet such challenges, the omics technologies serve as an advantageous toolbox. Despite the cost of equipment and the need for hands-on experts, omics technologies provide unprecedented high-throughput sample analysis (minute sample quantities) coupled to sophisticated data analysis. Notably, omics technologies enable a holistic study of health and disease statuses, which allow for a systems-level understanding even among individuals. In particular, when emphasis is given to translational medicine, such a systems-level understanding of interindividual variability becomes fundamental.

Aiming to implement precision medicine in the clinic, it became apparent that genotype-to-(clinical)phenotype associations need to be defined and, then, aligned with dosing guidelines and decision-making. To address the impact of environmental factors (such as life style choices, microbiome, diet), there is current interest in proteomics and metabolomics approaches, either as stand-alone or coupled to other omics strategies.

In this chapter, key features of proteomics and metabolomics will be summarized with an emphasis on data analysis. Specific examples will be introduced to comprehend current opportunities and challenges in translational medicine.

Human Genome Informatics. https://doi.org/10.1016/B978-0-12-809414-3.00005-X

5.2   THE NEED TO BRIDGE THE GAPS IN THE ERA OF PRECISION MEDICINE

When the principles of Mendelian genetics were incorporated into medicine, health professionals and scientists interpreted several disease states as monogenic and, by default, a specific clinical phenotype was attributed to every presenting patient showing symptoms of the monogenic disease in question. Indeed, the linkage of such monogenic diseases to their genetic alteration became routine; to name a few, the ΔF508 (*CFTR*) mutation is the most common pathogenic genomic variant leading to cystic fibrosis (Marson et al., 2017), hemoglobin-S is associated with sickle cell disease (Taher et al., 2017), α- and β-thalassemia are caused by mutations in the α- and β-globin gene locus, respectively (Taher et al., 2017), and a "fragile" site at the end of the long arm of the X-chromosome is responsible for the fragile X-syndrome (https://www.genome.gov/10001204/specific-genetic-disorders/) (Mila et al., 2018; Chokoshvili et al., 2018; Treff and Zimmerman, 2017). Interestingly enough, evidence from mRNA analysis and whole genome sequencing has indicated that pathogenic mutations can occur deep within the introns of over 75 disease-associated genes (Vaz-Drago et al., 2017), while monogenic diseases may also result in a wide range of clinical phenotypes (hemoglobinopathies serve as an example), for which population differences may also account (Papachatzopoulou et al., 2010). Such disease/clinical phenotype complex associations, together with the recognition of inherent uniqueness among the human population, empowered collaborative efforts and information technologies (Lederer et al., 2009; Giardine et al., 2011; Manolio et al., 2015; Weitzel et al., 2016; Rasmussen-Torvik et al., 2017; Sperber et al., 2017) and showed that the One Disease One Treatment paradigm needs to be enriched by inter-individual variability. Individual clinical outcomes range from full treatment efficacy to little or no benefit to severe toxicity and, for this, optimum disease management and patient stratification are of importance.

Human Genome Project paved the way to genomic medicine and the abundance of genetic variation profiling in disease susceptibility and treatment efficacy and/or toxicity in various populations (Consortium, 2005). The advent of technology has decreased the cost of sequencing analysis for the human genome from the $3 billion required by the Human Genome Project to just $1500 (https://www.genome.gov/27565109/the-cost-of-sequencing-a-human-genome/), allowing the implementation of large-scale genomic projects worldwide (Consortium, 2015; Sudmant et al., 2015). In 2008, the Genetic Information Nondiscrimination Act (GINA) was passed to protect against discrimination, following the use of genetic data (Honey, 2008) to ease the implementation of genomic medicine into the clinic. Yet, the Human Genome Project revealed about ~21,000 protein coding genes (~3% of the genome), whereas 97% of the genome needs to be further explored, while numerous molecular interactions have been revealed at the genetic, epigenetic, and protein level (Pennisi, 2012).

Currently, genomic information cannot account for interindividual variability on its own or predict clinical outcomes (Manolio et al., 2017), especially if the full complexity of cellular (patho)physiology and environmental influences are to be taken into account. Proteomics (and metabolomics) is a valuable asset. The proteome (and the metabolome) are rather dynamic and diverse themselves, able to adapt not only to uniqueness per individual, but also per health/disease states. Indicatively, neither the proteome or the metabolome of an individual are the same before, after, or during a disease state or in the presence of xenobiotics and the analytical technology has evolved as much to allow the analysis of a single biomolecule (hypothesis-driven targeted analysis) or the concomitant analysis of a plethora of diverse analytes (nontargeted analysis, hypothesis-generating) by nuclear magnetic resonance (NMR) spectroscopy and/or mass spectrometry (Tolstikov, 2016; Zhong et al., 2017). Additionally, protein/peptide arrays have enriched further the diversity of analytes in question, with an emphasis on the interactome and kinome (Gupta et al., 2016), even though only targeted analysis is feasible. The same holds for flow cytometry, which has been also widely used, and several limitations of spectral separation have been partially overcome by mass cytometer (CyTOF) (Bandura et al., 2009). Table 5.1 summarizes all available proteomic and metabolomic platforms that are available today and complement other omics technologies.

In terms of data analytics, protein/peptide arrays and flow cytometry are accompanied by predefined and well-established modes of analysis that largely depends on the marker or subset of markers used. This is the exact reason why both methodologies can only perform hypothesis-driven targeted analysis. On the contrary, NMR and mass spectrometry may generate enormously large and highly diverse datasets, following both hypothesis-driven and hypothesis-generating analyses. The charming fact that a single platform can reveal numerous datasets asks for our attention in terms of data accuracy, precision, reproducibility, robustness as well as data management and storage. The era of next-generation proteomics and metabolomics has begun.

5.3 CLINICAL PROTEOMICS

5.3.1 The Power of the Proteome

Proteomics enable the identification of differential protein expression, post-translational modifications, protein-protein interactions, cellular and subcellular distribution as well as temporal patterns of expression. Thus, differential and functional proteomics aims to inform and empower the understanding

Table 5.1 Technology Platforms for Clinical Proteomics and Metabolomics

	NMR[a]	Mass Spectrometry[a]	Arrays[b]	Flow Cytometry[c]
Protein alterations	Only if peptide/protein sequence is affected	Only if peptide/protein sequence is affected	Only if peptide/protein sequence is affected	Feasible for both peptide/protein sequence and structural alterations, if recognized by an antibody
Posttranslational modifications	Yes	Yes	Only for predefined modifications	If recognized by antibody
Differential expression	Yes	Yes	Only for predefined alterations	Yes
Targeted analysis	Yes	Yes	Yes	Yes
Nontargeted analysis	Yes	Yes	No	No
Analytes per analytical run/test	Thousands	Thousands	Thousands[d]	15/40[e]
Serial sample analysis	Yes	Yes	Yes	Yes
Monitoring	Yes	Yes	Potential	If refers to a cellular state
Pharmacokinetics	Yes	Yes	Potential	Yes
Pharmacodynamics	Yes	Yes	Potential	Yes

[a]Including proteomic, metabolomic, and lipidomic assays and their derivatives.
[b]Protein, aptamer, antibody, and peptide arrays.
[c]Flow cytometry, CyTOF, and imaging in flow assays.
[d]Only, if predefined.
[e]CyTOF allows the analysis of up to 40 analytes.

of molecular pathways and networks and their interrelationships in cells and living organisms. Today, the proteomics field explores its power and applicability, serving health professionals for high-throughput and fidelity theranostics. For this, clinical proteomics is considered as a valuable asset towards optimum disease management and patient stratification (Lehmann et al., 2017), especially if technological advances and their fine tuning are considered. Today, fast and consistent protein identification is achieved with corresponding increases in the dynamic range of proteins that may be detected in the proteome of interest (Chugh et al., 2012).

Historically, one- and two-dimensional gel electrophoresis and high-performance liquid chromatography have been the most common and widely used separation methods for protein identification in complex matrices. Currently, mass spectrometry is considered the gold standard method. For mass spectrometry-based approaches, several options are available for every step of a proteomic workflow (target enrichment, ionization, detection, and quantitation), each one optimized and validated and, yet, ready to be customized, optimized, and validated once again, if needed. Same is for NMR-based approaches. Samples may be derived from tissue, plasma, serum or urine, cells, or even cell compartments and, for this, sample depletion or enrichment may occur prior to analysis. As technology advances further, sample preprocessing may become obsolete, reducing the time to analysis without compromising sensitivity and accuracy. For mass spectrometry-based approaches, even though several methodologies have been developed for clinical proteomics (Scherl, 2015), Matrix Assisted Laser Desorption Ionization-Time of Flight mass spectrometry (MALDI-TOF MS) and liquid chromatography electrospray ionization tandem mass spectrometry (LC-ESI-MS/MS) have survived translational medicine demands. Notably, MALDI imaging mass spectrometry allows direct imaging of protein expression in normal and disease tissues (Cornett et al., 2007). Even though immunoassays are the traditional way to measure protein biomarkers, at present targeted proteomics allows faster assay development, multiplexing capabilities, and analytical specificity and, thus, is widely used to test subsets of candidate biomarkers prior to clinical validation studies (Rifai et al., 2006; Cima et al., 2011; Surinova et al., 2015; Schubert et al., 2017). Recently, mass spectrometric immunoassays claim to combine the selectivity of targeted immunoassays with the specificity of mass spectrometric detection towards clinical and population proteomics (Nedelkov et al., 2018). Although still at its infancy, protein chip technology is emerging, allowing for the analysis of protein-protein, protein-DNA, or protein-RNA interactions, depending on the substrate cross-linked to the chip. Nevertheless, proteins are highly heterogeneous and, thus, only peptide/protein arrays for the analysis of specific analytes or groups of analytes have been developed. While the arraying technology is improving, traditional immunohistochemical detection of protein expression in tissue sections can now be adapted to a high-throughput array format (Verrills, 2006).

Recently, the human (clinical) proteome has been defined (Kim et al., 2014; Wilhelm et al., 2014; Uhlén et al., 2015), being the basis for comparative clinical proteomics and sharing the vision that ultimately health professionals will be able to interpret the clinical proteome of a patient based on archived records (Lindskog, 2015). This scenario is anticipated to empower cost-effective theranostics. Indeed, from a therapeutics perspective, the majority of drug targets are proteins and not nucleic acids (Verrills, 2006). Taking into account that the power of proteomics lies in discovering the meaningful unknown within a massive repertoire of unrelated molecules with high mass accuracy and sensitivity, clinical proteomics are of outmost importance towards (i) the diagnosis of rare diseases or those of unknown etiology, (ii) treatment efficacy/toxicity monitoring, and (iii) patient stratification. Against traditional biomarker analysis, which focuses on the identification of one marker of a particular disease, there is general agreement of the statistical argument that a subset of independent disease-related proteins considered in an aggregate should be less prone to genetic and environmental "noise," when compared to the level of a single marker protein (Anderson, 2005). No doubt, proteomics has the power to identify such protein subsets in a high-throughput manner.

5.3.2   Steps Prior to Routine Proteome Analyses

Clinical proteomics still suffers from technological complexity of the analyses involved as well as lack of standardization. Thus, current applications of clinical proteomics include the screening of specific subsets of protein biomarkers for certain diseases (to name a few, cancer and cardiovascular disorders), rather than large-scale full protein profiling.

In 2001, the international Human Proteome Organization (HUPO) (http://www.hupo.org) was launched to (i) consolidate national and regional proteome organizations into a worldwide organization, (ii) disseminate knowledge on proteomics approaches and human proteome and/or model organisms by engaging in scientific and educational activities, and (iii) coordinate public proteome initiatives towards the characterization of specific tissue and cell proteomes (Hanash, 2004). The Proteomics Standards Initiative by HUPO undertakes the development of standards for data generation and analysis, including standardized formats for proteomic databases and measurements, especially if proteome analyses become routine use in the clinical setting (Verrills, 2006). To ease standardization and ensure data reliability, proteomic (raw) data are being publicly available (data repositories) and, hence, analyses are validated by independent users and several technological platforms (Domon and Aebersold, 2006). Indeed, this has been a unique initiative towards data sharing, turning information growth to knowledge growth, not only in terms of translational medicine, but also technology advances.

The implementation of clinical proteomics depends upon the meaningful translation of data to clinical outcomes. Translation of proteomics-based biomarker discovery research into the clinic has been limited by failures in the validation phase in testing large patient cohorts. Indeed, proteomics analyses of large patient cohorts are still prohibitively time-consuming and expensive, especially if many analytes are to be quantified consistently across the cohort. Mertins and coworkers as well as Zhang and coworkers took the risk resulting in several months of instrument time (Mertins et al., 2016; Zhang et al., 2016). To obtain large-scale yet cost-effective clinical proteomics strategies, high-throughput analysis needs to be coupled to robustness, repeatability, and sensitivity. Liu and coworkers introduced DIA/SWATH-type approaches, which may be considered as promising alternatives (Liu et al., 2015). Furthermore, stringent filters should be employed during the discovery phase to allow for data quality and validation in preclinical studies (Scherl, 2015; Lehmann et al., 2017). Notably, merely because it is possible to generate a large amount of hypersensitive data with mass accuracy to four decimal places from a tiny amount of starting material does not necessarily imply their clinical utility (Duarte and Spencer, 2016). To determine whether the protein in question is, in fact, clinically useful: (i) protein measurements should be easy to perform at a reasonable cost, (ii) increased protein levels would provide such knowledge that is not present when the protein is absent, and (iii) knowledge should guide medical decision-making (Morrow and de Lemos, 2007).

5.4 CLINICAL METABOLOMICS

5.4.1 The Advances and Promises of Metabolome

Clinical metabolomics appears to become an essential tool in translational medicine (Tolstikov, 2016) and, hence, metabolomics-based approaches accompany modern clinical studies, shedding light on disease pathobiology and allowing novel theranostics (Everett, 2015; Rankin et al., 2016). Clinical metabolomics has been also incorporated into the pipeline of modern pharmaceutical industry as an independent assay in toxicology and disease pathology as well as in drug discovery and preclinical studies (pharmacokinetic and pharmacodynamic profiling) (Matthews et al., 2016; Wishart, 2016; Kohler et al., 2017). To many, clinical metabolomics also offers a layer of understanding to bridge socioeconomic barriers, and thus, identify the needs among populations empowering therapeutic interventions and outcomes on the basis of metabolic phenotypes (metabotypes) (Everett, 2015; Rankin et al., 2016; Wishart, 2016). Dunn and coworkers have reported key demographic selection and association with metabolites enhancing predictive associations, when healthy UK individuals were considered (Dunn et al., 2015). Moreover, urinary metabolite differential excretion patterns were obtained among East Asian populations and their Western counterparts, Japanese individuals living in Japan or in the United States as well as Chinese participants living in the northern parts of China, when compared to Chinese individuals living in southern China (Kochhar and Martin, 2015). Such findings depict the high complexity of population studies and omics or even multiomics approaches towards clinical and disease phenotype assessment.

Such metabolomics applications are not only to be attributed to technological advances, but also to the creation of sophisticated software and databases, which allow data mapping and interpretation as well the centralization and distribution of clinical data (Wishart, 2010). At present, metabolomic analyses are expanded to understand the systems-level effects of metabolites. The power of metabolomics lies not only on its inherent sensitivity, which allows the detection of subtle alterations in biological networks and health/disease states, but also on common metabolites among species that ease pathway analysis and network construction (Johnson et al., 2016).

5.4.2 Needs Prior to Routine Metabolome Analyses

Today, even though (information) technology resources and tools are growing at a rapid rate, metabolomic analyses remain time-consuming and expensive, while metabolite identification is still a limiting factor. To overcome such hurdles, computational workflows promise the acceleration of data uploading and processing, coupled to automated or in silico metabolite identification. Obviously, biological interpretation remains an unmet need.

For metabolomics analyses, standardization becomes a gigantic issue, taking into account the dynamic nature of the metabolome itself coupled to the obstacles that the field needs to overcome towards metabolite identification. For this, specialization in informatics is highly required. At present, the US National Institutes of Health (NIH) Common Fund Metabolomics Program offers several resources, which enable hands-on and online training in data processing and interpretation (Sud et al., 2015). Same for the Coordination of Standards in Metabolomics (COSMOS), which promotes standardization in metabolomics via experimental and data sharing (Salek et al., 2015).

5.5 COMPUTATIONAL AND CHEMOINFORMATIC TOOLS

For NMR- and mass spectrometry-based analyses, computational and chemoinformatic tools are essential to effectively support experimental data upload, processing, statistical analyses, and analyte identification. Furthermore, data information needs to be translated into biological knowledge, and for this, bioinformatic tools have been coupled to computational and chemoinformatic tools. Such (information) technology resources and tools have been extensively reviewed elsewhere. Herein, we highlight some paradigms with an emphasis on their applicability.

In general, proteomic and metabolomic datasets contain information in the form of numerous atomic nuclei (NMR analyses) or ions (mass spectrometry analyses) that are generated upon the analysis of each sample. Therefore, computational tools are required for reducing redundancy in such datasets of high complexity and identifying the analytes of interest. Truly, computational resources are growing rapidly at present (Johnson et al., 2014). Yet, especially the metabolomics field is missing the means to (i) accelerate data upload and processing, (ii) enable metabolite identification, and (iii) translate such information into a biological context.

According to the current trend in proteomics and metabolomics, there are several tools that are user-friendly to satisfy the needs of the multidisciplinary omics laboratories and, at the same time, have advanced parameters for power/expert users, being an asset for all levels of expertise. Indicative examples are the MZmine2 (Pluskal et al., 2010), XCMS online (Tautenhahn et al., 2012a), and MetaboAnalyst (Xia et al., 2012) for metabolomics data analysis. Although several tools are available as part of the NMR/mass spectrometry vendor software, free and open-source software tools have been developed and widely employed allowing data sharing and empowering data reproducibility. For proteomics, this is fully evident (Deutsch et al., 2008).

In proteomics and metabolomics fields, greater sample sizes increase further data validity. Yet, greater sample sizes increase the complexity of the data handling process (collection, validation, and interpretation), while the issue of data storage remains a challenge (Taylor et al., 2007; Rosenling et al., 2011).

5.6 STRATEGIES TO ADDRESS DATA COMPLEXITY

Especially for NMR- and/or mass spectrometry-based metabolomics, high-throughput qualitative and quantitative information are delivered (Bingol and Brüschweiler, 2017), aiming to explore the high complexity of population studies and omics or even multiomics approaches towards clinical and disease phenotype assessment. This emerging complexity is at the same time a great opportunity and liability. Such datasets enable the translation of biological outcomes into viable theranostics towards optimum patient stratification and disease management. Yet, coupling proteomic data to metabolomic, transcriptomic, and genomic data can be rather challenging. Fig. 5.1 depicts such opportunities and challenges.

Today, multiple strategies are employed for such data integration. Several prefer to perform data acquisition and analysis per omics layer, so that the omics data (layer 1) that survive guide subsequent omics data acquisition and analysis (layer 2) and so on. Ji and coworkers explored citalopram/escitalopram treatment biomarkers, following a metabolomics analysis in plasma, according to which (i) glycine was reported to be negatively associated with treatment outcome leading to tag SNP genotyping for genes encoding glycine synthesis and (ii) rs10975641 (GLDC) was defined as a response biomarker in major depressive disorder patients (Ji et al., 2011). In a reverse context, traditional tag SNP genotyping was replaced by genotype imputation to determine genomic variants of interest in pathways identified during pharmacometabolomics studies (Suhre et al., 2011; Abo et al., 2012). Williams and coworkers implemented a transomics approach to analyze genetic and environmental variation in metabolic and mitochondrial phenotypes in a diverse population of BXD mice (Williams et al., 2016). For this, five distinct omics layers were first analyzed and then combined. The authors suggest that such a combined analysis of all layers together provided additional information, which would not be yielded by any single omics approach. Others propose an integrated analysis

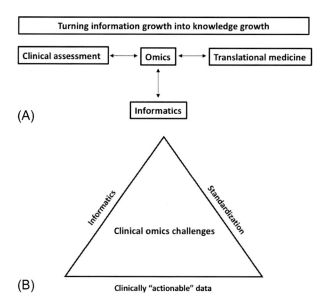

FIG. 5.1

Clinical proteomics and metabolomics approaches hold great promise, if implemented in the clinic, enabling powerful technologies coupled to informatics to serve decision-making in the clinic (A). Yet, several challenges need to be met if information growth turns into knowledge growth, since (i) databases and tools, (ii) automation, (iii) data visualization, and (iv) harmonization and regulatory compliance are still at their infancy (B).

and highlight the complementarity of such a multi-layered (multi-omics) strategy. Balasopoulou and coworkers select two layers of systems-scale molecular measurements, the pharmacometabolome (layer 1) and the virome (layer 2) (Balasopoulou et al., 2016), and multilayered (multiomics) datasets are generated that consist of pharmacometabolomics (layer 1) and viromics (layer 2) data. In this strategy, information technologies provide the means for data analysis, filtering, and argumentation and systems-level dynamic parameters from fewer samples across broad molecular interaction networks.

To name a few information technologies tools, PIUMet is a network-based approach, prize-collecting Steiner forest algorithm for integrative analysis of untargeted metabolomics (Pirhaji et al., 2016). PIUMet performs an integrative analysis of metabolite features to infer molecular pathways and components, without requiring their identification. The Dicode platform and services may facilitate data mining, analysis, collaboration, and decision-making in such diverse data-intensive and cognitively complex settings (Karacapilidis, 2014; Tsiliki et al., 2014). Two of the distinct features of the Dicode platform are: (i) users can customize the Dicode workbench via a proper assembly of tools that suit their needs, so that properly structured data can lead to more informed decisions and (ii) supports the synergy between artificial and human intelligence. We feel that this synergy is of fundamental importance since humans can detect patterns, while computer algorithms may fail to do so, yet data-intensive and cognitively complex processes limit human ability (Agrawal et al., 2012).

If emphasis is given to network construction and data interpretation, both can be based on current data warehouses and databases. There are several well-curated proteomics and metabolomics data repositories and databases (Wishart et al., 2007; Kamath et al., 2011; Paik et al., 2012; Tautenhahn et al., 2012b; Vizcaíno et al., 2012). Recent efforts aim to develop causal networks built with multiomics data with the guidance of Bayesian artificial intelligence (Boccard and Rudaz, 2014; Narain et al., 2015; Vemulapalli et al., 2016; Tolstikov et al., 2017). When Bayesian networks inference is employed, algorithms may operate on the basis of prior relationships with existing knowledge (metabolic pathways, networks) or in a strictly data-driven and unbiased fashion.

5.7   FROM TRANSLATIONAL MEDICINE DATA TO THERANOSTICS

To shape translational medicine data into companion diagnostics and next, theranostics, multidimensional molecular and clinical data need to be integrated into patient-centric models. Indeed, family and clinical history data are indispensable and such integration will allow for a robust biological narrative, when coupled to proteomics and metabolomics datasets. No doubt, the challenges are many, including adherence to protocol standardization and data accuracy towards predictive outcomes. Clinico-genomic prediction suites and tools can be of use and further enriched to serve multiomics data or serve as models to build new more versatile ones. This paradigm will account for the interplay among the genome and the so-called environment in light of inter- and intraindividual variability (genome*proteome*metabolome). Yet, for patient-centric models to be devoid as much as possible of confounding factors and biases, sample size and population differences need to be addressed. For this, collaborative platforms, argumentation, and data sharing among international networks and consortia become necessary.

RD-Connect (http://rd-connect.eu/about/) is a global infrastructure project that links up databases, registries, biobanks, and clinical bioinformatics data used in rare disease research into a central resource for researchers worldwide. In this initiative, harmonization and development of common standards have been of paramount importance, especially when collaboration with other projects, such as Neuromics (http://rd-neuromics.eu/), EURenOmics (https://

eurenomics.eu/), and IRDiRC (http://www.irdirc.org/), is taken into account. Same for ethical practices to balance patient-related interests associated with rare disease research using databases/registries, biobanks and omics databases, engaging with relevant stakeholders, including patient organizations, clinical and research networks, legislators and policymakers as well as the pharmaceutical industry.

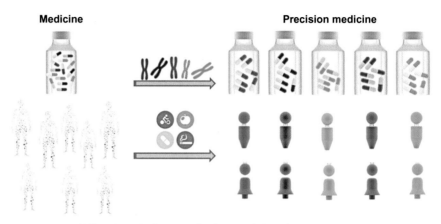

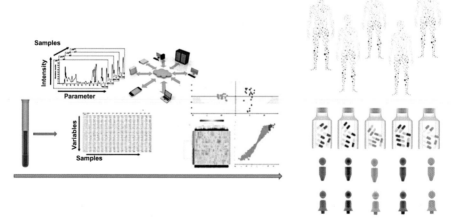

FIG. 5.2

Precision medicine is considered as the net outcome of genomic medicine and environmental influences (diet, polypharmacy, life style). Proteomics and metabolomics shed light on such environmental influences towards biomarker discovery accompanied by several success stories. To translate such knowledge into the clinic, proteomics and metabolomics data coupled to informatics and other omics approaches hold the promise for tailored-made diagnostics and therapeutics, accounting for interindividual variability towards optimum disease management and patient stratification. Data integration, curation, and sharing will ensure data quality and reproducibility against biases.

Similarly, the Implementing GeNomics In pracTicE (IGNITE) network (Weitzel et al., 2016; Sperber et al., 2017) aimed to synthesize data on challenges identified by six diverse projects, as part of a National Human Genome Research Institute (NHGRI)-funded network on the implementation of genomics into practice. The network explored viable strategies to facilitate genomic data integration into existing electronic health records and educate stakeholders about the value of genomic services, as they were defined as prerequisites for the clinical implementation of genomic medicine.

To determine the clinical utility of implementing a subset of pharmacogenomic biomarkers into routine care, the Ubiquitous Pharmacogenomics (U-PGx) Consortium carries out a prospective, block-randomized, controlled clinical study (PREemptive Pharmacogenomic testing for prevention of Adverse drug REactions [PREPARE]), in which preemptive genotyping of a subset of clinically relevant pharmacogenomic biomarkers, for which guidelines are available, will be implemented across healthcare institutions in seven European countries (van der Wouden et al., 2017). Genotyping data will be translated to clinical phenotypes and their impact on patient outcomes and cost-effectiveness will be estimated. Data management and privacy needs will be addressed in the context of pharmacogenomic-clinical patient cards and reports (Fig. 5.2).

5.8 CONCLUSIONS

Proteomics and metabolomics have pioneered our holistic understanding of organisms and their next big challenge is to translate such knowledge into clinically meaningful outcomes. Today, both fields have been flourishing with success stories, which demonstrate their analytical power to address biological questions and delineate the mechanisms that underlie phenotypes. The input of informatics has been fundamental.

Next, proteomics and metabolomics need to overcome their limitations to address the challenge of decision-making in the clinic, towards optimum disease management and tailored-made therapeutics in a cost-effective manner.

Acknowledgments

Most of our own work has been supported by funds from the European Commission grants (GEN2PHEN FP7-200754, RD-Connect FP7-305444 and UPGx H2020-668353) to GPP.

References

Abo, R., Hebbring, S., Ji, Y., Zhu, H., Zeng, Z.-B., Batzler, A., et al., 2012. Merging pharmacometabolomics with pharmacogenomics using "1000 genomes" SNP imputation: selective serotonin reuptake inhibitor response pharmacogenomics. Pharmacogenet. Genomics 22, 247.

Agrawal, D., Bernstein, P., Bertino, E., Davidson, S., Dayal, U., Franklin, M., et al., 2012. Challenges and Opportunities With Big Data—a Community White Paper Developed by Leading Researchers Across the United States.

Anderson, L., 2005. Candidate-based proteomics in the search for biomarkers of cardiovascular disease. J. Physiol. 563, 23–60.

Balasopoulou, A., Patrinos, G.P., Katsila, T., 2016. Pharmacometabolomics informs viromics toward precision medicine. Front. Pharmacol. 7.

Bandura, D.R., Baranov, V.I., Ornatsky, O.I., Antonov, A., Kinach, R., Lou, X., et al., 2009. Mass cytometry: technique for real time single cell multitarget immunoassay based on inductively coupled plasma time-of-flight mass spectrometry. Anal. Chem. 81, 6813–6822.

Bingol, K., Brüschweiler, R., 2017. Knowns and unknowns in metabolomics identified by multidimensional NMR and hybrid MS/NMR methods. Curr. Opin. Biotechnol. 43, 17–24.

Boccard, J., Rudaz, S., 2014. Harnessing the complexity of metabolomic data with chemometrics. J. Chemom. 28, 1–9.

Chokoshvili, D., Vears, D., Borry, P., 2018. Expanded carrier screening for monogenic disorders: where are we now? Prenat. Diagn. 38, 59–66.

Chugh, S., Sharma, P., Kislinger, T., Gramolini, A.O., 2012. Clinical proteomics. Circ. Cardiovasc. Genet. 5, 377.

Cima, I., Schiess, R., Wild, P., Kaelin, M., Schüffler, P., Lange, V., et al., 2011. Cancer genetics-guided discovery of serum biomarker signatures for diagnosis and prognosis of prostate cancer. Proc. Natl. Acad. Sci. 108, 3342–3347.

Consortium, G.P., 2015. A global reference for human genetic variation. Nature 526, 68.

Consortium, I.H., 2005. A haplotype map of the human genome. Nature 437, 1299.

Cornett, D.S., Reyzer, M.L., Chaurand, P., Caprioli, R.M., 2007. MALDI imaging mass spectrometry: molecular snapshots of biochemical systems. Nat. Methods 4, 828.

Deutsch, E.W., Lam, H., Aebersold, R., 2008. Data analysis and bioinformatics tools for tandem mass spectrometry in proteomics. Physiol. Genomics 33, 18–25.

Domon, B., Aebersold, R., 2006. Challenges and opportunities in proteomics data analysis. Mol. Cell. Proteomics 5, 1921–1926.

Duarte, T.T., Spencer, C.T., 2016. Personalized proteomics: the future of precision medicine. Proteomes 4, 29.

Dunn, W.B., Lin, W., Broadhurst, D., Begley, P., Brown, M., Zelena, E., et al., 2015. Molecular phenotyping of a UK population: defining the human serum metabolome. Metabolomics 11, 9–26.

Everett, J.R., 2015. Pharmacometabonomics in humans: a new tool for personalized medicine. Pharmacogenomics 16, 737–754.

Giardine, B., Borg, J., Higgs, D.R., Peterson, K.R., Philipsen, S., Maglott, D., et al., 2011. Systematic documentation and analysis of human genetic variation in hemoglobinopathies using the microattribution approach. Nat. Genet. 43, 295–301.

Gupta, S., Manubhai, K., Kulkarni, V., Srivastava, S., 2016. An overview of innovations and industrial solutions in protein microarray technology. Proteomics 16, 1297–1308.

Hanash, S., 2004. HUPO initiatives relevant to clinical proteomics. Mol. Cell. Proteomics 3, 298–301.

Honey, K., 2008. GINA: making it safe to know what's in your genes. J. Clin. Invest. 118, 2369.

Ji, Y., Hebbring, S., Zhu, H., Jenkins, G.D., Biernacka, J., Snyder, K., et al., 2011. Glycine and a glycine dehydrogenase (GLDC) SNP as citalopram/escitalopram response biomarkers in depression: pharmacometabolomics-informed pharmacogenomics. Clin. Pharmacol. Ther. 89, 97–104.

Johnson, C.H., Ivanisevic, J., Benton, H.P., Siuzdak, G., 2014. Bioinformatics: the next frontier of metabolomics. Anal. Chem. 87, 147–156.

Johnson, C.H., Ivanisevic, J., Siuzdak, G., 2016. Metabolomics: beyond biomarkers and towards mechanisms. Nat. Rev. Mol. Cell Biol. 17, 451–459.

Kamath, K.S., Vasavada, M.S., Srivastava, S., 2011. Proteomic databases and tools to decipher post-translational modifications. J. Proteome 75, 127–144.

Karacapilidis, N.I., 2014. Mastering data-intensive collaboration and decision making. In: KDIR/KMIS. Springer International Publishing, Switzerland.

Kim, M.-S., Pinto, S.M., Getnet, D., Nirujogi, R.S., Manda, S.S., Chaerkady, R., et al., 2014. A draft map of the human proteome. Nature 509, 575.

Kochhar, S., Martin, F.-P., 2015. Metabonomics and Gut Microbiota in Nutrition and Disease. Springer-Verlag, London.

Kohler, I., Hankemeier, T., van der Graaf, P.H., Knibbe, C.A., van Hasselt, J.C., 2017. Integrating clinical metabolomics-based biomarker discovery and clinical pharmacology to enable precision medicine. Eur. J. Pharm. Sci. 109S, S15–S21.

Lederer, C.W., Basak, A.N., Aydinok, Y., Christou, S., El-Beshlawy, A., Eleftheriou, A., et al., 2009. An electronic infrastructure for research and treatment of the thalassemias and other hemoglobin-opathies: the euro-mediterranean ITHANET project. Hemoglobin 33, 163–176.

Lehmann, S., Brede, C., Lescuyer, P., Cocho, J.A., Vialaret, J., Bros, P., et al., 2017. Clinical mass spectrometry proteomics (cMSP) for medical laboratory: what does the future hold? Clin. Chim. Acta 467, 51–58.

Lindskog, C., 2015. The potential clinical impact of the tissue-based map of the human proteome. Expert Rev. Proteomics 12, 213–215.

Liu, Y., Buil, A., Collins, B.C., Gillet, L.C., Blum, L.C., Cheng, L.Y., et al., 2015. Quantitative variability of 342 plasma proteins in a human twin population. Mol. Syst. Biol. 11, 786.

Manolio, T.A., Abramowicz, M., Al-Mulla, F., Anderson, W., Balling, R., Berger, A.C., et al., 2015. Global implementation of genomic medicine: we are not alone. Sci. Transl. Med. 7. 290ps213-290ps213.

Manolio, T.A., Fowler, D.M., Starita, L.M., Haendel, M.A., MacArthur, D.G., Biesecker, L.G., et al., 2017. Bedside back to bench: building bridges between basic and clinical genomic research. Cell 169, 6–12.

Marson, F.A., Bertuzzo, C.S., Ribeiro, J.D., 2017. Personalized or precision medicine? The example of cystic fibrosis. Front. Pharmacol. 8.

Matthews, H., Hanison, J., Nirmalan, N., 2016. "Omics"-informed drug and biomarker discovery: opportunities, challenges and future perspectives. Proteomes 4, 28.

Mertins, P., Mani, D., Ruggles, K.V., Gillette, M.A., Clauser, K.R., Wang, P., et al., 2016. Proteogenomics connects somatic mutations to signaling in breast cancer. Nature 534, 55.

Mila, M., Alvarez-Mora, M.I., Rodriguez-Revenga, L., 2018. Fragile X syndrome: an overview and update of the FMR1 gene. Clin. Genet. 93, 197–205.

Morrow, D.A., de Lemos, J.A., 2007. Benchmarks for the assessment of novel cardiovascular biomarkers. Circulation 115, 949–952.

Narain, N.R., Akmaev, V.R., Vemulapalli, V., 2015. Bayesian Causal Relationship Network Models for Healthcare Diagnosis and Treatment Based on Patient Data. In. Google Patents.

Nedelkov, D., Niederkofler, E.E., Oran, P.E., Peterman, S., Nelson, R.W., 2018. Top-down mass spectrometric immunoassay for human insulin and its therapeutic analogs. J. Proteome 175, 27–33.

Paik, Y.-K., Jeong, S.-K., Omenn, G.S., Uhlen, M., Hanash, S., Cho, S.Y., et al., 2012. The chromosome-centric human proteome project for cataloging proteins encoded in the genome. Nat. Biotechnol. 30, 221–223.

Papachatzopoulou, A., Kourakli, A., Stavrou, E.F., Fragou, E., Vantarakis, A., Patrinos, G.P., et al., 2010. Region-specific genetic heterogeneity of HBB mutation distribution in south-western Greece. Hemoglobin 34, 333–342.

Pennisi, E., 2012. ENCODE project writes eulogy for junk DNA. Science 337, 1159–1161.

Pirhaji, L., Milani, P., Leidl, M., Curran, T., Avila-Pacheco, J., Clish, C.B., et al., 2016. Revealing disease-associated pathways by network integration of untargeted metabolomics. Nat. Methods 13, 770–776.

Pluskal, T., Castillo, S., Villar-Briones, A., Orešič, M., 2010. MZmine 2: modular framework for processing, visualizing, and analyzing mass spectrometry-based molecular profile data. BMC Bioinformatics 11, 395.

Rankin, N.J., Preiss, D., Welsh, P., Sattar, N., 2016. Applying metabolomics to cardiometabolic intervention studies and trials: past experiences and a roadmap for the future. Int. J. Epidemiol. 45, 1351–1371.

Rasmussen-Torvik, L.J., Almoguera, B., Doheny, K.F., Freimuth, R.R., Gordon, A.S., Hakonarson, H., et al., 2017. Concordance between research sequencing and clinical pharmacogenetic genotyping in the eMERGE-PGx study. J. Mol. Diagn. 19, 561–566.

Rifai, N., Gillette, M.A., Carr, S.A., 2006. Protein biomarker discovery and validation: the long and uncertain path to clinical utility. Nat. Biotechnol. 24, 971.

Rosenling, T., Stoop, M.P., Smolinska, A., Muilwijk, B., Coulier, L., Shi, S., et al., 2011. The impact of delayed storage on the measured proteome and metabolome of human cerebrospinal fluid. Clin. Chem. 57, 1703–1711.

Salek, R.M., Neumann, S., Schober, D., Hummel, J., Billiau, K., Kopka, J., et al., 2015. COordination of standards in MetabOlomicS (COSMOS): facilitating integrated metabolomics data access. Metabolomics 11, 1587–1597.

Scherl, A., 2015. Clinical protein mass spectrometry. Methods 81, 3–14.

Schubert, O.T., Röst, H.L., Collins, B.C., Rosenberger, G., Aebersold, R., 2017. Quantitative proteomics: challenges and opportunities in basic and applied research. Nat. Protoc. 12, 1289–1294.

Sperber, N.R., Carpenter, J.S., Cavallari, L.H., Damschroder, L., Cooper-DeHoff, R.M., Denny, J.C., et al., 2017. Challenges and strategies for implementing genomic services in diverse settings: experiences from the implementing GeNomics in pracTicE (IGNITE) network. BMC Med. Genet. 10, 35.

Sud, M., Fahy, E., Cotter, D., Azam, K., Vadivelu, I., Burant, C., et al., 2015. Metabolomics workbench: an international repository for metabolomics data and metadata, metabolite standards, protocols, tutorials and training, and analysis tools. Nucleic Acids Res. 44, D463–D470.

Sudmant, P.H., Rausch, T., Gardner, E.J., Handsaker, R.E., Abyzov, A., Huddleston, J., et al., 2015. An integrated map of structural variation in 2,504 human genomes. Nature 526, 75.

Suhre, K., Shin, S.-Y., Petersen, A.-K., Mohney, R.P., Meredith, D., Wägele, B., et al., 2011. Human metabolic individuality in biomedical and pharmaceutical research. Nature 477, 54–60.

Surinova, S., Radová, L., Choi, M., Srovnal, J., Brenner, H., Vitek, O., et al., 2015. Non-invasive prognostic protein biomarker signatures associated with colorectal cancer. EMBO Mol. Med. 7, 1153–1165.

Taher, A.T., Weatherall, D.J., Cappellini, M.D., 2017. Thalassaemia. Lancet 391, 155–167.

Tautenhahn, R., Cho, K., Uritboonthai, W., Zhu, Z., Patti, G.J., Siuzdak, G., 2012a. An accelerated workflow for untargeted metabolomics using the METLIN database. Nat. Biotechnol. 30, 826–828.

Tautenhahn, R., Patti, G.J., Rinehart, D., Siuzdak, G., 2012b. XCMS online: a web-based platform to process untargeted metabolomic data. Anal. Chem. 84, 5035–5039.

Taylor, C.F., Paton, N.W., Lilley, K.S., Binz, P.-A., Julian, R.K., Jones, A.R., et al., 2007. The minimum information about a proteomics experiment (MIAPE). Nat. Biotechnol. 25, 887–893.

Tolstikov, V., 2016. Metabolomics: bridging the gap between pharmaceutical development and population health. Metabolites 6, 20.

Tolstikov, V., Akmaev, V.R., Sarangarajan, R., Narain, N.R., Kiebish, M.A., 2017. Clinical metabolomics: a pivotal tool for companion diagnostic development and precision medicine. Expert Rev. Mol. Diagn. 17, 411–413.

Treff, N.R., Zimmerman, R.S., 2017. Advances in preimplantation genetic testing for monogenic disease and aneuploidy. Annu. Rev. Genomics Hum. Genet. 18, 189–200.

Tsiliki, G., Karacapilidis, N., Christodoulou, S., Tzagarakis, M., 2014. Collaborative mining and interpretation of large-scale data for biomedical research insights. PLoS One. 9.

Uhlén, M., Fagerberg, L., Hallström, B.M., Lindskog, C., Oksvold, P., Mardinoglu, A., et al., 2015. Tissue-based map of the human proteome. Science 347.

van der Wouden, C., Cambon-Thomsen, A., Cecchin, E., Cheung, K., Dávila-Fajardo, C., Deneer, V., et al., 2017. Implementing pharmacogenomics in Europe: design and implementation strategy of the ubiquitous pharmacogenomics consortium. Clin. Pharmacol. Ther. 101, 341–358.

Vaz-Drago, R., Custódio, N., Carmo-Fonseca, M., 2017. Deep intronic mutations and human disease. Hum. Genet. 1–19.

Vemulapalli, V., Qu, J., Garren, J.M., Rodrigues, L.O., Kiebish, M.A., Sarangarajan, R., et al., 2016. Non-obvious correlations to disease management unraveled by Bayesian artificial intelligence analyses of CMS data. Artif. Intell. Med. 74, 1–8.

Verrills, N.M., 2006. Clinical proteomics: present and future prospects. Clin. Biochem. Rev. 27, 99.

Vizcaíno, J.A., Côté, R.G., Csordas, A., Dianes, J.A., Fabregat, A., Foster, J.M., et al., 2012. The PRoteomics IDEntifications (PRIDE) database and associated tools: status in 2013. Nucleic Acids Res. 41, D1063–D1069.

Weitzel, K.W., Alexander, M., Bernhardt, B.A., Calman, N., Carey, D.J., Cavallari, L.H., et al., 2016. The IGNITE network: a model for genomic medicine implementation and research. BMC Med. Genet. 9(1).

Wilhelm, M., Schlegl, J., Hahne, H., Gholami, A.M., Lieberenz, M., Savitski, M.M., et al., 2014. Mass-spectrometry-based draft of the human proteome. Nature 509, 582.

Williams, E.C., Wu, Y., Jha, P., Dubuis, S., Blattmann, P., Argmann, C.A., et al., 2016. Systems proteomics of liver mitochondria function. Science 352.

Wishart, D.S., 2010. Computational approaches to metabolomics. Methods Mol. Biol. 593, 283–313.

Wishart, D.S., 2016. Emerging applications of metabolomics in drug discovery and precision medicine. Nat. Rev. Drug Discov. 15, 473–485.

Wishart, D.S., Tzur, D., Knox, C., Eisner, R., Guo, A.C., Young, N., et al., 2007. HMDB: the human metabolome database. Nucleic Acids Res. 35, D521–D526.

Xia, J., Mandal, R., Sinelnikov, I.V., Broadhurst, D., Wishart, D.S., 2012. MetaboAnalyst 2.0—a comprehensive server for metabolomic data analysis. Nucleic Acids Res. 40, W127–W133.

Zhang, H., Liu, T., Zhang, Z., Payne, S.H., Zhang, B., McDermott, J.E., et al., 2016. Integrated proteogenomic characterization of human high-grade serous ovarian cancer. Cell 166, 755–765.

Zhong, Q., Guo, T., Rechsteiner, M., Rüschoff, J.H., Rupp, N., Fankhauser, C., et al., 2017. A curated collection of tissue microarray images and clinical outcome data of prostate cancer patients. Sci. Data. 4.

CHAPTER 6

Incentives for Human Genome Variation Data Sharing

George P. Patrinos

*Department of Pharmacy, School of Health Sciences, University of Patras, Patras, Greece;
Department of Pathology—Bioinformatics Unit, Faculty of Medicine and Health Sciences,
Erasmus University Medical Center, Rotterdam, The Netherlands; Department of Pathology,
College of medicine and Health Sciences, United Arab Emirates University, Al-Ain,
United Arab Emirates*

6.1 INTRODUCTION

The postgenomic revolution, characterized by the recent advances of high-throughput genomic approaches, such as next-generation sequencing and genome-wide association studies (Cooper et al., 2010), has led to the elucidation of the genetic basis of a considerable number of human-inherited disorders and the correlation of specific genomic variants with disease predisposition and progression. This has in turn resulted in an exponential increase in the generation of human genome variation data both from research centers and diagnostic laboratories, which enriches our knowledge of the genetic heterogeneity underlying both monogenic and multifactorial (i.e., complex) diseases.

To date, only few genetic databases have been able to successfully document either of the plethora of reported clinically relevant genomic variants, which is somewhat surprising considering their enormous potential utility in the field of molecular genetics diagnostics. As indicated in the previous chapter, genomic databases are online repositories of human genome variation data, with central variation databases (CVDBs), locus-specific variation databases (LSDBs), and National/Ethnic Genomic databases (NEGDBs) being the most popular types of genomic database available to date (Thorisson et al., 2009). Even if some of these genomic databases are individually quite small, their being well-curated, accurate, comprehensive, and up-to-date makes them invaluable resources for both molecular diagnostics and genetic counseling, while they have the potential to be of utmost importance for the establishment of genotype-phenotype correlations in any inherited disorder. Expansion of these data collections in order to maximize their practical utility represents a major challenge; although their potential value is obvious to everyone in the

Human Genome Informatics. https://doi.org/10.1016/B978-0-12-809414-3.00006-1

field, it is striking that, once these resources are established, subsequently identified variants are only rarely submitted, most likely not only due to the lack of motivation from the part of the database coordinator, but also due to the lack of sufficient benefit for the potential data submitter. So far, the scientific community has failed to establish a reliable and effective system to collect and systematically document this valuable information as it emerges. As such, a meaningful and ongoing reward system that appropriately incentivizes both clinicians and academic researchers to place their findings in the public domain at the earliest possible stage is more than ever urging. As it stands, virtually no credit is received for contributing genome variation data to or even curating genomic databases, and therefore, only the most open-access minded actually do so in the absence of journal-mediated compulsion. Without any appropriate attribution system, it is likely that much genomic variation data will stay unpublished, as scientists and clinicians often face major difficulties in publishing novel variants identified in exhaustively studied genes, such as the human globin genes, even if a newly appreciated genotype-phenotype correlation and multiple novel causative variants are presented in the same manuscript. As a result, a significant proportion of identified genome variation data, corresponding to both benign and clinically relevant genomic variants, fails to be deposited in the public domain and, therefore, fails to contribute to a broader understanding of the genetic basis of inherited disease, a trend that will most likely worsen over time. It is, therefore, imperative that new and innovative initiatives are undertaken to both encourage and promote the submission of genotype-phenotype data to genomic databases.

This chapter aims to summarize the various initiatives that have so far been undertaken to tackle this deficiency, the shortcomings of these undeniably well-intentioned efforts, as well as their future potential to be adopted by the research community as novel publication modalities that could serve to promote the capture of genome variation data and make possible their eventual placement in the public domain.

6.2 DATABASE PROJECTS LINKED TO SCIENTIFIC JOURNALS

There is an urgent need for systems and methodologies to facilitate the collection and deposition of hitherto unpublished genome variation, particularly since peer-reviewed journals are becoming increasingly reluctant to publish narrative reports of single (or even a few) causative variants in already well-characterized genes. The collation of all published causative variants in genomic databases is currently almost exclusively accomplished by (automated) literature data mining, while only a small proportion of entries listed in LSDBs are unpublished variants that have been directly contributed to the LSDBs by the authors.

In 2003, the Human Genome Variation Society (http://www.hgvs.org) launched the Human Genome Variation System (HGVSYS; Horaitis and Cotton, 2004), comprising three stages: (a) The WayStation (http://www.centralmutations.org), designed to be a "…central point with a consistent interface and format for the submission and collection of human genetic variation data" (Horaitis and Cotton, 2004), (b) the Office, comprising Gene Editors, experts in specific genes who made up the WayStation Review Board whose role was to peer-review and approve data submissions, and (c) the Central Database, the data warehouse to store the submitted variants and their corresponding phenotypes approved by the WayStation Review Board. This system would also have allowed dissemination to any other database; some entries, published in the journal *Human Mutation* as Genome Variation Reports, were indexed in PubMed and received a PubMed ID. Although the WayStation was well-publicized, it never reached the impact that had been hoped for. Indeed, in 4 years, it only received 77 submissions (encompassing 129 variants in 103 genes); as a result, this project was silently abandoned in 2007.

At the same time, the journal *Human Genetics* proposed a new approach, taking advantage of its informal affiliation with Human Gene Mutation Database (HGMD; http://www.hgmd.org). Thus, many electronically submitted genomic variants were published by *Human Genetics* and included in HGMD, together with very brief phenotypic descriptions. Hence, authors received scientific credit by obtaining a PubMed ID for their submissions. This scheme ran for a total of 12 years from February 1998, but was discontinued in April 2010 owing to the increasing workload for the publisher (Springer-Verlag) and editors, the decreasing benefit to the journal itself, and a potential detrimental effect on the journal's impact factor. A similar publication modality was also implemented for the HbVar database for hemoglobin variants and thalassemia mutations (http://globin.bx.psu.edu/hbvar), which invited submissions of novel pathogenic variants in the human globin genes, population-specific globin gene variation allele frequencies (Patrinos et al., 2004), and experimental protocols for globin-gene variant screening (Giardine et al., 2007). In 2004, the journal *Hemoglobin* introduced a requirement to place globin-gene variation data into HbVar prior to submission of a related manuscript to the journal (Patrinos and Wajcman, 2004), which led to a modest increase in data submission to HbVar. Such an approach has also latterly been adopted by *Human Mutation*, which now requires all novel genome variation data to be submitted to (and stored in) an LSDB prior to acceptance of the corresponding manuscript for publication.

Population-specific mutation datasets are considered to be more complete, as compared to the characterization of a single or a few novel genomic variants, and as such, these datasets may be more attractive to peer-reviewed journals. Based on this assumption, *Human Genomics and Proteomics* has affiliated itself with FINDbase worldwide database for clinically relevant genomic variation

Table 6.1 International Scientific Peer-Reviewed Journals That are (or Were) Related or Linked to a Genetic Database, Namely Central (CVDBs), Locus-Specific (LSDBs) or National/Ethnic Genetic Databases (NEGDBs)

Journal	Database	Description	Link	References
CVDBs				
Human Mutation	WayStation	An attempt to connect a major genetics journal with a central data repository[a]. Intended to incentivize genome variation data submission from the scientific community	http://www.hgvs.org/centraldb.html	Horaitis and Cotton (2004)
Human Genetics	Human Gene Mutation Database (HGMD)	Long standing connection[a] between a human genetics journal and the resource which acts as the de facto central depository for human gene mutations associated with inherited disease	http://www.hgmd.org	N/A
LSDBs				
Hemoglobin	HbVar	Connection of one of the main forums for the publication of hemoglobin research with the globin gene-specific database	http://globin.bx.psu.edu/hbvar	Patrinos and Wajcman (2004)
NEGDBs				
Human Genomics and Proteomics	FINDbase	Interconnection of a genomics journal with an open-access database that documents clinically relevant genome variation allele frequencies in various populations worldwide	http://www.findbase.org	Patrinos and Petricoin (2009)

[a]*No longer operational (see text for explanation).*

allele frequency (http://www.findbase.org), an international freely available database (Viennas et al., 2017). *Human Genomics and Proteomics* aimed to publish short descriptions of genetic datasets pertaining to population/ethnic-specific clinically relevant genome variation allele frequencies, namely causative variants and pharmacogenomic biomarkers (Georgitsi et al., 2011). The affiliation of *Human Genomics and Proteomics* with FINDbase allows authors to present their research on the characterization of causative genomic variation and/or pharmacogenomic marker frequency spectra to a broad audience (Patrinos and Petricoin, 2009).

The database-journal efforts that have been implemented to date are summarized in Table 6.1.

6.3 MICROATTRIBUTION AND NANOPUBLICATION: AN INNOVATIVE PUBLICATION MODALITY

One of the traditional measures used to evaluate a researcher's scientific career is: (a) their publication track record in international peer-reviewed scientific journals, assessed on the basis of the impact of the journal, related to its ISI Impact Factor, (b) the number of citations that each article attracts, and (c) the H-index, an aggregate measure of the total number of citations that each of the author's publications attracts (http://www.scopus.com). However, there are many other ways, apart from publishing scientific articles, in which researchers can contribute to the scientific community, one of them being submission to, and curation of, genomic databases.

6.3.1 The Concept of Microattribution

In 2008, the scientific journal *Nature Genetics* introduced the concept of "microattribution," in an attempt to introduce an alternative reward system for scientific contributions, such as database entries and records. The principle of microattribution is "… to produce a publication workflow that is open to all journals and that draws on the expertise of all those with a stake in understanding variation at a particular region in the human genome" (Anonymous, 2008).

Each Human Variome microattribution review would consist of the following components:

(a) The Public Genome Browser, to display the actual number of database entries and related articles. In this way, microattribution introduces the concept of curation of individual genome variants by gene, in order that data submitters and database curators will obtain all due credit. A prerequisite for this is that genomic variants should be deposited in stable, publicly available and well-maintained central repositories (e.g., NCBI, EBI) that would run independent microattribution services, based on an individual researcher's unique identity, such as is aimed for by the ResearcherID (http://www.researcherid.com), Open Researcher and Contributor ID consortium (ORCID; http://orcid.org), OpenID (http://openid.net), etc. Depositing genome data as "micro- or nanopublications" (Mons and Velterop, 2009), contributed by an individual researcher or research group, in a stable and accessible format in open repositories, would allow mining for citations associated with this/these authors' unique IDs. In other words, a micro- or nanopublication is the smallest unit of publishable information that can be linked to its contributor via their unique scientific identity and can be cited and evaluated in terms of its impact upon the research community.

(b) The Microattribution Analysis article, a comprehensive high-profile article commissioned by the journal's editor and stringently peer-reviewed, which would summarize the features of all variants at a particular locus, such as phenotypes, clinical findings, allele frequencies, and so on. In this article, all genome variant contributors who are authors of the micro- or nanopublication collection constituting the Microattribution Analysis article would be either considered as coauthors or would in the future receive citations on their individual micro- or nanopublications.

There are several advantages to this approach. First of all, the nature of the micro- or nanopublication is such that it does not need to be related to a full scientific article (although this may also be an option), but it can also refer to a genomic database from which it is derived. Moreover, micro- or nanopublications can be encoded in the Resource Description Framework (RDF) to become part of the Semantic Web and transferred between computers using Extensible Markup Language (XML). This way, micro- or nanopublications can be more easily mined, queried, and retrieved through the Internet and be subject to computer reasoning, something that is not feasible for regular articles. Most importantly, micro- or nanopublications have the potential to incentivize potential data contributors to place their data in the public domain for others to freely access and optimally exploit, since micro- or nanopublications can be attributed and cited in the same way as regular articles (Mons et al., 2011; Patrinos et al., 2012). This can have a significant impact upon future scientific publication modalities, since it facilitates data mining, enables data sharing, and therefore increases the likelihood of an article being retrieved and cited.

In the genetics/genomics world, a micro- or nanopublication would consist, at least, of the following elements:

(a) The *Assertion*, which is the minimal unit of information that needs to be published. It contains at least one statement with three elements, as a regular sentence: (i) the *Subject*, for instance the gene or a genomic variant, (ii) the *Predicate* that (for instance) relates the gene or the variant to a specific phenotype and/or clinical condition, as far as a GVD and/or LSDB is concerned or a specific allele frequency for a NEGDB, and (iii) the *Object* of the relation, that either links the Subject to another concept (for instance a phenotype) or to a feature or value. Finally, *Provenance* includes attribution information pertaining to the micro- or nanopublication, referring to who made the assertion (the contributing author(s)), when it was made by a date/time stamp, or how the assertion can be (re)used (copyright). Given that the attribution can be linked to each researcher through their unique identification, this can provide the basis for the necessary metrics for each scientific micro-contribution, in addition to the traditional publication metrics, such as article number, article citation index, H-index, etc. In addition, a micro- or nanopublication can include supporting information such as the nature of the data source, experimental conditions, and other contextual or "credibility" features that the authors consider essential evidence for their assertion.

Fig. 6.1 illustrates the minimal micro- or nanopublication and two representative examples. The first example, derived from an NEGDB, describes a genome variant (*SLCO1B1:c.521T>C; Subject*) that has (*Predicate*) a "minor allele frequency" (*Object*) with a value associated in a specific population (*Supporting*) and which has been contributed by a specific researcher

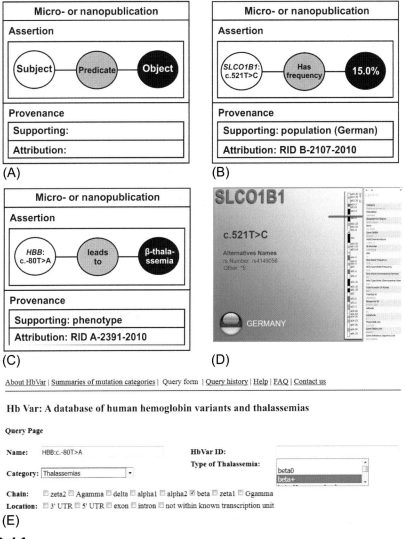

FIG. 6.1

(A) Outline of the micro- or nanopublication (see also text for details). (B, C) Indicative examples of micro- or nanopublications pertaining to national/ethnic genomic databases (NEGDBs) and locus-specific mutation databases (LSDBs), respectively. (D, E) The entries corresponding to the micro- or nanopublications, as outlined in B and C, are deposited in FINDbase and the HbVar databases respectively.

(*Attribution;* Fig. 6.1B). The second example, derived from a LSDB, describes a genome variant (*HBB*:c.-80T > A; *Subject*) that leads to (*Predicate*) β-thalassemia (*Object*), a specific phenotype (*Supporting*), and was contributed by a research group (*Attribution;* Fig. 6.1C). According to the latest micro- or nanopublication model, the *Supporting* and *Attribution* elements consist of the *Provenance*, while the *Subject, Predicate,* and *Object* consist of the *Assertion* of the micro- or nanopublication (Fig. 6.1A). In both cases, the corresponding entries have been hard-linked to, and hence are available for, further detailed study in the corresponding source databases, e.g., FINDbase (Fig. 6.1D) and HbVar (Fig. 6.1E), respectively. Since micro- or nanopublications are made such that they are fully computer readable, this format allows performing meta-analyses of many micro- or nanopublications pertaining to certain topics and linkage of such data to, for instance, genome browsers. Data contribution from multinational genome variation projects, such as the EuroPGx project (http://www.goldenhelix.org/activities/research/pharmacogenomics; Mizzi et al., 2016), could serve to accelerate the development of Semantic Web environments to support biomedical research areas.

Another example that demonstrates the utility of the micro- or nanopublication approach is querying a specific variant. Currently, one has to employ a genome browser to identify this variant, by activating all possible tracks, an approach that is not particularly user-friendly. Once this information becomes available in a micro- or nanopublication format, querying for this particular variant becomes much easier and more straightforward.

6.3.2 The Microattribution Process

In principle, the process of microattribution would begin with the deposition of genome variation data in LSDBs and/or NEGDBs (recorded by PubMedIDs or DoIs) or through direct submission by individual researchers, attributed to them through a unique ID, for instance, ResearcherID (www.researcherid.com) or ORCID (www.orcid.org). Database contents can then be additionally published as micro- or nanopublications, which would contain minimally four microattribution-enabling provenance elements:

(a) Submission of the genomic variants from the source database (CVDBs, LSDB, or NEGDB) to the central public repository,
(b) Microattribution details of the author(s),
(c) Association with a specific phenotype and/or clinical condition, and
(d) (Rare variant) allele frequencies.

Indicative provenance elements are outlined in Fig. 6.2.

These four microattribution elements could be submitted to the appropriate microattribution tables in the publicly available data repository and get

Source dbID HGVS name			
Element 1	**Element 2**	**Element 3**	**Element 4**
Central database ID	Disease ID type	Phenotype description	Population
Traditional name	Disease ID value	Zygosity	Ethnic group
OMIM ID	Clinical interpretation	Laboratory features	Allele Frequency
MicroattributionID[a]	PubMed ID(s)	Assay ID	
PMID	Local ID	Assay value	
DoI	OMIM allelic variant ID	Units	
	Germline/somatic	Clinical features	
	Other ID(s)	Other features	

FIG. 6.2

Possible contents of the 4 microattribution elements. Microattribution element 3 may vary between LSDBs and NEGDBs as well as between different LSDBs, depending on the disease for which the underlying genetic defect is described. [a]ResearcherID, OpenID, ORCID, etc. Source dbID and variant HGVS name should be common among all 4 microattribution elements.

automatically published in a micro- or nanopublication format, thereby also linking to the relevant microattribution article(s) submitted to the scientific journal for peer-review and, upon acceptance, being integrally linked to that publication. In a microattribution article, all data submitters are coauthors and should receive full credit for their contribution, whereas for micro- or nanopublications submitted to the central public database, these can be tracked in an aggregated manner, using their unique identifiers in the provenance part of the RDF-named graph to measure the credit due to any given researcher with respect to these submissions as a whole.

The introduction of microattribution should allow the measurement of a researcher's own contributions in publicly available central genomic data repositories by the unambiguous linking of their contributions to their unique data contributor ID. This would in turn facilitate data mining and the provision of citation metrics as per regular publications. Moreover, enabling data sharing by microattribution is likely to increase the likelihood of an article being retrieved and cited. Ultimately, micro- or nanopublication metrics can serve as a measure to provide academic credit to individual researchers, thereby contributing to academic advancement.

6.3.3 Implementation of Microattribution

The first demonstration of microattribution working in practice was achieved by Giardine et al. (2011) using HbVar, the globin-gene LSDB, as a model. At first, all causative variations, their associated phenotypes, and allele frequencies, where applicable, were comprehensively documented in 37 inter-related LSDBs for the globin genes as well as those genomic loci that, when mutated, lead to a hemoglobinopathy-related phenotype, e.g., *ATRX*, *KLF1*, and so on. Each genomic variant in these LSDBs was listed in four separate Microattribution Tables (Giardine et al., 2011), which include either published variants stored against the PubMed IDs, or unpublished variants contributed by individual researchers or research groups involved in hemoglobin research. In the latter case, unpublished variants were documented against contributors' own Researcher IDs. Each variant was linked to its LSDB accession number and unique IDs for the contributor(s) of the variant. Subsequently, the Micro-attribution Tables were deposited in the National Center for Biotechnology Information (NCBI; http://www.ncbi.nlm.nih.gov) public repository in an effort to measure microcitations for every data contributor or data unit centrally. In addition, the microattribution article, comprising 51 authors from 35 institutions, was published in *Nature Genetics* (Fig. 6.3).

The microattribution article itself could not contain all variant-phenotype relationships found in the 37 LSDBs. However, it was shown that the analysis article contained around 600 assertions in the text that could be individually captured as meaningful nanopublications contrary to the underlying databases

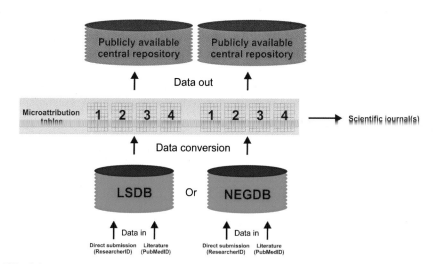

FIG. 6.3

Outline of the proposed microattribution approach. Data contributors deposit data into a LSDB or NEGDB either by directly contacting data curators or directly to the LSDB NEGDB, respectively and receive credit in the form of micro-citations vis-à-vis their data contributor ID (e.g., ResearcherIDs). Data are thereafter forwarded to the publicly available central data repository in the format provided in the microattribution tables (see also Fig. 6.2), while genome variation data are published as a scientific article in peer-reviewed scientific journals.

that contained well over 44,000 potentially meaningful nanopublications, thereby greatly enhancing the computer-enabled citation potential of the article itself as well as each nanopublication individually (Mons et al., 2011).

A number of useful conclusions were also drawn from this approach, derived from the clustering of *ATRX* and *KLF1* gene variants in the exons of the genes, from where new causative variation patterns have emerged (Giardine et al., 2011). Such conclusions would not have been possible without such an approach, further demonstrating the value of the immediate sharing of novel genome variants to databases even though they would not warrant classical narrative publication on their own.

Microattribution has been found to contribute significantly to the increase in data submission rate with respect to HbVar and related databases, showing a more than eightfold increase of data submission rate as compared to previous years in which HbVar was active, even compared to the official HbVar launch year (2001), emphasizing the potential impact of microattribution to genome variation data submission. Similarly, microattribution was successfully implemented to incentivize data sharing in NEGDBs (Georgitsi et al., 2011) and clinical genetics databases (Sosnay et al., 2013; Thompson et al., 2014).

6.4 MICROATTRIBUTION: HURDLES FROM CONCEPT TO IMPLEMENTATION

The examples presented in the previous paragraphs show how microattribution can be theoretically implemented, but prior to that there are several issues that should be addressed to ensure their successfully implementation in practice. The single most important contribution of microattribution to increase the rate of data submission is incentivizing individual researchers or research groups to submit their newly acquired and unpublished genome variation data to a public repository or database in return for appropriate credit and attribution.

First of all, microattribution presupposes the existence of well-curated and stable data repositories, such as LSDBs and/or NEGDBs, and underlying genome variation datasets as skeleton structures. Researchers will clearly be somewhat less motivated to submit their genome variation data to databases, however well-curated, if these databases lack substantial amounts of previously published data. One should bear in mind that the success of the pilot microattribution project involving the globin-gene datasets may be largely attributed to the following:

(a) Well-curated HbVar database, highly appreciated in the globin research community, long before the microattribution project was initiated (over 700 citations for the original and 3 subsequent update articles in the well-respected scientific journals *Human Mutation* and *Nucleic Acids Research*), providing a well-defined structure through which the new submissions could be shared,

(b) Novelty of the concept, derived from the pilot nature of the project itself, and

(c) Enthusiastic backing of a high-profile journal, such as *Nature Genetics*.

All these elements probably served as serious attractors for data submitters. However, it is likely that replication or even scaling up this project to hundreds or even thousands of genes, for which there are currently no well-structured databases and significantly less interest, is likely to be a slow and painstaking process and would probably have only very limited rewards for the huge number of participants required.

Furthermore, one would have to envisage that the successful adoption of the microattribution concept would require quality evaluation of data submission by international experts, possibly coordinated by committees to oversee the smooth implementation of the microattribution process. This requires their members to communicate with data submitters, review and curate the submitted data where necessary, and chase data contributors for any missing information. In the context of human genome variation data, a sensible approach would be to base the whole process around one or more preexisting and freely available high-quality centralized databases or database-journals, and possibly be coupled with publication of microattribution-type articles at regular intervals in genome research journals (possibly online only), such that the individual contributions of the data submitters in a consortium are recognized by their coauthorship. An example is the microattribution article by Giardine et al. (2011). Such articles further incentivize researchers to submit data to a central repository knowing that they will receive all due attribution, not only by the micro- or nanopublications itself, but also by the publication describing the database and the genome variation datasets that the database accommodates.

In relation to this approach, serious consideration must be given to the scale of funding required to run such a venture. As with any large database project, such a centralized approach would not only require significant funding, but would ultimately need to be financially self-sustaining. One possible solution would be partnering of the centralized data repository with a major publishing group(s) along the lines of a database-journal-like model (see above). Such venture already exists, in the form of a scientific journal, *Scientific Data*, published by Nature Publishing Group, which is derived from this very concept. *Scientific Data* (www.nature.com/sdata) is an open-access peer-reviewed scientific journal for descriptions of scientifically valuable datasets which advances the scientific data sharing and reuse. *Scientific Data* primarily publishes Data Descriptors, a new type of publication that focuses on helping others reuse data and crediting those who share. *Scientific Data* welcomes not only genomic data submissions, but also data from a broad range of research disciplines, including descriptions of big or small datasets, from major consortiums to single research groups. All datasets are hosted on www.nature.com and scientific publications are indexed in PubMed, MEDLINE, and Google Scholar literature databases and are automatically deposited into PubMed Central.

Another important issue that must be considered to implement microattribution is that the bulk of the genome variation data that are currently being generated are the product of whole-genome/exome sequencing studies. These are "raw" uncurated data, involving both pathogenic (clinically relevant) and benign variants, and as such, have to be handled in quite a different way from, for instance, curated data that involve allele frequencies and (endo)phenotypes. In many cases, these data have already been submitted to a central database (e.g., ClinVar, see Chapter 10), and hence, are freely available, but may still lack sufficient attribution and credit for the submitters. Such data, for instance hundreds of thousands of benign variants and perhaps 50–100 deleterious variants per genome (MacArthur et al., 2012), could in principle be extracted from publicly available whole-genome datasets, annotated and introduced into well-curated or even orphan LSDBs prior to announcing a call for genome variation data submission or for candidate curators to adopt an LSDB. The latter is particularly important for genome variation data that are no longer publishable on their own, e.g., by-products of whole exome or genome sequencing, and which, in the absence of any appropriate incentives, would in all likelihood remain outside the public domain. In this case, however, the workload involved in extracting, annotating, and introducing these data into LSDBs is substantial and every effort should be made to ensure that submission of these data would be made as easy and automated as possible and, at the same time, as rewarding as possible to the data submitter. Moreover, particular attention should be given to indicate the level of data confidence in a clear way, e.g., raw (and perhaps false positive) data versus well-curated data of unknown significance or with a clear genotype–phenotype correlation. The provenance and context elements of the micro- or nanopublication enable these elements.

Another important issue to consider when implementing microattribution is that those scientists who will be held responsible for generating most of the genome variation data, and by definition internationally renowned experts and leaders in the field, are expected to be among the least likely to be incentivized by this approach, since they are used to think exclusively in terms of high impact traditional publications. Indeed, the majority of LSDBs are not curated by the leaders in their respective fields, but they rather tend to be initiated by junior and/or transient workers who wish to maximize their gains in the short-term and often have little or no interest in the task of sustaining the LSDB into the long-term (Patrinos and Brookes, 2005). One of the underlying reasons for this may be the fact that submitting genome variation data to LSDBs is extremely laborious as opposed to the very little obvious reward one obtains at the end of it. Microattribution should, therefore, be implemented in such a way as to inspire busy high-profile researchers and their groups to populate LSDBs with genome variation data. Incentivization could, in principle, operate through assigning the responsibility for curating these data repositories to the more junior members of these research groups, who will be in turn rewarded for their contribution.

The above issues must be exhaustively and seriously considered and, if adequately addressed at an early stage, microattribution can be sustained beyond the initial excitement about the novelty of the concept. One way to address all these issues and, at the same time, to initiate the entire process would be to establish standards, set priorities, and discuss problems, shortcomings, and ways to bypass them. This can be accomplished as a result of an international meeting where all interested stakeholders would participate, similar to the way in which the Minimum Information About a Microarray Experiment (MIAME) standard was developed (Ball et al., 2004). Such a meeting, which might be placed under the auspices of a major bioinformatics center where the microattribution data would be gathered, could allow important deficiencies to be resolved in the early stages, which, in turn, will save time and effort during the full implementation of the microattribution process later on.

6.5 PROPOSED MEASURES AND STEPS FORWARD

There are numerous ongoing efforts by various groups worldwide whose aim is to comprehensively document all clinically relevant human genomic variants. The development and curation of genomic data repositories, such as LSDBs for a particular human gene/genetic disease, NEGDBs for a particular population, or even patient registries with microattribution, provide examples of how such repositories might be established and populated with data in the future. By adopting several different approaches to incentivizing data contribution, such as coupling databases to scientific journals or by implementing the microattribution process, the rate of accumulation of high-quality and accessible data may be significantly increased. At the same time, as with many novel initiatives, there are bottlenecks that remain to be tackled, such as which repository these variants must be deposited to, e.g., in central databases (e.g., ClinVar) or NEGDBs and/or LSDBs (Mitropoulou et al., 2010).

Usually, traditional publication modalities do not require submission of primary data to genomic databases, unless specifically requested to do so by the journal or the funding agencies. These examples include submission of

DNA sequences and transcription profiling datasets into public databases (e.g., GenBank, MIAME, dbSNP) against their submission accession numbers. Since several years, *Human Mutation* requires novel gene-specific variation data submission to LSDBs prior to manuscript acceptance. As such, a way forward would be genetics and genomics journals to adapt their Instructions to Authors accordingly with the aim of enhancing genome variation data capture. Further, in several journals, where submission of 'mutation updates' is already encouraged, such articles could be coupled to microattribution so as to increase unpublished genome variation data capture. On top of that, every paper that has been annotated for microattribution could be marked as such in PubMed, adding to the incentive to publish data (e.g., a clearly visible acronym such as *"mA"* could be shown). This acronym could also be placed, e.g., in the top right hand corner of the print and online versions of the article to indicate that microattribution data are available for a given manuscript. This approach might also inspire the authors to get their papers marked with such a label.

Other important parameters that must be considered in the context of the abovementioned novel publication modalities, which is either database-journals or microattribution, are data privacy protection and ownership. In principle, existing recommendations for the curation of human genome data repositories (Povey et al., 2010; Celli et al., 2012) should be adhered to, including policies for data sources derived both from research and clinical genetic diagnostic settings, published or unpublished results, special provisions for the protection of certain cultural and minority population groups (e.g., Bedouins, Roma, etc.), and removal of sensitive identifying information from the submitted datasets, the latter being particularly important since it is clear that the availability of microarray data or even whole-genome sequencing data can no longer guarantee a person's anonymity/nonidentifiability (Homer et al., 2008).

A key hurdle in capturing all clinically ascertained genome variation data lies in the lack of incentivization of public and private genetic diagnostic laboratories as well as research groups to share their data. Some genomic variant data derived in a clinical diagnostic laboratory setting may never be revealed to the public or the scientific community, because they will probably never be published or documented in a public database (Cotton et al., 2009). Private laboratories, in particular, are not always prepared to place their data in the public domain, one of their reasons being fear of possible patient identification and of subsequent litigation, or in other cases, lack of consent to publicly share patient data; this could also lead to uncomfortable questions about the company's strategy for interpreting the data that they generate (Guttmacher et al., 2010). To overcome this bottleneck, the necessary consent could be requested from patients upon ordering a genetic test or, alternatively, a policy of concealing their personal details upon data submission could be adopted.

Certain database management systems have made provisions to contribute patients' genomic data without revealing any sensitive personal details, both in the context of individuals (Fokkema et al., 2011) and of culturally sensitive population minority groups (Zlotogora et al., 2007, 2009). Another practical step that could be implemented to tackle this bottleneck would be to include data contribution to public repositories as a necessary quality control requirement for the accreditation process of a genetic testing laboratory. Such a measure is already being implemented by the Israeli Ministry of Health (Zlotogora et al., 2009). Also, the EQA/accreditation system may be exploited, such that the EQA bodies require database submission of all novel pathogenic variants identified by molecular diagnostic laboratories, which should be listed in the report as a reference number.

As far as the microattribution process is concerned, it may be possible not only to track a researcher's contribution as a means to provide publication credit, but also to establish a microcredit measure for database development, maintenance, and curation efforts, involving, e.g., manual data content curation and expansion by adding related information from the literature, verification of data correctness, cross-linking with other databases, and so on (as already performed by existing data repositories, e.g., HGMD). At present, such efforts go largely unnoticed, since they cannot be measured by traditional publications metrics. In the postgenomic era, in which the submission and curation of large datasets is one of the most pressing issues, attribution of both data submission and curation is going to be of the utmost importance in providing the research community with access to these datasets. The microattribution approach described in the previous paragraphs enables both data attribution and citation. It is important to clearly distinguish between these two factors. "Assertional" data, such as the association between genomic variants and their phenotypes, when deposited in databases, can be linked and attributed to their creators/submitters via the provenance described above. Contributors to the data quality control and curation process, including the curators of LSDBs and NEGDBs, can also be recognized. This is particularly important since many databases do not annotate their data appropriately (i.e., by reference to genomic sequence numbering) and, consequently, the quality of the submitted, uncurated data in many LSDBs is relatively poor. A mechanism to value and assign these contributions, as well as formally acknowledging mechanisms to use them as part of the scientific accreditation process, is currently lacking. Therefore, the actual citation of data elements after publication as micro- or nanopublications can provide significant added value. First of all, established mechanisms for citation metrics already exist and, hence, it would be less cumbersome to implement than developing new citation metrics methods to assess the value and extent of individual depositions in those databases and the latter as a whole. Thus, for example, a single database collating globin-gene variants has received more than 700 citations of its original report and subsequent updates (HbVar; http://globin.bx.psu.edu/hbvar) since its establishment in 2001 (an average of ~50 citations per year; Giardine B, personal communication). Especially when a database has been published in a formal scientific journal, citations of individual micro- or nanopublications in that database can be linked to the paper describing the resource. Citations to microattribution papers could, therefore, be significantly enhanced by rendering individual micro- or nanopublications from the underlying databases visible in computer-readable format. For this to work in practice, these should become an integral part of modern *enhanced* Science (e-Science) approaches, such as computer-aided reasoning over gene and protein interaction maps and other in silico knowledge discovery methods. When individual associations are discovered to be part of a wider interaction or causative network, they can receive citations, but these will also reflect the relevant article(s) describing the databases from which the micro- or nanopublications were derived. Thus, microattribution articles and micro- or nanopublications should mutually reinforce the proper citation and attribution of individual and batch contributions to the human variome (Fig. 6.4).

Lastly, it should be clarified that the technical ability to cite and attribute small yet meaningful—omics contributions is necessary, but not sufficient to change the culture of attribution and reward modalities in the biological sciences. Even though the value of incremental microcontributions to the scientific knowledgebase is widely recognized by the research community itself, the reward systems used by funders and faculty bodies to decide upon issues of tenure and promotion are not necessarily ready for this dramatic shift. It is, therefore, imperative that, alongside technical changes, social changes are also required to make the scientific reward system ready for the next decades, taking into account the crucial role that curated genomic data resources play in modern genomics and translational research. This may, in turn, facilitate the fundraising efforts of database curators to ensure the long-term maintenance and sustainability (and hence viability) of genomic databases, since this would be a significantly more robust and objective measure of database quality for reviewers to assess than either the number of citations to the reports describing the database itself or the database traffic metrics.

6.6 CONCLUSIONS

In the previous paragraphs, the key elements of the microattribution process, as a means to incentivize genomic variation data sharing, were outlined. Developing an incentivization process for placing human genome variation data into the public domain, such as by microattribution or through database-journals, is nowadays entirely feasible in technical terms and the first efforts to build such

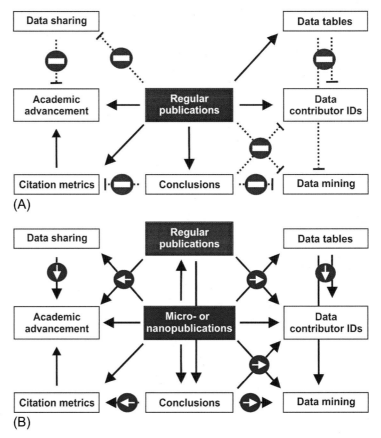

FIG. 6.4

Schematic depiction of the envisaged impact of microattribution in future scientific publishing. (A) Regular narrative publications often include large data tables which allow citation metrics by author name and hence can contribute to academic advancement. However, data mining and data sharing are both limited, while linking contents of the data tables back to the data contributor is trivial and often impossible, since a unique author ID is missing. (B) The introduction of micro- or nanopublications allows the unambiguous linking of data tables to data contributor IDs, which in turn can facilitate data mining. Moreover, data sharing is enabled which increases the likelihood of an article being retrieved and cited. Hence, citation metrics are also possible for micro- or nanopublications, whilst academic credit by microattribution is also possible, thereby contributing to academic advancement.

a process have already been materialized, while others are currently underway. However, the widespread adoption of these efforts will rely on innovative publication approaches, the means to uniquely identify individual data contributors as well as the enthusiasm and motivation of specific research laboratories, funders, and academic institutions to support these initiatives for the benefit of the wider research community and, most importantly, the patients and their families. The unfortunate reality is that persuading busy scientists to contribute their valuable time to the public good is always going to be problematic. However, the value of those data currently being gathered by genomic data resources is such that fundamental changes to the culture of data sharing are necessary for scientific progress to be made.

References

Anonymous, 2008. Human variome microattribution reviews. Nat. Genet. 40(1).

Ball, C., Brazma, A., Causton, H., Chervitz, S., Edgar, R., Hingamp, P., Matese, J.C., Parkinson, H., Quackenbush, J., Ringwald, M., Sansone, S.A., Sherlock, G., Spellman, P., Stoeckert, C., Tateno, Y., Taylor, R., White, J., Winegarden, N., 2004. Standards for microarray data: an open letter. Environ. Health Perspect. 112, A666–A667.

Celli, J., Dalgleish, R., Vihinen, M., Taschner, P.E., den Dunnen, J.T., 2012. Curating gene variant databases (LSDBs): toward a universal standard. Hum. Mutat. 33, 291–297.

Cooper, D.N., Chen, J.M., Ball, E.V., Howells, K., Mort, M., Phillips, A.D., Chuzhanova, N., Krawczak, M., Kehrer-Sawatzki, H., Stenson, P.D., 2010. Genes, mutations, and human inherited disease at the dawn of the age of personalized genomics. Hum. Mutat. 31, 631–655.

Cotton, R.G., Al Aqeel, A.I., Al-Mulla, F., Carrera, P., Claustres, M., Ekong, R., Hyland, V.J., Macrae, F.A., Marafie, M.J., Paalman, M.H., Patrinos, G.P., Qi, M., Ramesar, R.S., Scott, R.J., Sijmons, R.H., Sobrido, M.J., Vihinen, M., members of the Human Variome Project Data Collection from Clinics, Data Collection from Laboratories and Publication, Credit and Incentives Working Groups, 2009. Capturing all disease-causing mutations for clinical and research use: toward an effortless system for the human Variome project. Genet. Med. 11, 843–849.

Fokkema, I.F., Taschner, P.E., Schaafsma, G.C., Celli, J., Laros, J.F., den Dunnen, J.T., 2011. LOVD v.2.0: the next generation in gene variant databases. Hum. Mutat. 32, 557–563.

Georgitsi, M., Viennas, E., Gkantouna, V., Christodoulopoulou, E., Zagoriti, Z., Tafrali, C., Ntellos, F., Giannakopoulou, O., Boulakou, A., Vlahopoulou, P., Kyriacou, E., Tsaknakis, J., Tsakalidis, A., Poulas, K., Tzimas, G., Patrinos, G.P., 2011. Population-specific documentation of pharmacogenomic markers and their allelic frequencies in FINDbase. Pharmacogenomics 12, 49–58.

Giardine, B., van Baal, S., Kaimakis, P., Riemer, C., Miller, W., Samara, M., Kollia, P., Anagnou, N.P., Chui, D.H., Wajcman, H., Hardison, R.C., Patrinos, G.P., 2007. HbVar database of human hemoglobin variants and thalassemia mutations: 2007 update. Hum. Mutat. 28, 206.

Giardine, B., Borg, J., Higgs, D.R., Peterson, K.R., Philipsen, S., Maglott, D., Singleton, B.K., Anstee, D.J., Basak, A.N., Clark, B., Costa, F.C., Faustino, P., Fedosyuk, H., Felice, A.E., Francina, A., Galanello, R., Gallivan, M.V., Georgitsi, M., Gibbons, R.J., Giordano, P.C., Harteveld, C.L., Hoyer, J.D., Jarvis, M., Joly, P., Kanavakis, E., Kollia, P., Menzel, S., Miller, W., Moradkhani, K., Old, J., Papachatzopoulou, A., Papadakis, M.N., Papadopoulos, P., Pavlovic, S., Perseu, L., Radmilovic, M., Riemer, C., Satta, S., Schrijver, I., Stojiljkovic, M., Thein, S.L., Traeger-Synodinos, J., Tully, R., Wada, T., Waye, J.S., Wiemann, C., Zukic, B., Chui, D.H., Wajcman, H., Hardison, R.C., Patrinos, G.P., 2011. Systematic documentation and analysis of human genetic variation in hemoglobinopathies using the microattribution approach. Nat. Genet. 43, 295–301.

Guttmacher, A.E., McGuire, A.L., Ponder, B., Stefánsson, K., 2010. Personalized genomic information: preparing for the future of genetic medicine. Nat. Rev. Genet. 11, 161–165.

Homer, N., Szelinger, S., Redman, M., Duggan, D., Tembe, W., Muehling, J., Pearson, J.V., Stephan, D.A., Nelson, S.F., Craig, D.W., 2008. Resolving individuals contributing trace amounts of DNA to highly complex mixtures using high-density SNP genotyping microarrays. PLoS Genet. 4.

Horaitis, O., Cotton, R.G., 2004. The challenge of documenting mutation across the genome: the human genome variation society approach. Hum. Mutat. 23, 447–452.

MacArthur, D.G., Balasubramanian, S., Frankish, A., Huang, N., Morris, J., Walter, K., Jostins, L., Habegger, L., Pickrell, J.K., Montgomery, S.B., Albers, C.A., Zhang, Z.D., Conrad, D.F., Lunter, G., Zheng, H., Ayub, Q., DePristo, M.A., Banks, E., Hu, M., Handsaker, R.E., Rosenfeld, J.A., Fromer, M., Jin, M., Mu, X.J., Khurana, E., Ye, K., Kay, M., Saunders, G.I., Suner, M.M., Hunt, T., Barnes, I.H., Amid, C., Carvalho-Silva, D.R., Bignell, A.H., Snow, C., Yngvadottir, B., Bumpstead, S., Cooper, D.N., Xue, Y., Romero, I.G., 1000 Genomes Project Consortium, Wang, J., Li, Y., Gibbs, R.A., McCarroll, S.A., Dermitzakis, E.T., Pritchard, J.K., Barrett, J.C., Harrow, J., Hurles, M.E., Gerstein, M.B., Tyler-Smith, C., 2012. A systematic survey of loss-of-function variants in human protein-coding genes. Science 335, 823–828.

Mitropoulou, C., Webb, A.J., Mitropoulos, K., Brookes, A.J., Patrinos, G.P., 2010. Locus-specific database domain and data content analysis: evolution and content maturation toward clinical use. Hum. Mutat. 31, 1109–1116.

Mizzi, C., Dalabira, E., Kumuthini, J., Dzimiri, N., Balogh, I., Başak, N., Böhm, R., Borg, J., Borgiani, P., Bozina, N., Bruckmueller, H., Burzynska, B., Carracedo, A., Cascorbi, I., Deltas, C., Dolzan, V., Fenech, A., Grech, G., Kasiulevicius, V., Kádaši, Ľ., Kučinskas, V., Khusnutdinova, E., Loukas, Y.L., Macek Jr., M., Makukh, H., Mathijssen, R., Mitropoulos, K., Mitropoulou, C., Novelli, G., Papantoni, I., Pavlovic, S., Saglio, G., Setric, J., Stojiljkovic, M., Stubbs, A.P., Squassina, A., Torres, M., Turnovec, M., van Schaik, R.H., Voskarides, K., Wakil, S.M., Werk, A., Del Zompo, M., Zukic, B., Katsila, T., Lee, M.T., Motsinger-Rief, A., Mc Leod, H.L., van der Spek, P.J., Patrinos, G.P., 2016. A European spectrum of pharmacogenomic biomarkers: implications for clinical pharmacogenomics. PLoS One 11(9).

Mons, B., Velterop, J., 2009. Nano-Publication in the e-Science Era. Workshop on Semantic Web Applications in Scientific Discourse (SWASD 2009), Washington, DC, USA.

Mons, B., van Haagen, H., Chichester, C., Hoen, P.B., den Dunnen, J.T., van Ommen, G., van Mulligen, E., Singh, B., Hooft, R., Roos, M., Hammond, J., Kiesel, B., Giardine, B., Velterop, J., Groth, P., Schultes, E., 2011. The value of data. Nat. Genet. 43, 281–283.

Patrinos, G.P., Brookes, A.J., 2005. DNA, disease and databases: disastrously deficient. Trends Genet. 21, 333–338.

Patrinos, G.P., Petricoin, E.F., 2009. A new scientific journal linked to a genetic database: towards a novel publication modality. Hum. Genomics Proteomics 1.

Patrinos, G.P., Wajcman, H., 2004. Recording human globin gene variation. Hemoglobin 28, v–vii.

Patrinos, G.P., Giardine, B., Riemer, C., Miller, W., Chui, D.H., Anagnou, N.P., Wajcman, H., Hardison, R.C., 2004. Improvements in the HbVar database of human hemoglobin variants and thalassemia mutations for population and sequence variation studies. Nucleic Acids Res. 32, D537–D541.

Patrinos, G.P., Cooper, D.N., van Mulligen, E., Gkantouna, V., Tzimas, G., Tatum, Z., Schultes, E., Roos, M., Mons, B., 2012. Microattribution and nanopublication as means to incentivize the placement of human genome variation data into the public domain. Hum. Mutat. 33, 1503–1512.

Povey, S., Al Aqeel, A.I., Cambon-Thomsen, A., Dalgleish, R., den Dunnen, J.T., Firth, H.V., Greenblatt, M.S., Barash, C.I., Parker, M., Patrinos, G.P., Savige, J., Sobrido, M.J., Winship, I., Cotton, R.G., Ethics Committee of the Human Genome Organization (HUGO), 2010. Practical guidelines addressing ethical issues pertaining to the curation of human locus-specific variation databases (LSDBs). Hum. Mutat. 31, 1179–1184.

Sosnay, P.R., Siklosi, K.R., Van Goor, F., Kaniecki, K., Yu, H., Sharma, N., Ramalho, A.S., Amaral, M.D., Dorfman, R., Zielenski, J., Masica, D.L., Karchin, R., Millen, L., Thomas, P.J., Patrinos, G.P., Corey, M., Lewis, M.H., Rommens, J.M., Castellani, C., Penland, C.M., Cutting, G.R., 2013. Defining the disease-liability of mutations in the cystic fibrosis transmembrane conductance regulator gene. Nat. Genet. 45, 1160–1167.

Thompson, B.A., Spurdle, A.B., Plazzer, J.P., Greenblatt, M.S., Akagi, K., Al-Mulla, F., Bapat, B., Bernstein, I., Capellá, G., den Dunnen, J.T., du Sart, D., Fabre, A., Farrell, M.P., Farrington, S.M., Frayling, I.M., Frebourg, T., Goldgar, D.E., Heinen, C.D., Holinski-Feder, E., Kohonen-Corish, M., Robinson, K.L., Leung, S.Y., Martins, A., Moller, P., Morak, M., Nystrom, M., Peltomaki, P., Pineda, M., Qi, M., Ramesar, R., Rasmussen, L.J., Royer-Pokora, B., Scott, R.J., Sijmons, R., Tavtigian, S.V., Tops, C.M., Weber, T., Wijnen, J., Woods, M.O., Macrae, F., Genuardi, M., InSiGHT, 2014. Application of a 5-tiered scheme for standardized classification of 2,360 unique mismatch repair gene variants in the InSiGHT locus-specific database. Nat. Genet. 46, 107–115.

Thorisson, G.A., Muilu, J., Brookes, A.J., 2009. Genotype-phenotype databases: challenges and solutions for the post-genomic era. Nat. Rev. Genet. 10, 9–18.

Viennas, E., Komianou, A., Mizzi, C., Stojiljkovic, M., Mitropoulou, C., Muilu, J., Vihinen, M., Grypioti, P., Papadaki, S., Pavlidis, C., Zukic, B., Katsila, T., van der Spek, P.J., Pavlovic, S., Tzimas, G., Patrinos, G.P., 2017. Expanded national database collection and data coverage in the FINDbase database worldwide database for clinically relevant genomic variation allele frequencies. Nucleic Acids Res. 45, D846–D853.

Zlotogora, J., van Baal, S., Patrinos, G.P., 2007. Documentation of inherited disorders and mutation frequencies in the different religious communities in Israel in the Israeli National Genetic Database. Hum. Mutat. 28, 944–949.

Zlotogora, J., van Baal, J., Patrinos, G.P., 2009. The Israeli National Genetics database. Isr. Med. Assoc. J. 11, 373–375.

PART 2

Human Genome Informatics Tools and Related Resources

A Review of Tools to Automatically Infer Chromosomal Positions From dbSNP and HGVS Genetic Variants

**Alexandros Kanterakis*,†, Theodora Katsila†, George Potamias*,
George P. Patrinos†, Morris A. Swertz‡**

**Institute of Computer Science, Foundation for Research and Technology Hellas (FORTH),
Heraklion, Greece, †Department of Pharmacy, School of Health Sciences, University of Patras,
Patras, Greece, ‡Genomics Coordination Center, University Medical Center Groningen,
University of Groningen, Groningen, The Netherlands*

7.1 INTRODUCTION

A large proportion of scientific publications in the field of genomics focuses on discovering novel genomic variations in the human genome and investigating known ones. Moreover, there are hundreds of online databases with known variants. For example, LOVD (Fokkema et al., 2011), one of the most popular databases for variants, is installed in 89 different instances reporting more than 10 million variants, whereas dbSNP contains more than 100 million validated SNPs. DbSNP assigns a unique ID (rs#) for validated variants, in which the specifics of a variant—like the type of mutation and the position—are unambiguously defined. For new or under-investigation variants, the most common nomenclature for reporting genomic variants and their positions is HGVS. According to HGVS, variations are reported in terms of their relative position and basepair differences in a reference sequence. This sequence can be a transcript, a chromosome, or a database accession number. Official HGVS directives were introduced in 2001 and the last update was published in 2016 (den Dunnen et al., 2016). Nevertheless, HGVS-like variation reporting was common even in publications in the 1990s[1] (Tack et al., 2016).

Although HGVS supports the definition of a variant according to a complete reference genome (e.g., chr1:12345678A>G) this is rarely used in scientific reports.[2] One reason for this is that a complete reference genome was only introduced years after the HGVS was being used in practice. The hg1 human genome reference was introduced in 2000, as part of the Human Genome Project and has

---

[1]See http://www.hgvs.org/mutnomen/history.html.

[2]See also practical problems genomic reference sequence: http://www.hgvs.org/mutnomen/refseq.html.

Human Genome Informatics. https://doi.org/10.1016/B978-0-12-809414-3.00007-3

undergone 18 major updates since then. Another reason is that HGVS allows the position to be stated according to the location on the mRNA of a gene transcript. According to HGVS this is called coding location (denoted by "c." prefix, e.g., NM_001104.3:c.1729C>T) in contrast with the genomic location (denoted by "g." prefix, e.g., NG_013304.2:g.18705C>T). The practical value of reporting the coding location is that it is obvious if a variant is in an exonic or intronic region and sometimes it reveals the effect in the translation (e.g., a variant in a splice site). This makes the description of the variant more informative because it provides the exact position of the mRNA where the mutation occurred. Another advantage of the mRNA level notation is that it is shorter. For example, a report that investigates a mutation of the BRCA1 gene is more coherent when stating "c.211A>G" rather than "chr17:g.41258474T>C of hg19."

To address potential confusion and ambiguity stemming from multiple reference sequences that describe the same variant, HGVS nomenclature recommends the use of Locus Reference Genomic (LRG) sequence ID for this purpose. LRGs are a manually curated record of reference sequences that guarantees an un-versioned (i.e., unchanged) coding scheme (MacArthur et al., 2014). Nevertheless, because LRG has not been widely adopted, the HGVS consortium also recommends the use of RefSeq or Ensembl transcript. RefSeq sequences are a collection of genomic DNA, gene transcripts, and proteins that exhibit certain qualitative criteria for a variety of well-studied organisms maintained by NCBI.[3] Ensembl reference sequences are based partly on RefSeq but include additional resources like UniProt as well as manually curated entries.[4] It is known that the differences between these two resources have effects on certain analysis pipelines (Zhao and Zhang, 2015) including variant annotation (McCarthy et al., 2014). Moreover, as we will present later, certain HGVS analysis tools have preferences for specific sources, so that the choice of a reference sequence source in an HGVS variant can affect the readability and subsequently the significance of the variant.

Recently there is an increasing desire to identify already reported variants in existing genotyped or sequenced samples, with a strong driver being the clinical genetics community. An example of a relevant service is Promethease,[5] which uses a community curated variant database, SNPedia, to locate variants in a set of genotypes provided by the user (Cariaso and Lennon, 2012). These efforts belong to the "translational bioinformatics" field, which tries to apply genetics knowledge in clinical and medical practice (Overby and Tarczy-Hornoch, 2013) in order to bring the vision of personalized medicine closer to reality.

---

[3]See http://www.ncbi.nlm.nih.gov/books/NBK21105/#ch1.Appendix_GenBank_RefSeq_TPA_and_UniP.

[4]See http://www.ensembl.org/info/genome/genebuild/index.html.

[5]See https://www.promethease.com/.

These services analyze individual genetic profiles that have been generated either by whole genome genotyping techniques or whole genome sequencing after the application of a variant calling pipeline. In both cases the matching between the individual genetic profile and the set of reported variants available is based on the genetic location. Therefore, the process of identifying the chromosomal location of an HGVS variant is becoming increasingly important in clinical genetics.

In theory this variant identification process should be trivial, because an HGVS variant reports a unique genomic position regardless of the chosen coding scheme. However, recent studies have identified several difficulties that severely impede this process. For example, in one study (Deans et al., 2016), 26 different laboratories were given tumor samples and asked to locate and report genetic variants. Only six laboratories reported all the correct variants in the right HGVS format. Among other discrepancies, four laboratories reported incorrect positions of indels, while one laboratory reported a correct position but an incorrect reference. In another study (Pandey et al., 2012), the authors examined the validity of reported variants in LSMDs and reported common mistakes; among the most prominent errors were numbering issues on the cDNA level as well as on the genetic level of reported transcripts. For example, although the HGVS directives state that the numbering should start from the translation initiation site, some papers report the position according to the transcription initiation site. Many LSMDs copy this information directly from papers without performing basic quality and uniformity checks. Finally, another study (Tack et al., 2016) focused on mistakes that involved the incorrect use of the reference sequence in HGVS. These are the omission of the transcript version or the use of a gene name instead of a transcript name. They also discussed the consequences of these mistakes for research and even in health care. Interestingly, errors on variant reporting can occur even in well-designed studies that have submitted their findings to dbSNP (Sand, 2007).

Various tools have been developed to remedy these issues by performing quality checks on reported variants in HGVS. Perhaps the most prominent tools are Mutalyzer (Wildeman et al., 2008), BioCommons/HGVS (Hart et al., 2015), and Counsyl[6] ; they offer methods for HGVS parsing and conversion. Other tools for variant annotation and clinical effect prediction (e.g., variant effect predictor [VEP] (McLaren et al., 2010), Variation Reporter,[7] and TransVar (Zhou et al., 2015)) also accept input in HGVS and perform this quality check as a secondary task. In contrast to some well-known annotation tools that do not accept HGVS variants as input like ANNOVAR[8] (Wang et al., 2010),

---

[6]See https://github.com/counsyl/hgvs.

[7]See http://www.ncbi.nlm.nih.gov/variation/tools/reporter.

[8]See http://annovar.openbioinformatics.org/en/latest/user-guide/input/.

GEMINI (Paila et al., 2013), and SNPnexus (Dayem Ullah et al., 2013). Additionally, existing databases that list variants in HGVS format are LOVD (Fokkema et al., 2011) and ClinVar (Landrum et al., 2014).

Here we assess the ability of HGVS-compatible tools to successfully infer the genomic positions of variants. For the remaining of this report, we will call this task "chromosomal position resolution of HGVS variants." Here, we discuss the effectiveness and limitations of each tool to perform this task and we also suggest improvements. We also introduce a new tool, MutationInfo, which offers a unified access to Mutalyzer, Biocommons/HGVS, and Counsyl for the purpose of locating the genomic position of an HGVS variant. In case all these methods fail (or if explicitly instructed), it performs a high precision, but slow "low-level" technique to infer this information. In short, MutationInfo parses the HGVS variants and extracts the reference sequence. Then it downloads this sequence in FASTA format and performs a BLAT search (Kent, 2002) in the reference genome to locate the starting position of this sequence. Finally, it calculates the offset of the reported position of the HGVS variant in the downloaded sequence and reports the genomic position.

As a testing set we used PharmGKB (Medina et al., 2011), which is a prominent database in the field of pharmacogenetics. The four main benefits of using PharmGKB as a testing platform are: (1) it contains variants with direct clinical significance; (2) it does not have any standardized method for variant reporting as it contains variants in both dbSNP, HGVS, and arbitrary formats; (3) it is manually curated; and (4) it contains variants published over a long time (1995 onward). These characteristics make PharmGKB both very important as an LSDB and, at the same time, error-prone with regard to the quality and validity of the variants it contains. The evaluation of PharmGKB variants from MutationInfo revealed that certain classes of HGVS variants could not be correctly located in a reference genome by existing tools. Consequently, we applied MutationInfo as an alternatively pipeline that attempts to use a variety of existing tools and if they all fail then a BLAT search is used. Using this pipeline we located the genetic location of all HGVS variants in PharmGKB. Additionally, we identified errors that most probably came from careless transfer of variants from papers to the database. We also identified errors that could be attributed to the incorrect use of HGVS notation in the original papers.

MutationInfo is implemented in python with a very simple programming interface. It offers a simple function `get_info` that accepts variants in both dbSNP and HGVS. This format agnostic interface makes MutationInfo ideal for fast genomic location and quality check of a set of variants that belong either in dbSNP or HGVS nomenclature. For dbSNP variants it sequentially uses CruzDB (Pedersen et al., 2013), VEP and MyVariant.info (Xin et al., 2015, 2016) to determine the genetic position of the variant. As we will demonstrate, we also need a combination of tools in order to get the genetic position of all (880) dbSNP variants in PharmGKB.

7.2 EXISTING TOOLS FOR HGVS POSITION RESOLUTION

In this section we present how to use existing online services and tools that offer position resolution of HGVS variants. All presented tools offer position resolution as a small part of their functionality, because there is not any tool solely targeted to this purpose. Therefore, this section does not cover the many useful tasks that these tools also perform. Moreover, we do not focus on the ability of these tools to process complex and uncommon mutation types that HGVS supports.

Being able to automatically perform position resolution of HGVS variants through a scripting language relieves us from a repetitive manual work and also contributes to the reproducibility of our analysis. We will use the python programming language (version 3) and we also assume that the packages requests and pandas are installed. Requests (Reitz, 2011) is a package that simplifies access to online resources and pandas (McKinney, 2010) offers convenient methods for managing tabular data. We will start by importing these packages:

```
import requests
import pandas as pd
```

7.2.1 Mutalyzer

Mutalyzer (Wildeman et al., 2008) offers an online Application Programming Interface (API) to automatically access most of its functionality. Assuming that we want to infer the chromosomal position of the variant NM_001276506.1: c.204C>T in the hg38 version of the human genome, we will type:

```
r = requests . get (
   ' https :// mutalyzer . nl / json / numberConversion ' ,
   {
   ' build ' : ' hg38 ' ,
   ' variant ' : ' NM_001276506 .1: c .204 C > T '
   })
data = r . json ()
print ( data )
OUTPUT : [ ' NC_000011 .10: g .112088901 C > T ' ]
```

Unfortunately Mutalyzer offers this functionality only in HGVS variants that are using NM_, NC_, or LRG_ accession numbers. For variants that are using other NCBI reference sequences, Mutalyzer can still be useful by performing a "c. to g." conversion. For example, assuming the variant NG_008377.1:c.-1289G>A, we can apply the following commands:

```
r = requests . get (
    ' https :// mutalyzer . nl / json / runMutalyzer ' ,
    {
    ' variant ' :  ' NG_008377 .1: c .144 G > A '
    })
data  =  r . json ()
print  ( data [ ' summary ' ])
OUTPUT :  0  Errors ,  0  Warnings .

print  ( data [ ' genomicDescription ' ])
OUTPUT :  NG_008377 .1: g .5165 G > A
```

Knowing the genomic position of a variant can help us to recover the chromosomal position as we will present later.

7.2.2   The HGVS Python Package

The HGVS python package (Hart et al., 2015) offers methods to parse variants, perform "c. to g." conversions, and map a variant to arbitrary transcripts. To perform these operations HGVS maintains a connection with a database that contains transcript information. By default this database is Universal Transcript Archive (UTA). To perform position resolution with HGVS, we initially have to import the appropriate HGVS modules:

```
import  hgvs . parser
import  hgvs . dataproviders . uta
import  hgvs . assemblymapper

hdp  =  hgvs . dataproviders . uta . connect ()
hgvs_parser  =  hgvs . parser . Parser () . parse_hgvs_variant
assembly_mapper  =  hgvs . assemblymapper . AssemblyMapper (
                hdp ,  assembly_name = " GRCh38 " )
transcript_mapper  =  hgvs . variantmapper . VariantMapper ( hdp )
```

In this example, the object hgvs_parser validates and parses HGVS variants, the object assembly_mapper maps a variant to a reference genome and the object transcript_mapper maps a variant to another transcript. To locate the chromosomal position of the variant NM_001276506.1:c.204C>T that we also used in the Mutalyzer example, we will have to use the following commands:

```
var  =  ' NM_001276506 .1: c .204 C > T '
var_parsed  =  hgvs_parser ( var )
var_mapped  =  assembly_mapper . c_to_g ( var_parsed )
print  ( var_mapped )
OUTPUT :  ' NC_000011 .10: g .112088901 C > T '
```

As we can see the result is the same with the one obtained from Mutalyzer. Mutalyzer managed to perform "c. to g." conversion for variant NG_008377.1:
c.-1289G>A but failed to locate a chromosomal position. Interestingly, HGVS
shows a inverse functionality: fails to perform "c. to g." on this variant, but succeeds in locating a chromosomal position. For this reason we will take the output
of the "c. to g." operation from Mutalyzer, which is NG_008377.1:g.5165G>A
and we will locate a chromosomal position with HGVS:

```
var  =  ' NG_008377 .1: g .5165 G > A '
var_parsed  =  hgvs_parser ( var )
transcripts  =  assembly_mapper . relevant_transcripts ( var_parsed )
print  ( transcripts )
OUTPUT :  [ ' NM_000762 .5 ' ]

var_NM  =  transcript_mapper . g_to_c ( var_parsed ,  ' NM_000762 .5 ' )
print  ( var_NM )
OUTPUT :  NM_000762 .5: c .144 G > A

print  ( assembly_mapper . c_to_g ( var_NM ) )
OUTPUT :  NC_000019 .10: g .40850283 C > T
```

This is also a demonstration of the synergy that can be achieved by combining
more than one tools.

7.2.3   Variant Effect Predictor

The primary use of VEP (McLaren et al., 2010) is to annotate a variant with
effect information. In addition, VEP also provides useful location information
including possible affected transcripts. Since VEP belongs to EMBL's large family of bioinformatics tools, it also supports Ensembl transcripts. Accessing VEP
is easy through the requests library. Initially we define VEP's URL and the
enquired HTML headers:

```
headers ={  " Content - Type "  :  " application / json " }
vep_url  =  " https :// rest . ensembl . org / vep / human / hgvs /{ var }? "

def  vep ( var ) :
```

```
url = vep_url . format ( var = var )
responce = requests . get ( url , headers = headers )
if responce . ok :
  return responce . json ()
else :
  print ( ' Error : ' )
  print ( responce . text )
```

Now we can request location information for a specific HGVS variant:

```
var = ' NM_001276506 .1: c .204 C > T '
data = vep ( var )

print ( data [0] [ ' assembly_name ' ])
OUTPUT : GRCh38

print ( data [0] [ ' seq_region_name ' ])
OUTPUT : 11

print ( data [0] [ ' start ' ])
OUTPUT : 112088901
```

Again we notice that the output has the same information with the previous methods. VEP fails to locate position information for both `NG_008377.1: c.-1289G>A` and `NG_008377.1:g.5165G>A` variants.

7.2.4 Variation Reporter

Variation reporter belongs to the NCBI family of bioinformatics tools and provides similar functionality to VEP. As with VEP, it focuses on providing clinical effect of variants and it also performs position resolution. On the following lines we define the function `Variation_Reporter` which handles the requests to this service.

```
url = ' https :// www . ncbi . nlm . nih . gov / projects / SNP / VariantAnalyzer
  / var_rep . cgi '

def Variation_Reporter ( var ) :
  r = requests . post ( url , { ' annot1 ' : var , })

  if r . ok :
    return r . text
  else :
    print ( ' ERROR : ' )
    print ( r . text )
```

Variation reporter generates data in simple text format as opposed to other tools that export data in a structured JSON format. This means that users have to write custom functions to parse the output. The following listing is a simple function that converts output from Variation Reporter into a pandas object (McKinney, 2010).

```
from io import StringIO

def parse_vr_results ( results ) :

   lines = results . split ( ' \n ' )
   lines = filter ( lambda x : x [0:2] not in
      [ ' Su ' , ' . ' , ' .. ' , ' ## ' ], lines )
   lines = ' \n ' . join ( lines )

   data_f = StringIO ( lines )
   return pd . read_csv ( data_f , sep = ' \t ' )
```

We can demonstrate the functionality of the above with the following commands:

```
var = ' NM_001276506 .1: c .204 C > T '
vr_results = Variation_Reporter ( var )
vr_parsed = parse_vr_results ( vr_results )

print ( vr_parsed [ ' Hgvs_g ' ] [0])
OUTPUT : ' NC_000011 .9: g .111959625 C > T '
```

We notice that Variation Reporter mapped the variant in an older genome assembly (hg19) compared with other tools. Fortunately, apart from VEP, the EMBL also includes a service that converts the coordinates of a variant from one assembly to another. This tool is Assembly Converter and can be accessed through the EMBL's REST API with the following code:

```
AC_url = ' https :// rest . ensembl . org / map / human / GRCh37 '
AC_url += ' /{ chr }:{ pos }..{ pos }:1/ GRCh38 '

def from_19_to_38 ( chromosome , position ) :
    url = AC_url . format (
        chr = chromosome ,
        pos = position
    )

   headers = { ' content – type ' : ' application / json ' }

   request = requests . get ( url , headers = headers )

   if request . ok :
```

```
data = request . json ()
return data [ ' mappings ' ] [0] [ ' mapped ' ] [ ' start ' ]
else :
print ( ' ERROR : ' )
print ( request . text )
```

Now we can convert the variant `NC_000011.9:g.111959625C>T` from hg19 to hg38:

```
new_location = from_19_to_38 (11, 111959625)
print ( new_location )
OUTPUT : 112088901
```

Again we notice that the location is the same as the one inferred from the previous tools. Also Variation Reporter succeeded to locate a chromosomal position for `NG_008377.1:g.5165G>A` but failed for `NG_008377.1:c.-1289G>A`. Converting a genomic loci from one assembly to another is called "liftover." NCBI also has a tool for this task, which is called Remap. Nevertheless, this tool is available either through a web interface or through a Perl script.

7.2.5   Transvar

Transvar (Zhou et al., 2015) is a tool that applies various annotations and conversions for genomic loci that are defined in HGVS. Assuming a common Linux or OSX command-line environment, Transvar can be installed with the following commands:

```
pip install transvar
pip install pytabix
transvar config - download_anno - refversion hg38
transvar config - download_ref - refversion hg38
```

The last two commands download locally a set of annotation databases that Transvar consults. These databases include RefSeq (NCBI), Ensembl, UCSC, and AceView (Thierry-Mieg and Thierry-Mieg, 2006). Transvar is a command-line tool, which means that we cannot access its functionality through python. Nevertheless we can execute the tool from the command line and then process the output. For example, the following commands consult the RefSeq database for the `'NM_001276506.1:c.204C>T'` variant:

```
transvar canno - i 'NM_001276506 .1: c .204' -- refseq -- refversion hg38
   > output . txt
```

```
data = pd . read_csv ('output . txt ', sep = '\t')
print data ['coordinates ( gDNA / cDNA / protein )'][0]
OUTPUT : chr11 : g .112088901 C / c .204 C /.
```

Again we can verify that this is the same position that was inferred with all the previous tools. Transvar failed to process variants having the sequence NG_008377.1 since it could not locate it in any of the databases that it supports.

7.2.6 BioPython

BioPython (Cock et al., 2009) is Python's most known library for biological computation. It contains a large variety of methods for accessing online genetic databases, sequence annotation, statistic inference, machine learning, and clustering. Here, we will show how we can use Biopython to infer the chromosomal position of the HGVS variant: NG_008377.1:g.5165G>A. Initially we will download the reference sequence (NG_008377.1) in FASTA format and then we will apply BLAT in order to identify the location in the human genome where this sequence is aligned. Assuming we have BioPython installed, initially we import the necessary libraries:

```
from Bio import Entrez
from Bio . Blast import NCBIWWW
from Bio . Blast import NCBIXML

Entrez . email = 'anonymous@gmail . com'
```

The Entrez library (Maglott et al., 2004) allows access to NCBI's large collection of databases. With the following commands, we download the reference sequence NG_008377.1 of the variant in FASTA format:

```
handle = Entrez . efetch (
  db = 'nuccore ',
  id = 'NG_008377 .1 ',
  retmode = 'text ',
  rettype = 'fasta ')

fasta_data = handle . read ()
```

We can inspect the `fasta_data` variable to confirm that it contains a FASTA record. The next step is to perform a BLAT search. BioPython offers many options regarding the BLAT algorithm, target database, and organisms. A full list of available target databases (e.g., 16S ribosomal RNA sequences or expressed sequence tags) for NCBI is available in NCBI's FTP site documentation (Tao et al., 2008). The following commands perform a BLAT search in the human reference genome and then parse the results.

```
result_handle = NCBIWWW.qblast("blastn", "refseq_genomic_human",
    fasta_data)
blast_record = NCBIXML.read(result_handle)
```

A BLAT search results in various alignments whose number and quality depends on the choice of the target database. These alignments are usually called "hits." Each alignment has many high-scoring pair (HSP), which are fragments (or frames) of queried-reference sequence pairs accompanied with statistics produced from the search algorithm. For our example, we will take the first HSP from the first alignment of the BLAT search:

```
first_sequence = blast_record.alignments[0]
print (first_sequence.title)
OUTPUT: '.. Homo sapiens chromosome 19, GRCh38.p7 Primary Assembly'

first_match = first_sequence.hsps[0]
print (first_match)
OUTPUT:
Score 27820 (25086 bits), expectation 0.0e+00, alignment length
  13318
Query: 1 CTTGTCATCAATCTGAAATGTGAAACACAGACATTTCACTCTCTG ... TCT 13910
         |||||||||||||||||||||||||||||||||||||||||||||| ... |||
Sbjct:40855447 CTTGTCATCAATCTGAAATGTGAAACACAGACATTTCACTCTCTG ... TCT
  40841538
```

Here we have the chromosome and reference genome information. Also, the "expectation" is a metric of the quality of the match. The 0.0 value of the "expectation" in our case indicates a perfect match. With the following commands we get the absolute position of the aligned sequence in the reference genome:

```
print (first_match.sbjct_start)
OUTPUT: 40855447

print (first_match.sbjct_end)
OUTPUT: 40841538

print (first_match.query_start)
OUTPUT: 1
```

Here we notice the start is lower than the end. This indicates that the sequence that we queried was aligned in the negative strand of the reference. Therefore, in order to locate the position of the `NG_008377.1:g.5165G>A` variant, we need to count 5165 nucleotides before the start position. If the alignment was in the positive strand then we should count forward (plus) instead of backwards (minus). The final calculations are as follows:

```
result = first_match.query_start + first_match.sbjct_start - 5165
print ( result )
OUTPUT : 40850283
```

For validation, we confirm that the result is the same as with the HGVS python package.

7.3 THE MUTATIONINFO PIPELINE

From the previous section it is eminent that existing tools vary on their ability to perform position resolution. Moreover, this task often requires the combination of more than one tool. To activate this problem we present MutationInfo. MutationInfo is foremost a frontend to existing tools for recovering the chromosomal position of HGVS and dbSNP formatted variants. It offers a simple function named `get_info`, which accepts variants in both formats. The methods are summarized in Fig. 7.1.

If a variant with a dbSNP name is detected (the format should be "-rs#") then MutationInfo applies the following pipeline: it queries VEP, MyVariant.info, and CruzDB in this order. If any of them succeeds in recovering a chromosomal position then it returns this position and stops, otherwise it moves to the next service in the list. The order of the tool list was constructed according to their efficiency to locate variants in PharmGKB. VEP is a service from Ensembl that determines the effect of a variant. It is basically an aggregator of useful tools like SIFT (Kumar et al., 2009) and PolyPhen (Adzhubei et al., 2010), along with basic location and reference/alternative information. Although VEP also accepts variants in HGVS, it does not recognize reference accession numbers from RefSeq (it recognizes gene names and Ensembl transcripts). Because the vast majority of reported HGVS variants are using a RefSeq accession number, MutationInfo does not use this tool for HGVS processing in the default settings. The second tool is MyVariant.info, which is a variant annotation service that has collected information from various resources.[9] The documentation of this tool (Xin et al., 2015) states that it offers variant annotation for HGVS

---

[9]For a complete list see: http://myvariant.info/metadata.

variants that state explicitly the chromosomal position (i.e., chr1: g.35367G>A). Although it mentions that alternative reference sequences (like NM_003002.2:c.274G>T) are also supported, we could not verify this by running the tool. For this reason MutationInfo by default uses this tool only for dbSNP variants. Finally CruzDB (Pedersen et al., 2013) is a python interface for the database tables of UCSC's genome browser and it does not support HGVS.

If a variant is not in dbSNP format ("rs#"), then MutationInfo assumes that it is formatted as HGVS. Initially, it checks if the variant contains any common format error. MutationInfo recognizes and corrects four common mistakes in HGVS formatting (see Table 7.1) that made all the available tools fail. Once potential format errors have been corrected, MutationInfo runs a pipeline for recovering a chromosomal position for the variant. The first step is to use the "spline" and "BLAT" method of BioCommons (Hart et al., 2015). If both of these methods fail, then it tries the Counsyl tool. Although these tools do not have the highest efficiency, according to our experiments, they do perform the complete computation locally without

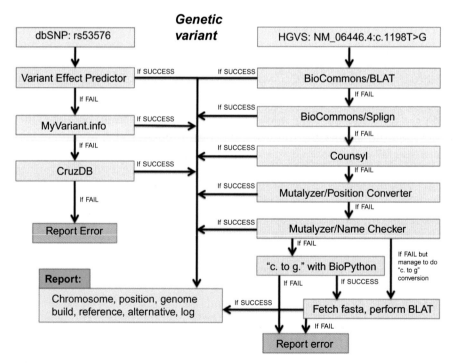

FIG. 7.1

MutationInfo's default pipeline for recovering the chromosomal position of a genetic variant.

Table 7.1 Parsing Problems of HGVS Variants in PharmGKB

Problem	Example		PharmGKB Cases
	Erroneous	Correct	
"->" instead of "-"	1048G->C	1048G>C	121
Multiple alternatives	844G>C/T	844G>C, 844G>T	8
Parentheses	−1126(C>T)	−1126C>T	75
Invalid substitution	1160CC>GT	1160_1161delinsGT	2
			Total: 206

Notes: *MutationInfo corrects these problems.*

requiring Internet access to external services. Therefore, for the sake of speed, MutationInfo prioritizes use of these tools. If these tools fail, MutationInfo makes a request to Mutalyzer's Position Converter tool.[10] This is a tool dedicated to locating a chromosomal position in HGVS variants. Position Converter usually fails if the HGVS reference sequence is not an NM_ RefSeq transcript. However, the Name Checker tool in Mutalyzer, although its purpose is not solely to locate a chromosomal position, is very successful in making "c. to g." conversions. This is the task of converting the position of a mutation from coding DNA to genomic coordinates.[11] Whenever Position Converter fails then MutationInfo makes a request to the Name Checker tool from Mutalyzer. This task can have three possible outcomes: the first is that Name Checker recovers the chromosomal position of the variant and performs a "c. to g." conversion; the second is that it succeeds only in the "c. to g." conversion; and the third is that it returns an error. If the first scenario happens, then MutationInfo returns the chromosomal position and stops. For the second, MutationInfo takes the genomic coordinates and performs a BLAT search to find the chromosomal position. For the third scenario, MutationInfo downloads the reference sequence's GenBank record from NCBI. The GenBank record has information on the relative genomic positions of the CDS (if any) and the start and end positions of the sequence's exons (if any). MutationInfo feeds this information to a BioPython (Cock et al., 2009) external module[12] that performs a "c. to g." conversion. Subsequently, based on the genomic position of the variant, MutationInfo performs a BLAT search to recover the chromosomal position.

---

[10]See https://mutalyzer.nl/position-converter.

[11]See http://www.hgvs.org/mutnomen/recs.html\#general.

[12]In GIT terms, this is a "fork" of BioPython that has not been "merged" to the official release: https://github.com/lennax/biopython/tree/f_loc5/Bio/SeqUtils/Mapper.

Whenever needed, the BLAT method in MutationInfo works as follows: MutationInfo downloads the FASTA file of the reference sequence of the variant. Given that the genomic position (g.) of the variant is known, it first checks if the reference sequence in FASTA is the same as the reference in the HGVS variant. The next step is to isolate a subsequence starting 20,000 nucleotides upstream to 20,000 downstream from that genomic position (or shorter if the start or end of the reference sequence is reached). Subsequently it accesses UCSC's BLAT search service (Kent, 2002; Bhagwat et al., 2012) and performs a sequence alignment of the isolated sequence in the human reference genome. Finally, it reports the position with the highest BLAT score (Bhagwat et al., 2012).

Fig. 7.1 shows the default pipeline that MutationInfo follows to recover the chromosomal position of a genetic variant. For more customized tasks, the user can provide additional options to use only specific tools. One limitation of the current implementation is that although the desired genome assembly is an optional parameter, MutationInfo does not guarantee that the returned position will belong on this assembly. The reason for this is that certain tools are still not fully adapted to the hg38 version; therefore, an older assembly is used. Of course the returned item always contains the version of the genome assembly used. In our future work, we plan to embed liftover methods in order to offer seamless conversion between genome assemblies.

Finally some additional implementation notes: MutationInfo keeps a local storage of all data generated from requests on external services in order to minimize Internet access and possible disruption to these services from repetitive calls that contain common variants. Moreover, it includes a setup script, which installs all software dependencies (BioCommons, Counsyl, CruzDB, BioPython) along with their required datasets (i.e., reference genomes). Finally, MutationInfo contains a web application implemented with the Django web framework[13] that is easily embeddable in other web services. The purpose of this application is to offer a lightweight web service that infers (or validates) the chromosomal position of HGVS (or dbSNP) variants as an extra component in existing LSMD installations.

7.4   RESULTS

To evaluate the ability of MutationInfo to locate the chromosomal positions of HGVS variants, we analyzed the highly curated content of PharmGKB (Medina et al., 2011). PharmGKB is an LSMD with variants and haplotypes of pharmacogenomics interest that are collected manually from publications.

---

[13]See https://www.djangoproject.com/.

The content of PharmGKB can be divided into four categories: variant annotations, clinical annotations, dosing guidelines, and haplotypes. The difference between clinical annotations and dosing guidelines is that the latter are supported by official or professional societies, whereas the former contain information with lower scientific support. Both are referred to either in individual variants or in haplotypes that contain many variants. Although the clinical interpretation of single variants is relatively straightforward, haplotypes are more challenging since they are continuously updated (Crews et al., 2012). Since haplotypes contain more dynamic and thus error-prone content, in this study we focused on variants that comprise the haplotypes listed in PharmGKB.

PharmGKB's website[14] was accessed in May 2016. It contained 89 haplotypes from 78 different genes.[15] These haplotypes contained 1554 different variants in total; 880 were in dbSNP format, 634 are in HGVS format, and 40 were free text descriptions of the variants (e.g., "TAA ins between 1065 and 1090"). The reference sequences were not directly available for 542 of these HGVS variants. To locate this we used either the reference sequence suggested by the official nomenclature of the gene (e.g., cypalleles for the CYP gene family) or the more frequently occurring transcript of the gene according to NCBI's gene search service.[16] For our analysis, we tested all the tools available in MutationInfo's pipeline (VEP, MyVariant.info, CruzDB, BioCommons, Counsyl, Mutalyzer, BLAT) along with two external tools (Variant Reporter, Transvar) and one database (LOVD). It should be noted that we used the latest versions of tools and services as of May 2016. Also, we used the dbSNP and HGVS formatted variants for the evaluation, (described following) and excluded the free text descriptions from our analysis.

7.4.1 Analysis of dbSNP Variants

PharmGKB contained 1014 SNPs in RS# format of which 880 were unique. All three tools were able to infer a genetic position for 781 SNPs, 91 SNPs were located only by VEP and CruzDB, 6 SNPs were located only by VEP, and 1 SNP was located only by MyVariant.info. There were no SNP located only by CruzDB (see Venn diagram in Fig. 7.2). Finally, one SNP (rs2839940) was not identified by any of the available tools. After closer examination it was revealed to be an erroneously reported SNP in PharmGKB, the correct name was rs28399440. We also located the following problems: rs72549375 from CYP1B1 does not exist in dbSNP although it is referred to in the relevant cypalleles page; rs35044222, rs67944833, rs67878449 from SULT1A3 gene are listed in dbSNP without any position information; and rs72558184 from CYP2C9

---

[14]See https://www.pharmgkb.org/.

[15]See https://www.pharmgkb.org/search/browse.action?browseKey=haplotypeGenes.

[16]See http://www.ncbi.nlm.nih.gov/gene/.

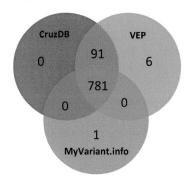

FIG. 7.2
Venn diagram of the number of rs# SNPs identified by CruzDB, VEP, and MyVariantinfo.

does not exist in cypalleles but is referred to in PharmGKB. This SNP does not have position information in dbSNP. For these five SNPs although the dbSNP ID (rs#) was declared in PharmGKB we used instead the HGVS notation that was also available.

All tools (CruzDB, VEP, and MyVariant.info) returned the same location for the same SNPs. Nevertheless we noticed that for indels, the tools returned different start and end positions. Although these values were consistent for each tool, the semantics of "start" and "end" positions differed. In Table 7.2, we present an example of the different start and end positions returned by all tools for an insertion and a deletion. For insertions, CruzDB returned the same position as start and end. By contrast, in VEP and in MyVariant.info, the start position differed from the end position by the length of the insertion. Surprisingly, VEP's start position was higher than the end position. For deletions, all tools returned the same end position. Nevertheless, the start position of CruzDB was lower than the end position by a value equal to the size of the deletion. In general, the only tool that provided values in accordance with the HGVS equivalent

Table 7.2 Differences in Start and End Positions Returned by Different Tools for Insertions and Deletions

Type	Insertion		Deletion	
rs#	rs72549385		rs3215983	
HGVS	NC_000002.12:g.38074887_38074888insA		NC_000001.11:g.46815075_46815076delAT	
	Start	End	Start	End
CruzDB	38074887	38074887	46815074	46815076
VEP	38074888	38074887	46815075	46815076
MyVariant.info	38074887	38074888	46815075	46815076

of the dbSNP variant was MyVariant.info. These differences can confound an analysis pipeline or researcher that uses more than one tool. MutationInfo returns consistent location information regardless of the underlying tool that has been used. This location is the same with the HGVS start position as if a chromosomal reference sequence had been used (e.g., chr1:123456_123457insA).

7.4.2 Analysis of HGVS Variants

PharmGKB contains 634 variants in HGVS format. These variants are using multiple types of RefSeq sequences as well as GenBank sequences. We applied all aforementioned tools individually and we also applied the MutationInfo pipeline. Table 7.3 contains the number of variants in PharmGKB that were correctly located as well as the mistakes that were found during this procedure.

Assuming that researchers have access to and knowledge of all the tools presented in this chapter, they could directly infer chromosomal positions for 502 out of the 634 HGVS variants that are present in PharmGKB, or else 79%. This percentage also assumes that the researchers identify and correct the parsing problems that we located for 206 variants (see Table 7.1). With MutationInfo's pipeline they could directly get the positions for 628 (99%) of these variants without performing any quality check or preprocessing.

Table 7.3 Summary of the Number of HGVS Variants From PharmGKB Processed Correctly by Various Tools to Infer a Chromosomal Position

| Method | HGVS Reference Sequence Type | | | | | Total |
	NT_	NG_	NM_	NC_	GenBank	
Total in PharmGKB	118	187	207	59	63	634
Mutalyzer	0	0	207	0	0	207
Mutalyzer + BLAT	116	183	0	0	21	320
Mutalyzer + BLAT after error correction	117	185	0	0	21	323
Biocommons/Splign	0	124	203	59	0	386
Biocommons/Blat	0	115	40	59	0	214
Counsyl	0	0	207	0	0	207
LOVD	0	0	155	0	0	155
MutationInfo's BLAT	0	172	203	59	42	476
Variation Reporter	0	143	205	59	1	408
Transvar	90	7	203	0	3	303
Mistakes found	2	4	0	0	0	6
Mistakes corrected	1	2	0	0	0	3

The six variants that fail MutationInfo's pipeline are erroneous entries, three of them can be easily corrected.

7.5 DISCUSSION

The difference between MutationInfo and other tools stems mainly from the fact that existing tools focus only on specific types of HGVS variants. Mutalyzer, BioCommons, and Counsyl perform very well when HGVS variants have reference sequences with NM_ accession numbers. Yet, this is the case for only 32% of the variants present in PharmGKB. As we described, the consortia responsible for cataloguing and curating the variants and haplotypes for genes with known pharmacogenetics implications list NT_, NG_, NC_, and arbitrary GenBank accession numbers as their reference sequences (covering in total 68% of variants in PharmGKB). If we omit NC_ variants (9% of total) where the conversion is simple then there are 368 variants (58%) and existing tools can infer a position for only 144 of them (39%).

Among all existing tools, Mutalyzer was the most successful at parsing and inferring a genetic position (c. to g. conversion) since it performed this task for 527 variants (83%). Nevertheless, Mutalyzer failed to infer a chromosomal position for 320 of these variants and a BLAT search was required. It is interesting that in Mutalyzer's initial report (see Table 2 in Wildeman et al., 2008), more than half of the variants failed parsing in three different databases (HbVar, BRCA2, and PAH). They also mention: "Mutalyzer can only be effective when variant descriptions in HGVS-format are combined with a well-annotated genomic reference sequence." This underlines the necessity of a tool like MutationInfo that applies alternative methods (BLAT search) when similar tools fail.

The most successful tool that inferred chromosomal positions without the use of a BLAT search was Variation Reported, which completed this task for 408 variants (65%). Nevertheless if we use MutationInfo's BLAT service without any other tools, we will be able to infer positions for 476 out of the 634 variants (75%).

7.6 CONCLUSIONS

We expect the quality of reported genetic variants to improve as the process of identifying genetic variants becomes more automated using high throughput techniques like Next-Generation Sequencing (Richards et al., 2015). Issues like ambiguity of the location or the alternative sequence of a mutation will be tackled sufficiently by existing pipelines for variant discovery

Table 7.4 List of Guidelines for Correct HGVS Processing

Guidelines for Correct HGVS Processing
Support for many RefSeq accession numbers (NM_, NC, NT_, NM_, NG_)
Support for arbitrary GenBank accession numbers
Support for Ensembl and RefSeq accession numbers
Support for LRG accession numbers (officially recommended by HGVS)
Support for gene names as reference transcripts (although this is not recommended, it is commonly used)
Support for explicit HGVS forms like chr1:123456A>C
Support for multiple position identifiers (c., g., r., p.)
Support for many variant kinds (Substitutions, Indels, Inversions)
Make visible which version of HGVS guidelines the tool adheres to
Provide basic quality control indicators, like correctness of reference sequence
Offer conversion between reference sequences that include chromosome sequences

Note: *According to our experiments no tool adheres to all these guidelines.*

(Duitama et al., 2014). Yet there is still a large and valuable legacy of identified variants discovered with less automated techniques, obsolete technology or simply by following incorrect reporting guidelines. Since, the impact of variant discovery in clinical practice is increasing, the availability of software tools for performing uniform quality checks and extracting metainformation from reported variants is also becoming more important. MutationInfo is a tool that belongs in this category for the purpose of quickly recovering (or validating) the chromosomal position, the reference and alternative sequence of a variant. It is mainly a frontend for existing tools that tries to combine them for this purpose while offering a "low-level" BLAT search method if all other tools fail. The application of MutationInfo in PharmGKB, which is a typical human curated LSMD of high significance, demonstrated that existing tools exhibit substantial limitations in identifying positional information and revealing errors in the database entries. For this reason, we have compiled a list of guidelines (Table 7.4) that future tools should follow in order to successfully parse and process HGVS variants. Since at present, no individual tool follows all the guidelines on this list, we propose a pipeline that combines most of the existing tools for this task; MutationInfo is an open source tool that tries to fill this gap. It targets bioinformaticians with minimum knowledge of the Python programming language. It contains an installation script that manages all dependencies and offers a single function (`get_info`) for accessing its methods. Finally, MutationInfo offers a web service that can be easily accessed by individually researchers or embedded in existing online services.

In our future work we plan to augment MutationInfo's pipeline with a liftover method as well as with additional tools (our first priorities are Transvar and Variation Reporter). We also plan to make MutationInfo compatible with the latest version of HGVS nomenclature guidelines, including the extensions suggested by the authors of Mutalyzer, that it can support more complex structural changes (Taschner and den Dunnen, 2011).

References

Adzhubei, I.A., Schmidt, S., Peshkin, L., Ramensky, V.E., Gerasimova, A., Bork, P., Kondrashov, A.S., Sunyaev, S.R., 2010. A method and server for predicting damaging missense mutations. Nat. Methods 1548-7105. 7 (4), 248–249. https://doi.org/10.1038/nmeth0410-248.

Bhagwat, M., Young, L., Robison, R.R., 2012. Using BLAT to find sequence similarity in closely related genomes. Curr. Protoc. Bioinform. 37, 10–18.

Cariaso, M., Lennon, G., 2012. SNPedia: a wiki supporting personal genome annotation, interpretation and analysis. Nucleic Acids Res. 1362-4962. 40, D1308–D1312. https://doi.org/10.1093/nar/gkr798. http://nar.oxfordjournals.org/content/40/D1/D1308.short.

Cock, P.J.A., Antao, T., Chang, J.T., Chapman, B.A., Cox, C.J., Dalke, A., Friedberg, I., Hamelryck, T., Kauff, F., Wilczynski, B., et al., 2009. Biopython: freely available python tools for computational molecular biology and bioinformatics. Bioinformatics 25 (11), 1422–1423.

Crews, K.R., Hicks, J.K., Pui, C.-H., Relling, M.V., Evans, W.E., 2012. Pharmacogenomics and individualized medicine: translating science into practice. Clin. Pharmacol. Ther. 92 (4), 467–475.

Dayem Ullah, A.Z., Lemoine, N.R., Chelala, C., 2013. A practical guide for the functional annotation of genetic variations using SNPnexus. Brief. Bioinform. 1477-4054. 14 (4), 437–447. https://doi.org/10.1093/bib/bbt004. http://bib.oxfordjournals.org/content/14/4/437.

Deans, Z., Fairley, J.A., den Dunnen, J.T., Clark, C., 2016. HGVS nomenclature in practice: an example from the United Kingdom National External Quality Assessment Scheme. Hum. Mutat. 1098-1004. https://doi.org/10.1002/humu.22978. http://www.ncbi.nlm.nih.gov/pubmed/26919400.

den Dunnen, J.T., Dalgleish, R., Maglott, D.R., Hart, R.K., Greenblatt, M.S., McGowan-Jordan, J., Roux, A.-F., Smith, T., Antonarakis, S.E., Taschner, P.E.M., 2016. HGVS recommendations for the description of sequence variants: 2016 update. Hum. Mutat. 37 (6), 564–569.

Duitama, J., Quintero, J.C., Cruz, D.F., Quintero, C., Hubmann, G., Foulquié-Moreno, M.R., Verstrepen, K.J., Thevelein, J.M., Tohme, J., 2014. An integrated framework for discovery and genotyping of genomic variants from high-throughput sequencing experiments. Nucleic Acids Res. 42 (6), e44.

Fokkema, I.F.A.C., Taschner, P.E.M., Schaafsma, G.C.P., Celli, J., Laros, J.F.J., den Dunnen, J.T., 2011. LOVD v. 2.0: the next generation in gene variant databases. Hum. Mutat. 32 (5), 557–563.

Hart, R.K., Rico, R., Hare, E., Garcia, J., Westbrook, J., Fusaro, V.A., 2015. A Python package for parsing, validating, mapping and formatting sequence variants using HGVS nomenclature. Bioinformatics (Oxford, England) 1367-4811. 31 (2), 268–270. https://doi.org/10.1093/bioinformatics/btu630. http://www.pubmedcentral.nih.gov/articlerender.fcgi?artid=4287946tool=pmcentrezrendertype=abstract.

Kent, W.J., 2002. BLAT-the BLAST-like alignment tool. Genome Res. 1088-9051. 12 (4), 656–664. https://doi.org/10.1101/gr.229202. http://www.pubmedcentral.nih.gov/articlerender.fcgi?artid=187518tool=pmcentrezrendertype=abstract.

Kumar, P., Henikoff, S., Ng, P.C., 2009. Predicting the effects of coding non-synonymous variants on protein function using the SIFT algorithm. Nat. Protoc. 1750 2799. 4 (7), 1073–1081. https://doi.org/10.1038/nprot.2009.86. http://www.ncbi.nlm.nih.gov/pubmed/19561590.

Landrum, M.J., Lee, J.M., Riley, G.R., Jang, W., Rubinstein, W.S., Church, D.M., Maglott, D.R., 2014. ClinVar: public archive of relationships among sequence variation and human phenotype. Nucleic Acids Res. 1362-4962. 42, D980–D985. https://doi.org/10.1093/nar/gkt1113. http://nar.oxfordjournals.org/content/early/2013/11/14/nar.gkt1113.short.

MacArthur, J.A.L., Morales, J., Tully, R.E., et al., 2014. Locus reference genomic: reference sequences for the reporting of clinically relevant sequence variants. Nucleic Acids Res. 1362-4962. 42, D873–D878. https://doi.org/10.1093/nar/gkt1198. http://nar.oxfordjournals.org/content/42/D1/D873.

Maglott, D., Ostell, J., Pruitt, K.D., Tatusova, T., 2004. Entrez gene: gene-centered information at NCBI. Nucleic Acids Res. 1362-4962. 33, D54–D58. https://doi.org/10.1093/nar/gki031. https://academic.oup.com/nar/article-lookup/doi/10.1093/nar/gki031.

McCarthy, D.J., Humburg, P., Kanapin, A., Rivas, M.A., et al., 2014. Choice of transcripts and software has a large effect on variant annotation. Genome Med. 1756-994X. 6 (3), 26. https://doi.org/10.1186/gm543. http://genomemedicine.biomedcentral.com/articles/10.1186/gm543.

McKinney, W., 2010. Data structures for statistical computing in Python. In: van der Walt, S., Millman, J. (Eds.), Proceedings of the 9th Python in Science Conference. pp. 51–56.

McLaren, W., Pritchard, B., Rios, D., Chen, Y., Flicek, P., Cunningham, F., 2010. Deriving the consequences of genomic variants with the Ensembl API and SNP Effect Predictor. Bioinformatics (Oxford, England) 1367-4811. 26 (16), 2069–2070. https://doi.org/10.1093/bioinformatics/btq330. http://bioinformatics.oxfordjournals.org/content/26/16/2069.

Medina, M.W., Sangkuhl, K., Klein, T.E., Altman, R.B., 2011. PharmGKB: very important pharmacogene-HMGCR. Pharmacogenet. Genomics 1744-6880. 21 (2), 98–101. https://doi.org/10.1097/FPC.0b013e328336c81b. http://www.pubmedcentral.nih.gov/articlerender.fcgi?artid=3098759tool=pmcentrezrendertype=abstract.

Overby, C.L., Tarczy-Hornoch, P., 2013. Personalized medicine: challenges and opportunities for translational bioinformatics. Pers. Med. 1741-0541. 10 (5), 453–462. http://www.pubmedcentral.nih.gov/articlerender.fcgi?artid=3770190tool=pmcentrezrendertype=abstract.

Paila, U., Chapman, B.A., Kirchner, R., Quinlan, A.R., 2013. GEMINI: integrative exploration of genetic variation and genome annotations. PLoS Comput. Biol. 1553-7358. 9 (7), e1003153. https://doi.org/10.1371/journal.pcbi.1003153. http://journals.plos.org/ploscompbiol/article?id=10.1371/journal.pcbi.1003153.

Pandey, K.R., Maden, N., Poudel, B., Pradhananga, S., Sharma, A.K., 2012. The curation of genetic variants: difficulties and possible solutions. Genomics Proteomics Bioinformatics 2210-3244. 10 (6), 317–325. https://doi.org/10.1016/j.gpb.2012.06.006. http://www.sciencedirect.com/science/article/pii/S1672022912000927.

Pedersen, B.S., Yang, I.V., De, S., 2013. CruzDB: software for annotation of genomic intervals with UCSC genome-browser database. Bioinformatics (Oxford, England) 1367-4811. 29 (23), 3003–3006. https://doi.org/10.1093/bioinformatics/btt534. http://bioinformatics.oxfordjournals.org/content/29/23/3003.full.

Reitz, K., 2011. Requests: http for humans (online). http://docs.python-requests.org/en/master/. (Accessed 25 January 2018).

Richards, S., Aziz, N., Bale, S., Bick, D., Das, S., Gastier-Foster, J., Grody, W.W., Hegde, M., Lyon, E., Spector, E., Voelkerding, K., Rehm, H.L., 2015. Standards and guidelines for the interpretation of sequence variants: a joint consensus recommendation of the American College of Medical Genetics and Genomics and the Association for Molecular Pathology. Genet. Med. 1098-3600. 17 (5), 405–423. https://doi.org/10.1038/gim.2015.30.

Sand, P.G., 2007. A lesson not learned: allele misassignment. Behav. Brain Funct. 1744-9081. 3, 65. https://doi.org/10.1186/1744-9081-3-65. http://www.pubmedcentral.nih.gov/articlerender.fcgi?artid=2231368tool=pmcentrezrendertype=abstract.

Tack, V., Deans, Z.C., Wolstenholme, N., Patton, S., Dequeker, E., 2016. What's in a name? A co-ordinated approach towards the correct use of a uniform nomenclature to improve patient reports and databases. Hum. Mutat. 37, 570–575.

Tao, T., Madden, T., Christiam, C., 2008. BLAST FTP Site. https://www.ncbi.nlm.nih.gov/books/NBK62345/.

Taschner, P.E.M., den Dunnen, J.T., 2011. Describing structural changes by extending HGVS sequence variation nomenclature. Hum. Mutat. 1098-1004. 32 (5), 507–511. https://doi.org/10.1002/humu.21427. http://www.ncbi.nlm.nih.gov/pubmed/21309030.

Thierry-Mieg, D., Thierry-Mieg, J., 2006. AceView: a comprehensive cDNA-supported gene and transcripts annotation. Genome Biol. 1474-760X. 7 (suppl 1), S12.1–S12.14. https://doi.org/10.1186/gb-2006-7-s1-s12.

Wang, K., Li, M., Hakonarson, H., 2010. ANNOVAR: functional annotation of genetic variants from high-throughput sequencing data. Nucleic Acids Res. 1362-4962. 38 (16), e164. https://doi.org/10.1093/nar/gkq603. http://nar.oxfordjournals.org/content/38/16/e164.

Wildeman, M., van Ophuizen, E., den Dunnen, J.T., Taschner, P.E.M., 2008. Improving sequence variant descriptions in mutation databases and literature using the Mutalyzer sequence variation nomenclature checker. Hum. Mutat. 1098-1004. 29 (1), 6–13. https://doi.org/10.1002/humu.20654. http://www.ncbi.nlm.nih.gov/pubmed/18000842.

Xin, J., Mark, A., Afrasiabi, C., Tsueng, G., Juchler, M., Gopal, N., Stupp, G., Putman, T., Ainscough, B., Griffith, O., et al., 2015. Mygene.info and myvariant.info: gene and variant annotation query services. bioRxiv, 035667.

Xin, J., Mark, A., Afrasiabi, C., Tsueng, G., Juchler, M., Gopal, N., Stupp, G.S., Putman, T.E., Ainscough, B.J., Griffith, O.L., Torkamani, A., Whetzel, P.L., Mungall, C.J., Mooney, S.D., Su, A.I., Wu, C., 2016. High-performance web services for querying gene and variant annotation. Genome Biol. 1474-760X. 17 (1), 91. https://doi.org/10.1186/s13059-016-0953-9.

Zhao, S., Zhang, B., 2015. A comprehensive evaluation of ensembl, refseq, and ucsc annotations in the context of rna-seq read mapping and gene quantification. BMC Genomics 16 (1), 97.

Zhou, W., Chen, T., Chong, Z., Rohrdanz, M.A., Melott, J.M., Wakefield, C., Zeng, J., Weinstein, J.N., Meric-Bernstam, F., Mills, G.B., Chen, K., 2015. TransVar: a multilevel variant annotator for precision genomics. Nat. Methods 1548-7105. 12 (11), 1002–1003. https://doi.org/10.1038/nmeth.3622.

CHAPTER 8

Translating Genomic Information to Rationalize Drug Use

Alexandros Kanterakis*,†, Theodora Katsila†, George P. Patrinos†,‡,§

Institute of Computer Science, Foundation for Research and Technology Hellas (FORTH), Heraklion, Greece, †Department of Pharmacy, School of Health Sciences, University of Patras, Patras, Greece, ‡Department of Pathology—Bioinformatics Unit, Faculty of Medicine and Health Sciences, Erasmus University Medical Center, Rotterdam, The Netherlands, §Department of Pathology, College of medicine and Health Sciences, United Arab Emirates University, Al-Ain, United Arab Emirates

8.1 INTRODUCTION

Genomic medicine, though being there soon after the completion of the Human Genome Project, >15 years ago, is still far from being considered as commonplace in clinical practice. At the same time, it is obvious that the growth of genomic knowledge has not met with a reciprocal increase in clinical implementation. The concept of genomic (or personalized or precision) medicine, which is the tailoring of medical treatment modalities to the individual genetic characteristics, needs, and preferences of each patient, is not new. Indeed, it goes back to the era of Hippocrates, who quoted that "...*it is more important to know what person suffers from a disease than knowing which disease the person suffers.*" The discovery, made back in 1956, that the genetic basis for the selective toxicity of both fava beans and the antimalarial drug primaquine is a deficiency of the metabolic enzyme glucose-6-phosphate dehydrogenase (G6PD) (Luzzatto and Seneca, 2014) presents one of the earliest illustrations of the principle of personalized medicine.

Nowadays, it is almost taken for granted that the advances in genomics research will revolutionize the way in which genomic medicine is performed. Building on these advances, Pharmacogenomics (PGx), namely the collective impact that many genomic variants have on the response to medication, are currently driving discovery, analysis, and interpretation in the context of research into the genetic basis of interindividual variation in drug response (Qiang and Anthony, 2011). As an integral part of genomic medicine, PGx targets the delineation of the relationship between genomic variation/gene expression and drug

Human Genome Informatics. https://doi.org/10.1016/B978-0-12-809414-3.00008-5

efficacy and/or toxicity (Evans and Relling, 2004). To date, there are several genes, referred to as pharmacogenes, which play a key role in the absorption, distribution, metabolism, excretion, and toxicity (ADMET) of several drugs. The most important ADMET genes can be grouped into four main categories: (a) Phase-I and (b) Phase-II drug metabolism enzymes, (c) drug transporters, and (d) drug modifiers.

Interestingly, the rise of next-generation sequencing technology has created unprecedented opportunities to analyze whole genomes (Schuster, 2008). This approach promises to be extremely useful in PGx, since unlike conventional medium- or even high-throughput genetic screening approaches, such as microarray-based assays, the AmpliChip CYP450 (Roche Molecular Diagnostics, Basel, Switzerland) and the DMET Plus (Affymetrix, Santa Clara, CA, United States), it allows the complete scanning and determination of a full picture with respect to individual ADMET variome, from the frequent to the very rare ones. This is important because it is very likely that each individual harbors rare and/or novel variants of functional significance in well-established pharmacogenes, which may render an individual and/or patient intermediate or poor metabolizer to certain drugs and may go undetected when using a genetic screening assay (Katsila and Patrinos, 2015).

Over the recent years, whole genome sequencing (WGS) has been one of the drivers of PGx research, with many interesting findings (Mizzi et al., 2014). A search in PubMed (September 2016) for PGx-related papers shows that the proportion of papers related to "pharmacogenomics" (PGx) and "whole genome sequencing" (WGS) as search terms has risen significantly over the last 10 years.

However, most variants that impact drug response still remain to be identified (Katsila and Patrinos, 2015). As genome-wide association studies (GWAS) may not identify all risk biomarkers (Gurwitz and McLeod, 2013), since this a genetic screening rather than full scanning approach, the identification of unknown (possibly very rare or even unique) variants from WGS may provide indicative associations between specific genotypes and (rare) adverse drug reactions. Whole exome and/or WGS can now be easily performed using several commercially available or proprietary platforms to analyze genome variation comprehensively and with a high degree of accuracy at reasonable cost, as compared to the recent past (Pareek et al., 2011).

In this chapter, we discuss recent advancements in the application of WGS in PGx research and highlight the current shortcomings of PGx data integration, while we describe the design principles and the development of integrated PGx electronic services, as a potential approach to address the problem of PGx information management and overload, and the delivery of personalized PGx translation services.

8.2 PERSONALIZED PGx PROFILING USING WHOLE GENOME SEQUENCING

Recently, WGS analysis of almost 500 individuals identified a very large number of rare and potentially functional genomic variants in ADMET genes, which would not have been identified with a conventional high-throughput genetic screening approach. Mizzi et al. (2014) showed that the number of ADMET-related genomic variants identified by WGS per individual was significantly higher compared to those that would have been identified if the DMET Plus platform, the most comprehensive genotyping platform for pharmacogenomic biomarkers available to date (Sissung et al., 2010), was used. These authors reported the identification of over 400,000 genomic variants in the 231 ADMET-related genes, of which almost 10% were frequent, attaining population frequencies of >20%. On average, roughly 18,000 variants were found for each individual in these 231 ADMET-related genes (pharmacovariants), compared to an anticipated 250 variants in the same genes that would have been identified by the DMET Plus platform. This analysis also revealed almost 16,500 novel (not annotated in dbSNP) pharmacovariants within exons and regulatory regions, of which ~900 attained frequencies of over 1% and are likely to be functionally significant (pathogenic). The latter finding clearly demonstrates that any result from a currently available PGx screening assay, from medium- to even high-throughput, would not be indicative of a patient's ability to respond to certain drugs and, as such, should be interpreted with a degree of caution.

From the above, it seems plausible that, in the light of the continuous decrease of WGS costs and the gradual increase in WGS data accuracy, one would envisage that comprehensive WGS-based PGx testing could be readily applicable in a clinical setting (Gillis et al., 2014). By applying WGS analysis to two unrelated family members suffering from atrial fibrillation and presenting with differential response rates to anticoagulation treatment, Mizzi et al. (2014) were able to delineate the differential response rate to anticoagulation treatment of these family members, further demonstrating that the potential applicability of this approach for PGx testing in a clinical setting is not in the distant future.

Although the application of WGS in PGx is still in its infancy, one might envisage that the ultimate PGx test would involve, at the very least, the resequencing of the ADMET-related pharmacogenes, particularly those that have been acknowledged to be credible PGx biomarkers by regulatory agencies. Several PGx tests have been developed, representing tangible deliverables from the numerous genomic studies which have attempted to correlate genetic variation with variable drug response. The US Food and Drug Administration (FDA) established the Genomics and Targeted Therapy Group to advance the application of genomic technologies in the discovery, development, regulation, and use of medications and the first PGx testing device to get approval by

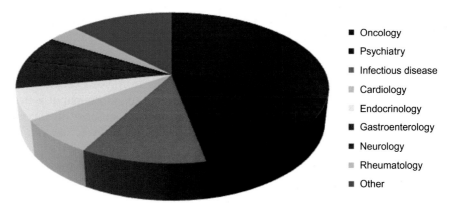

FIG. 8.1

Distribution of FDA-approved drugs with PGx information in their labels among the various medical specialties. "Other" specialties include Anesthesiology, Antidotal treatment, Autoimmune diseases, Dermatology, Hematology, Metabolic disorders, Pulmonology, Transplantation, and Urology.

the US FDA was Roche's AmpliChip in 2004, assessing *CYP2D6* and *CYP2C19* genomic biomarkers, linked to the function of these drug metabolizing enzymes. To date (October 2016), the FDA has relabelled >140 approved drugs to include genetic information (www.fda.gov/drugs/scienceresearch/researchareas/pharmacogenetics/ucm083378.htm), from which ~25% are metabolized by cytochrome CYP2D6 with variable rates of metabolism. Fig. 8.1 shows the distribution of these drugs among different medical specialties, with oncology and psychiatry being the dominating ones. However, these drug labels do not always provide, based on relevant genetic information, specific guidelines (e.g., in relation to putative adverse drug reactions) and recommendations about what actions should be taken (Amur et al., 2008).

The most challenging and perhaps crucial part of such an approach would be the accurate target enrichment of those ADMET-related pharmacogenes, followed by whole-pharmacogenome resequencing, capturing novel and putative deleterious variants in the known pharmacogenes. By contrast, the main disadvantage would be omitting important variants in (yet unknown) modifier genes involved in drug metabolism. Several Consortia, such as the Pharmacogenomics Research Network (PGRN; http://pgrn.org) and the eMERGE Consortium (http://emerge.mc.vanderbilt.edu), are developing such whole pharmacogenes resequencing platforms, focusing on the most significant fraction (actionable) of pharmacogenes.

For such an approach to be viable and readily applicable in clinical practice, it should be accompanied by the necessary genome informatics platforms and services to (a) potentiate accurate analysis of the resulting pharmacogenome resequencing, (b) address the secure storage of the sheer amount of genomic information resulting from pharmacogenome resequencing, and (c) provide a meaningful and clinician-friendly report so that it can be readily exploited in the clinic. To cope with this challenge, innovative approaches that derive meaningful insights and knowledge from large and complex PGx resources need to be explored and developed. The scope of the task is twofold: firstly, to facilitate and enhance identification and evidence-based documentation of (existing or newly reported) PGx gene/variant-drug-phenotype associations and, secondly, the translation and transfer of well-documented PGx knowledge to clinical implementation with the aim of both rationalizing and individualizing drug treatment modalities. The following paragraphs focus on this particular challenge.

8.3 TOWARDS PHARMACOGENOMIC DATA INTEGRATION

As previously mentioned, a crucial component of personalized medicine is the individualization of drug therapy. Understanding the complex interactions and detailed characterization of the functional variants of individual ADMET-related genes is needed to demonstrate clinical utility. In other words, associating pharmacogene variants with specific drug responses in individual patients improves clinical decision-making by informing the adjustment of the dosage or the selection of a different drug (Amur et al., 2008). However, true individualization of therapy, which would maximize drug efficacy and minimize drug toxicity, would need to consider genomic and phenotypic data, as well as any environmental factors that could influence the response to drug treatment, in the context of the specific individual and/or patient. Therefore, the design and development of IT solutions, which are (a) continually updatable by inclusion of newly generated PGx knowledge, (b) able to disseminate information in the form of guidelines, and (c) capable of linking the results of PGx tests to recommendations for therapeutic interventions with the aim of supporting drug-prescribing decision makers, is a prerequisite for incorporating PGx into routine clinical practice.

Once the Internet became an indispensable tool for biomedical researchers, genomic information overload was inevitable. Although there are numerous genomic and biological, in general, databases, which often create confusion for users in terms of which might be the most appropriate to investigate a given biological question, the number of the currently existing databases that are directly related to PGx is limited in number (Potamias et al., 2014).

The main problem regarding the exploitation of PGx knowledge and its utilization in clinical practice relates to the heterogeneity and low degree of connectivity between different PGx resources. In most cases, the amount of raw data is so overwhelming that PGx biomedical researchers and stakeholders are often confused and hence unable to capture all that is known and being discovered regarding genomic variation and its correlation with variable drug response (Gordon et al., 2012). As such, the design of an integrated web information system to interconnect and federate diverse PGx information resources into a single portal represents a formidable challenge.

8.3.1   The Concept of Integrated PGx Assistant Services

The PGx information overload challenge calls for specialized informatics services for the interpretation and integration of the increasingly large amounts of molecular and clinical data. Such a multifaceted endeavor entails both translational and clinical bioinformatics approaches that would not only offer analytical and interpretational methods to optimize the transformation of increasingly voluminous biomedical data into proactive, predictive, preventive and participatory ("4Ps") medicine (Squassina et al., 2010), but also enable the clinical application of discovery-driven bioinformatics methods to understand molecular mechanisms and prompt the search for potential therapies for human diseases (Wang and Liotta, 2011). In this respect, an electronic PGx assistant platform could act as the PGx *bench-to-bedside* enabling medium. The fundamental components that together underpin the novelty of such a platform revolve around its ability to provide translation services resulting from interrelating genotypic to phenotypic (metabolizer status) information as a valuable tool both to clinicians and biomedical researchers, by (a) supporting them in making informed decisions based on state-of-the-art PGx data, and (b) providing an "one-stop solution" where information can be found to facilitate an understanding of interindividual differences in drug efficacy, toxicity, and pharmacokinetics, as well as driving the discovery of new PGx variants.

To this end, the goal is to provide such a web-based platform to ease the processing, assimilation, and sharing of PGx knowledge and facilitate the aggregation of different PGx stakeholders' perspectives. The platform should take advantage of, and be designed around, interoperable and flexible bioinformatics and advanced information processing components that are able to offer personalized treatment recommendations, based on reliable genomic evidence and, at the same time, reduce healthcare costs by increasing drug efficacy and minimizing adverse drug reactions.

To develop such a system, one would first need to determine its functional requirements, from the user's perspective. Such requirements would include: (a) retrieval of PGx information regarding ADMET genes, their respective variants and drugs, (b) a format that is readily updatable with information on newly discovered PGx variants, and (c) the capability to receive personalized treatment recommendations based on personalized PGx profiles. At the same time, special emphasis should be given to specific users' roles, particularly the following four likely different types of potential users:

(a) *Individual and/or patient*: Any user who provides SNP genotype profiles with the aim of receiving corresponding clinical annotations and personalized PGx recommendations (as assessed and validated by healthcare professionals);

(b) *Healthcare professional*: Any healthcare professional (physician, geneticist, etc.) who needs to infer the phenotypic status of individuals (based on their genotype profiles, and by reference to genotype-phenotype translational tables), to review and supervise an individual patient's personalized treatment recommendations, assess them, and decide upon ensuing therapeutic protocols and treatment options;

(c) *Data submitter*: Any biomedical researcher who discovers and identifies a new gene variant and its putative PGx associations. The data submitter can either validate the findings and enrich the system's database, or request a (local) version of the PGx database to work with; and

(d) *Administrator*: Any user with administrative rights, responsible for maintaining and upgrading the electronic PGx assistant's database server (backups, versioning, restoration, etc.), managing application tools and services, assigning and authorizing user roles and privileges, and providing appropriate security and privacy-preserving services.

The reference architecture, in a multilayer level, and the basic components of the proposed electronic PGx assistant platform is shown in Fig. 8.2 and described below.

8.3.2 Development of an Electronic PGx Assistant

At first, several external data sources are leveraged to extract and integrate PGx information. To achieve this goal, one needs to adopt the notion of a data warehouse as the basic and perhaps the most appropriate data model to encompass the different requirements for database technology, compared to traditional database systems that support typical online transaction processing applications (Fig. 8.3).

For example, a medical professional might need to address the following questions: (a) *"Are there any specific recommendations for prescribing olanzapine to a psychiatric patient who is intermediate metabolizer of this drug?"* (b) *"Which alternative treatment modalities are available for the TPMT*2/*3A genotype?"* This is posed as an Online Analytical Processing Query (OLAP) to the data warehouse. The latter is centered around the gene/variant-drug-phenotype-

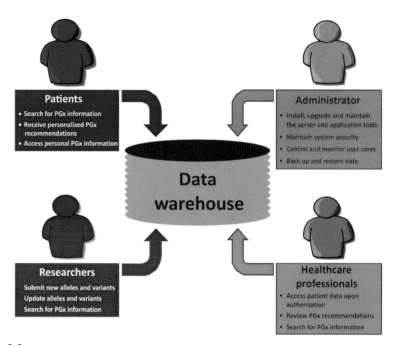

FIG. 8.2
Schematic drawing of the ePGA system depicting the different users and their roles and access rights.

recommendation concept, embodied in the fact table which, in turn, reference the dimensional tables around it, corresponding to the entities of: (i) gene, (ii) drug, (iii) diplotype, (iv) phenotype, and (iv) clinical annotations, guidelines, and recommendations. These entities correspond to the dimension attributes that act as foreign keys to the fact table. Different types of data extraction tools, such as APIs, Web-Services, JSON/XML, or text parsers, are utilized in order to fetch and transform data from the various heterogeneous data sources, such as PharmGKB, NCBI databases (e.g., dbSNP, PubMed, etc.), and genotyping platform(s) (e.g., Affymetrix) annotations, into the central data warehouse, following an extraction-transform-load (ETL) process. Standard ontologies and nomenclatures are utilized in an effort to uniformly represent the various data and information PGx items (e.g., gene-variant nomenclatures, gene ontology, ICD for disease classification and encoding, etc.).

With regard to the management of individuals' genotype/SNP profiles, an electronic healthcare record (EHR) solution can be adopted. To this end, state-of-the-art guidelines and data models related to the genetic tests and their interpretations can be employed (e.g., the HL7/CDA2 guide for genetic testing report (Butte, 2008)). Standard ontologies and data models could be utilized for the representation of genotype profiles (e.g., Genetic Variant Format (GVF),

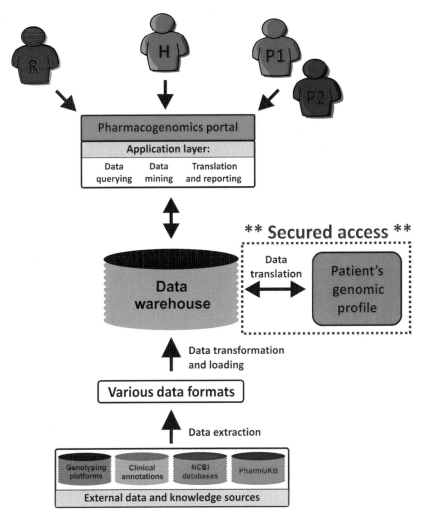

FIG. 8.3

Reference layered architecture of an electronic pharmacogenomics translation tool, including the various components, data/information flow, and services. Data are extracted from various external sources, such as genotyping platforms (such as Affymetrix DMET Plus, Illumina, etc.), in various formats (such as.csv, . xml, .xls, .txt, etc.) and then transformed and loaded into the main data warehouse. A patient's genomic profile is subsequently entered in a secured (e.g., password- or smart card-protected) manner into the data warehouse. These constitute the back office of the electronic pharmacogenomics translation tool, which communicates with the pharmacogenomics portal, which includes the various applications, namely data querying, mining, analysis, translation, and reporting. The pharmacogenomics portal can be accessed by researchers (R), healthcare professionals (H), and patients (P1, P2, and so on), each having different access rights to the portal.

www.sequenceontology.org/resources/gvf.html and LOINC, https://loinc.org; Wang and Liotta, 2011). The utility of linking genotype data to EHRs is crucial for the translation and transfer of PGx knowledge into the clinic. The approach is both cost-effective and time-efficient as there is no need to actively recruit and gather samples from a study population, as cases and controls are readily available and consistently identified from EHRs and the linked genetic samples. The eMERGE Consortium (http://emerge.mc.vanderbilt.edu/) has already successfully exploited this alternative (Shabo, 2012; Shoenbill et al., 2014).

Regarding the accumulation, storage, and management of individuals' genotype profiles, and taking into consideration the current debate about genetic tests and translational research (Denny et al., 2013), a number of key ethical issues are raised, namely public genetic awareness and genomic literacy, physicians' knowledge of genomics, handling of genomic information in and beyond the clinic, and online direct-to-consumer (DTC) (pharmaco)genomic testing, with the associated arguments to be highly polarized (Denny, 2013). In this respect, all the relevant ethical, privacy-preserving, and security issues should be employed and implemented, with special effort devoted to surveys and the assessment of guidelines, in order to critically appraise the impact of genetics and PGx on society and increase the level of awareness of the general public, healthcare professionals, and biomedical researchers to PGx and personalized medicine.

Lastly, from the researcher's perspective, the proposed PGx assistant will enhance PGx research by facilitating the discovery of new PGx variants. Methods for discovering genetic factors in drug response, including GWAS, expression analysis, and even WGS, already exist. However, more sophisticated knowledge-based tools to assign meaning to novel variants are required. The proposed PGx assistant brings reporting and analytical services to the end user, through a simple and user-friendly interface, and supports research by revealing hidden relations between genes, variants, and drugs, thereby driving the discovery of candidate genomic regions of interest. Moreover, as knowledge about drug-gene and drug-drug interactions accumulates, the proposed freely available system which is additionally coupled with advanced literature mining features and updatable components becomes even more beneficial to the research community and society.

8.4   PERSONALIZED PGx TRANSLATION SERVICES

Surely, the most important notion in the proposed PGx assistant platform is the idea of personalization. The inclusion of a personalized PGx translation component in such a platform is based on the assumption that *"clinical high-throughput and preemptive genotyping will eventually become common practice and clinicians will increasingly have patients' genotypes available before a prescription is written"* (Vayena and Prainsack, 2013). Several ongoing studies, such as PREDICT (http://www.mc.vanderbilt.edu/documents/predictpdx/files/PREDICT %20Fact%20Sheet%20for%20Providers%202-15-12.pdf) and PREPARE (www. upgx.eu) are built around this notion. The personalized translation component would serve the:

(i) Automated matching of patients' genotype profiles with established and/or newly discovered gene-variants/alleles, based on the customization of an elaborate allele-matching algorithm (Farmen et al., 2009);

(ii) Inference of respective phenotypes (e.g., metabolizer profiles), and

(iii) Delivery of relevant and updated clinical annotations and (drug) dosing recommendations.

Such component would be based upon the harmonization of PGx haplotype/ translation tables from the DMET Plus assay and PharmGKB knowledgebase (www.pharmgkb.org), but would also provide services for updating the haplotype/translation tables with newly discovered and validated gene-variants/ alleles, particularly those derived from WGS studies.

In order to produce accurate personalized PGx recommendations, the focus should be placed on the relationship between a variant and its related gene, drug(s), and phenotype(s). For example, a patient with bipolar disorder receives a genetic test that has the potential to determine the rs2032582 (A/C) variant in the *ABCB1* gene. How can this information be translated into clinical knowledge? The medical professional should be made aware that although the majority (70%–80%) of patients with bipolar disorder respond well to lithium treatment, the remaining 20%–30% of the patients present with patterns of partial or nonresponsiveness. Before prescribing this drug, the physician enters the patient's encrypted genotype data into the proposed PGx assistant, which then identifies that the specific variant is related to lithium response. More specifically, the medical professional receives the following information that the PGx assistant would bring to his/her attention: *"Patients with the AC genotype* (for the rs2032582 variant in the *ABCB1* gene) *and depression may have an increased risk of suicidal ideation when treated with clomipramine, lithium,..., or venlafaxine as compared to patients with the CC genotype"* (relevant text from PharmGKB). At the same time, the PGx assistant displays three new studies related to genetic variants associated with lithium response and provides links to the respective sources of information. By considering the patient's family and medical history, the medical professional may then decide to provide an alternative treatment and closely monitor this patient. The medical professional may also prescribe a new genetic test based on the findings of the recommended articles in the literature that associate additional genetic variants with response to lithium treatment.

A related protocol, called PG4KDS, is employed at St. Jude Children's Research Hospital (http://www.stjude.org/pg4kds), aiming to selectively migrate microarray-based genotypes for clinically relevant genes into each patient's electronic medical record preemptively. By leveraging "look up" translation tables created by the Translational Pharmacogenetics Project (TPP) (Schildcrout et al., 2012), a PGRN-led initiative with the goal of operationalizing the work of the Clinical Pharmacogenetics Implementation Consortium (CPIC) (Relling et al., 2012), they assigned phenotypes to each unique *CYP2D6* or *TPMT* diplotype based on assessments of functional allele activity (Relling and Klein, 2011).

Of course, the continued deposition of such data in the public domain is of paramount importance to maximize both their scientific and clinical utility, which dictate the incentivization of individual researchers or research groups to submit their newly acquired and unpublished mutation/variation data to public repositories or knowledge bases in return for appropriate credit and attribution (see also Chapter 6). As previously mentioned, microattribution is perhaps the most promising approach, where not only the overall contributions from individual scientists increased (Giardine et al., 2011; Patrinos et al., 2012), but most importantly, a number of useful conclusions were drawn in every case that the microattribution was implemented, derived from, e.g., variant clustering, clinical phenotype and/or PGx variant allele frequencies comparisons, and so on. Such conclusions would not have been possible without such an approach, further demonstrating the value of the immediate contribution and sharing of novel genome variants even though they would not warrant classical narrative publication on their own.

8.5   THE ELECTRONIC PGx ASSISTANT

Based on the concepts described in the previous paragraphs, a web-based electronic PGx Assistant (ePGA; www.epga.gr) that provides personalized genotype-to-phenotype translation, linked to state-of-the-art clinical guidelines, was developed (Lakiotaki et al., 2016). ePGA's translation service matches individual genotype profiles with PGx gene haplotypes and infers the corresponding diplotype and phenotype profiles, accompanied with summary statistics. Additional features include the ability (i) to customize translation based on subsets of interesting variants and (ii) to update the knowledge base with novel PGx findings. In its current implementation, ePGA utilizes PGx information and data mainly from PharmGKB (www.pharmgkb.org), which in concert with CPIC (Relling and Klein, 2011) and PGRN (PharmacoGenomics Research Network; www.pgrn.org) (Giacomini et al., 2007) presents the most comprehensive resource on genes related to drug response, their variations, their pharmacokinetics and pharmacodynamics pathways, and their effects on drug-related phenotypes. To the best of our knowledge, ePGA is currently the only web-based application that combines: (i) information retrieval on state-of-the-art PGx genes, variants, drugs, and their associations, (ii) matching of individual genotype profiles with PGx gene haplotypes and inference of corresponding diplotype profiles, accompanied by respective summary statistics, (iii) automated linkage of the inferred diplotypes with respective clinical annotations, recommendations, and dosing guidelines, and (iv) update services on newly discovered PGx variants and haplotypes (Fig. 8.4A).

ePGA was built in Django (www.djangobook.com), a high-level Python web-framework and a free, open-source platform which comes with an object-relational mapper in which database layout can be described in Python code. Different types of data extraction tools (APIs, Web-Services, JSON/XML, and text parsers) were developed to fetch and transform data from various heterogeneous data sources into the ePGA database. The main PGx information and data comes from public and private (licensed) PharmGKB sources. The core of the translation process is implemented in the open-source R environment (www.r-project.org) and uses R Studio's Shiny web-application framework (shiny.rstudio.com) to build its web-based interface.

ePGA features three distinct services, namely the (a) Explore, (b) Translate, and (c) Update services, which are analyzed further below.

8.5.1 Explore Service

The *Explore service* is simply a browser where the user can search for PGx associations, referring to: (i) pharmacogenes that relate to drug response, (ii) PGx variants and haplo/diplotypes, (iii) metabolizer status (e.g., extensive, intermediate, poor, and ultra-rapid), and (iv) drug dosing-guidelines and/or clinical annotations. The retrieved information and data are organized and presented in an expandable tree structure that can be easily browsed. PGx information was retrieved from JSON-formatted drug dosing guidelines and recommendation files (freely available for download from PharmGKB). PGx clinical annotations were manually retrieved from PharmGKB pages, as well as from specially devised parsers of private (licensed) PharmGKB files. ePGA also connects to DruGeVar database (Dalabira et al., 2014) to provide additional information on drug/gene combinations that have been approved by any or both of the two major regulatory agencies (FDA and EMA). In this respect, ePGA's explore service is considered as a flexible and efficient browser to state-of-the-art PGx knowledge, allowing a direct clinical exploitation of relevant gene-drug associations and respective guidelines. Explore service is currently offered for 579 (pharmaco)genes, related to 544 drugs and 2920 (pharmaco)variants.

When a user makes a selection in any of the dropdown menus, all other menus are narrowed down respectively (Fig. 8.5B). For example, if the user first selects

FIG. 8.4

(A) Snapshot of the front page of ePGA PGx assistant webservice, where the two main services, namely "Explore" and "Translate", are available for the user. (B) Overview of the ePGA "Explore" service, where the user can browse for PGx associations from corresponding drop-down menus, indicated at the right hand side of the figure (please see text for details), namely pharmacogenes that relate to drug response, PGx variants, metabolizer status, drug dosing guidelines, and/or clinical annotations. The example provided refers to Simvastatin and *SLCO1B1*, from where the user can access the respective recommendations and clinical annotations.

gene *CYP2C19*, then only the 34 drugs related to that gene become available. Subsequently, if the user selects a drug, i.e., amitryptiline, all 44 variations and 5 different metabolizer statuses related to *CYP2C19* and amitryptiline are activated for further exploration. Each variation (or combination of variations) and any related clinical annotations link to the source knowledge base; PharmGKB in this case.

8.5.2 Translation Service

The other service that ePGA envisions to provide, and perhaps the key feature of this tool, is the ***Translation service***, defined as the inference of diplotypes from individual genotype profiles and based on the utilization of the PGx haplotype tables (see below). The input to the ePGA translation service would be a.vcf file, which is a standard file format for storing the individual genotype profiles (www.1000genomes.org/wiki/Analysis/vcf4.0) and an optional file that contains a list of variants by their rsIDs (unique SNP cluster IDs as assigned by dbSNP, www.ncbi.nlm.nih.gov/projects/SNP), in order to perform customized translation. Particularly, the user will be able to upload a file with a list of rsIDs for the variants of interest and, in this case, the translation will only include this specific set of markers. The output of the translation service would be a five column table with the following features: (a) *Sample ID*, (b) *Gene symbol*, (c) *Assigned diplotype*, (d) *Phenotype* color indicator (green, yellow, and red; see below; Fig. 8.5A), and (e) *Recommendations*, hyperlinked (when available) to the explore ePGA service regarding the gene-diplotype-phenotype association (Fig. 8.5B). In ePGA, the assignment of a specific PGx phenotype (e.g., extensive, intermediate, poor, or ultra-rapid metabolizer status) is not provided. Instead, a color classification is employed to indicate the inferred diplotype category (green for "Wt/Wt" (wild-type/wild-type diplotype), yellow for "Wt/Var" (wild-type/variant diplotype), and red for "Var/Var" (variant/variant diplotype)). If for the inferred gene diplotype, respective PGx information is available (in the form of clinical annotations and/or dosing guidelines and recommendations), as indicated with the "Check" hyperlink, the user is transferred to the ePGA Explore service (see Section 8.5.1) that displays the relevant information (in the above described tree-based representation of ePGA's Explore service results). The ePGA translation service also provides gene summary statistics (e.g., percentages of the three diplotype categories assigned to samples and sample summary statistics). The input genotype profiles may also be displayed. The user will be able to download ePGA's translation results as a report in.pdf format (Fig. 8.5C).

8.5.3 Update Services

As mentioned in the previous paragraphs, most gene variations and their impact on drug response still remain to be identified, while identification of

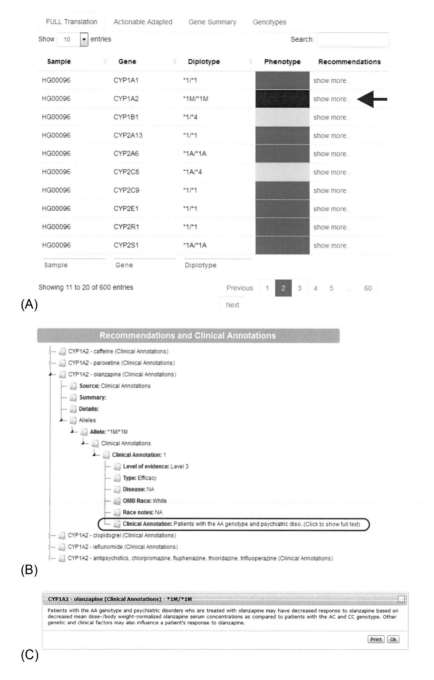

FIG. 8.5

Overview of the envisioned ePGA Translation service (A). The various phenotypes corresponding to each diplotype in various pharmacogenes are translated and displayed using a color code (please see text for details). The user can opt to access the recommendations for a homozygous variant diplotype (colored in red, depicted with a red arrow), which leads to a new window with the corresponding recommendations and clinical annotations (B). In this example, by selecting the *CYP1A2*/olanzapine gene drug combination, the menu unfolds, where the user can click on the clinical annotation (depicted in red circle), where the full clinical annotation for that particular diplotype is provided (C).

unknown and possibly rare variants with WGS presents a quite promising direction for the discovery of indicative associations between specific genotypes and adverse drug reactions (Pareek et al., 2011). Such an approach would be dynamic in the sense that it would allow enrichment and/or modification of the PGx knowledge (Potamias et al., 2014). This need is fully addressed in the ePGA system by an appropriately devised update service that allows the user to incorporate novel gene variants (as well as new haplotypes) and test their presence in individual genotype profiles. As such, one of the hallmarks of the ePGA tool is the *Update services,* particularly for PGx researchers, to update the haplotype table of a gene, when a new variant and/or haplotype is discovered, or to devise a new haplotype table for a gene discovered to be engaged with a particular gene-drug association. This can only be done once the user agrees with the ePGA's purpose and use (www.epga.gr/static/dosing2/ePGA_Purpose_and_Use.pdf). This can be particularly useful when a large number of putatively actionable PGx variants are identified through large WGS-based studies (Mizzi et al., 2014).

8.6 TRANSLATING PGx KNOWLEDGE INTO CLINICAL DECISION-MAKING

From the previous paragraphs, it is clear that once WGS-based PGx testing becomes widely available, it will require a substantial effort to translate this genomic information into clinically meaningful guidelines. As far as PGx is concerned, the PGx clinical scenarios are truly complex in the real life situation, which often, if not always, pose significant dilemmas to the medical professionals regarding the selection of a most appropriate treatment modality and drug dose. This complexity does not occur because of the inability to correlate genomic with clinical variables, as genomics research has already revealed and produced (as will continue to do so) valuable PGx associations and knowledge. This complexity arises mostly due to the large translation gap in moving PGx (as with the other—omics) scientific discoveries towards successful innovations applicable in the clinic. This gap occurs because of the lack of orientation to innovation that conceptualizes knowledge-based PGx innovation as an ecosystem of communicating innovation actors, such as pharmacology, PGx, molecular biology and genetics researchers, and innovation narrators (Georgitsi and Patrinos, 2013). Such ecosystem can be realized by the currently missing intermediate medium that facilitates communication and supports bench-to-bedside translation endeavors, by harnessing knowledge from basic PGx to produce treatment options for patients (Mitropoulos et al., 2011; Mizzi et al., 2016).

In such a setting, it is imperative to adopt a multidisciplinary approach based on a portfolio of interoperating translational or clinical genome informatics components and their alignment with contemporary information engineering and processing approaches. Such approaches should aim to devise: (a) a PGx knowledge assimilator that seamlessly (e.g., based on standard semantics and data-models) links diverse PGx knowledge sources, and (b) knowledge-extraction services able to identify useful genotype-to-phenotype associations and knowledge from these sources. Moreover, the identified PGx genotype-to-phenotype associations should be explored in relation to their pharmacokinetic and pharmacodynamic background. Such exploration could be served by the elaboration of the appropriate PK/PD simulation models that help to assess PGx association's covariance in virtually devised populations, e.g., following the approach of SimCYP (www.simcyp.com) and NONMEM (www.iconplc. com/technology/products/nonmem) virtual simulation commercial packages, as well as utilizing free open-source PK modeling s/w tools such as PKreport (cran.r-project.org/web/packages/PKreport) and WFN (wfn.sourceforge.net/ wfnxpose.htm) R-packages. In addition, and based on the coupling of web 2.0 and social networking technology, it would be essential to facilitate and support the engaged collaboration needs and try to fill in the missing communication medium between the diverse PGx knowledge sources, the simulated PGx genotype-to-phenotype associations, and the various PGx actors.

To accommodate these needs, one should incorporate (a) the linkage and seamless integration of established PGx resources (e.g., PharmGKB, CPIC, DruGeVar, etc.), literature, and other genomic databases (e.g., PubMed, dbSNP, dbGAP, ClinVar, FINDbase, etc.), to be based on the elaboration and operationalization of standard (pharmaco)genomic/clinical ontologies and data-models, (b) literature mining/natural language processing (NLP), to extract putative disease-drug-gene/variant-phenotype associations from PGx resources and the published literature, (c) a virtual population pharmacokinetic simulator, to test putative variant-phenotype associations and assess relevant genotype-to-phenotype covariance statistics in virtual populations, and (d) a collaborative environment, to enable communication and collaboration between various PGx players towards the formation, validation, and evidential assessment of such associations. In addition, two additional components and respective services are required to align and harmonize such a platform with a bench-to-bedside orientation and its utilization in a clinical decision-making setting. First, an Electronic Healthcare Genotype component that would be readily compatible with the general EHR, to service the management of patients' genotype profiles, and secondly, a Genotype-to-Phenotype translation component, to service the automated matching of patients' genotype profiles with established and/or newly discovered gene/variant alleles, inference of the respective phenotypes (e.g., metabolizer profiles), and delivery of up-to-date relevant clinical annotations and respective (drug) dosing guidelines. Finally, a portal will be required as a single-access-point PGx environment that embraces the aforementioned components and services.

Such a system, once operable, would facilitate the integration and translation of PGx knowledge into the clinical decision-making process and bring genomic medicine closer to the clinical reality. To this end, one would also need to circumvent three additional fundamental hurdles, namely (a) ensuring that all the necessary consents are provided by the patients, (b) safeguarding sensitive personal data to avoid the inappropriate leaking of genetic information which may lead to a person's stigmatization, and (c) enhancing the genetics awareness and genetics education of healthcare professionals (Kampourakis et al., 2014).

8.7 CONCLUSION AND FUTURE PERSPECTIVES

The postgenomic revolution, characterized by the rise of massively parallel next-generation sequencing technologies, has led to the correlation of specific genomic variants with disease predisposition and other clinical features, including response to some of the most commonly prescribed drugs. As personalized drug treatment and genomic medicine get closer to becoming a reality, the use of whole genome sequencing that spans all ethnicities and covers all possible genetic alterations is the most useful approach (Drmanac, 2012). Recent evidence, though limited at the present time, confirms that WGS can reveal a relatively large number of unique (or rare) pharmacogenomic markers that would otherwise go undetected by conventional genetic screening method (Mizzi et al., 2014; Kozyra et al., 2017).

An important aspect of the next-generation sequencing technology that would be critical for its early adoption in the clinic is its cost-effectiveness. In other words, it becomes clear that performing a comprehensive personalized PGx profile using WGS (currently ~1000 US$), which would include almost all of the germline and de novo genomic variants needed to manage all current and future treatment modalities, would be cost-effective when compared to the cost of testing for a single marker or several markers in a few pharmacogenes (from 300 US$ to up to 1500 US$, respectively). At present, setting up a (centralized) WGS facility and PGx data translation to clinicians are two of the most important hurdles to be overcome, but sample outsourcing for data analysis and interpretation might be the answer to surmounting this obstacle using an economy-of-scale model. Ultimately, as PGx testing costs using WGS and cost-effectiveness are well-documented, it should only be a matter of time until the cost of PGx testing reimbursement is adopted by national insurance bodies (Vozikis et al., 2016).

In light of the above, the design and development of advanced informatics solutions that ease to fill in the gap between PGx research findings and clinical practice surfaces as a major need. In this chapter, we presented the operational requirements and design specifications of an electronic PGx assistant that aims to act as the medium between the various PGx communities (biomedical researchers, geneticists, healthcare providers, and PGx regulatory bodies) equipping them with innovative services that enable PGx research findings to reach clinical implementation. The orchestration of the provisioned PGx assistant's services in an integrated platform empowers the capabilities of PGx communities to grasp, assess, and maximize the use of relevant biomedical and molecular PGx knowledge. Finally, the implementation of PGx assistant services should address and provide feasible solutions to challenges related to the PGx annotation of whole genomes (Altman et al., 2013) that concerns the accuracy of PGx markers across the genome, the ambiguity of gene-variants and PGx markers in relevant literature references, the effect of multi gene-variants and PGx markers on individual phenotypes, the combined effects of variants on multiple drugs, as well as the limited body of clear guidelines and recommendations.

Lastly, until now, drug development and clinical trials for most drugs on the market are performed mainly in developed countries and in a limited ethnic diversity (Mitropoulos et al., 2011). By identifying population groups that are likely to be more susceptible to a certain adverse drug reaction, drug development companies can eliminate the huge cost and length of clinical trials. ePGA's translation service can be proved extremely valuable in identifying population differences in drug response.

Acknowledgments

Part of our own work was supported by the Golden Helix Foundation (P3; RD-201201), European Commission (GEN2PHEN (FP7-200754), RD-CONNECT (FP7-305444), U-PGx (H2020-668353)), and Greek National Secretariat of Research and Technology (COOPERATION (11ΣΥΝ_415)) grants to GPP.

References

Altman, R.B., Whirl-Carrillo, M., Klein, T.E., 2013. Challenges in the pharmacogenomic annotation of whole genomes. Clin. Pharmacol. Ther. 94, 211–213.

Amur, S., Frueh, F.W., Lesko, L.J., Huang, S.M., 2008. Integration and use of biomarkers in drug development, regulation and clinical practice: a US regulatory perspective. Biomark. Med 2, 305–311.

Butte, A.J., 2008. Translational bioinformatics: coming of age. J. Am. Med. Inform. Assoc. 15, 709–714.

Dalabira, E., Viennas, E., Daki, E., Komianou, A., Bartsakoulia, M., Poulas, K., et al., 2014. DruGeVar: an online resource triangulating drugs with genes and genomic biomarkers for clinical pharmacogenomics. Public Health Genomics 17, 265–271.

Denny, J.C., 2013. Mining electronic health records in the genomics era. PLoS Comput. Biol. 8.

Denny, J.C., Bastarache, L., Ritchie, M.D., Carroll, R.J., Zink, R., Mosley, J.D., Field, J.R., Pulley, J.M., Ramirez, A.H., Bowton, E., et al., 2013. Systematic comparison of phenome-wide association study of electronic medical record data and genome-wide association study data. Nat. Biotechnol. 31, 1102–1110.

Drmanac, R., 2012. The ultimate genetic test. Science 336, 1110–1112.

Evans, W.E., Relling, M.V., 2004. Moving towards individualized medicine with pharmacogenomics. Nature 429, 464–468.

Farmen, M., Close, S., Koh, W., 2009. Translation of drug metabolic enzyme and transporter (DMET) genetic variants into star allele notation using SAS. PharmaSUG, 2009 Conference, Portland, Oregon, May 31–June 3, PR03-2009.

Georgitsi, M., Patrinos, G.P., 2013. Genetic databases in pharmacogenomics: the frequency of inherited disorders database (FINDbase). Methods Mol. Biol. 1015, 321–336.

Giacomini, K.M., Brett, C.M., Altman, R.B., Benowitz, N.L., Dolan, M.E., Flockhart, D., et al., 2007. The pharmacogenetics research network: from SNP discovery to clinical drug response. Clin. Pharmacol. Ther. 81, 328–345.

Giardine, B., Borg, J., Higgs, D.R., Peterson, K.R., Philipsen, S., Maglott, D., Singleton, B.K., Anstee, D.J., Basak, A.N., Clark, B., et al., 2011. Systematic documentation and analysis of human genetic variation in hemoglobinopathies using the microattribution approach. Nat. Genet. 43, 295–301.

Gillis, N.K., Patel, J.N., Innocenti, F., 2014. Clinical implementation of germ line cancer pharmacogenetic variants during the next-generation sequencing era. Clin. Pharmacol. Ther. 95, 269–280.

Gordon, A.S., Smith, J.D., Xiang, Q., Metzker, M.L., Gibbs, R.A., Mardis, E.R., Nickerson, D.A., Fulton, R.S., Scherer, S.E., 2012. PGRNseq: a new sequencing-based platform for high-throughput pharmacogenomic implementation and discovery. American Society of Human Genetics (ASHG) Annual Meeting, Session 46—Pharmacogenetics: From Discovery to Implementation, Boston, USA, November 6–12.

Gurwitz, D., McLeod, H.L., 2013. Genome-wide studies in pharmacogenomics: harnessing the power of extreme phenotypes. *Pharmacogenomic*s 14, 337–339.

Kampourakis, K., Vayena, E., Mitropoulou, C., van Schaik, R.H., Cooper, D.N., Borg, J., Patrinos, G.P., 2014. Key challenges for next-generation pharmacogenomics. EMBO Rep. 15, 472–476.

Katsila, T., Patrinos, G.P., 2015. Whole genome sequencing in pharmacogenomics. Front. Pharmacol. 6, 61.

Kozyra, M., Ingelman-Sundberg, M., Lauschke, V.M., 2017. Rare genetic variants in cellular transporters, metabolic enzymes, and nuclear receptors can be important determinants of interindividual differences in drug response. Genet. Med 19, 20–29.

Lakiotaki, K., Kartsaki, E., Kanterakis, A., Katsila, T., Patrinos, G.P., Potamias, G., 2016. ePGA: a web-based information system for translational pharmacogenomics. PLoS One 11.

Luzzatto, L., Seneca, E., 2014. G6PD deficiency: a classic example of pharmacogenetics with on-going clinical implications. Br. J. Haematol. 164, 469–480.

Mitropoulos, K., Johnson, L., Vozikis, A., Patrinos, G.P., 2011. Relevance of pharmacogenomics for developing countries in Europe. Drug Metabol. Drug Interact. 26, 143–146.

Mizzi, C., Peters, B., Mitropoulou, C., Mitropoulos, K., Katsila, T., Agarwal, M.R., van Schaik, R.H., Drmanac, R., Borg, J., Patrinos, G.P., 2014. Personalized pharmacogenomics profiling using whole genome sequencing. Pharmacogenomics 15, 1223–1234.

Mizzi, C., Dalabira, E., Kumuthini, J., Dzimiri, N., Balogh, I., Başak, N., Böhm, R., Borg, J., Borgiani, P., Bozina, N., Bruckmueller, H., Burzynska, B., Carracedo, A., Cascorbi, I., Deltas, C., Dolzan, V., Fenech, A., Grech, G., Kasiulevicius, V., Kádaši, Ĺ., Kučinskas, V., Khusnutdinova, E., Loukas, Y.L., Macek, M., Makukh, H., Mathijssen, R., Mitropoulos, K., Mitropoulou, C., Novelli, G., Papantoni, I., Pavlovic, S., Saglio, G., Setric, J., Stojiljkovic, M., Stubbs, A.P., Squassina, A., Torres, M., Turnovec, M., van Schaik, R.H., Voskarides, K., Wakil, S.M., Werk, A., Del Zompo, M., Zukic, B., Katsila, T., Lee, M.T., Motsinger-Rief, A., Mc Leod, H.L., van der Spek, P.J., Patrinos, G.P., 2016. A European spectrum of pharmacogenomic biomarkers: Implications for clinical pharmacogenomics. PLoS One 11.

Pareek, C.S., Smoczynski, R., Tretyn, A., 2011. Sequencing technologies and genome sequencing. J. Appl. Genet. 52, 413–435.

Patrinos, G.P., van Mulligen, E., Gkantouna, V., Tzimas, G., Tatum, Z., Schultes, E., Roos, M., Mons, B., 2012. Microattribution and nanopublication as means to incentivize the placement of human genome variation data into the public domain. Hum. Mutat. 3, 1503–1512.

Potamias, G., Lakiotaki, K., Katsila, T., Lee, M.T., Topouzis, S., Cooper, D.N., et al., 2014. Deciphering next-generation pharmacogenomics: an information technology perspective. Open Biol. 4.

Qiang, M., Anthony, Y.H., 2011. Pharmacogenetics, pharmacogenomics, and individualized medicine. Pharmacol. Rev. 63, 437–459.

Relling, M.V., Klein, T.E., 2011. CPIC: clinical pharmacogenetics implementation consortium of the pharmacogenomics research network. Clin. Pharmacol. Ther. 89, 464–467.

Relling, M.V., Johnson, J., Roden, D., Weinshilboum, R., Sadee, W., Klein, T., 2012. Translational Pharmacogenomics Project (TPP) Annual Progress Report. pgrn.org/download/attachments/131165/TPP_Report%20to%20ESP_Spring2012.pdf?version=1&modificationDate=1333398152000&api=v2.

Schildcrout, J.S., Denny, J.C., Bowton, E., Gregg, W., Pulley, J.M., Basford, M.A., Cowan, J.D., Xu, H., Ramirez, A.H., Crawford, D.C., et al., 2012. Optimizing drug outcomes through pharmacogenetics: a case for preemptive genotyping. Clin. Pharmacol. Ther. 92, 235–242.

Schuster, S.C., 2008. Next-generation sequencing transforms today's biology. Nat. Methods 5, 16–18.

Shabo, A., 2012. HL7/CDA Implementation Guide for Genetic Testing Report (GTR). www.hl7.org/implement/standards/product_brief.cfm?product_id=292.

Shoenbill, K., Fost, N., Tachinardi, U., Mendonca, E.A., 2014. Genetic data and electronic health records: a discussion of ethical, logistical and technological considerations. J. Am. Med. Inform. Assoc. 21, 171–180.

Sissung, T.M., English, B.C., Venzon, D., Figg, W.D., Deeken, J.F., 2010. Clinical pharmacology and pharmacogenetics in a genomics era: the DMET platform. Pharmacogenomics 11, 89–103.

Squassina, A., Manchia, M., Manolopoulos, V.G., Artac, M., Lappa-Manakou, C., Karkabouna, S., Mitropoulos, K., Del Zompo, M., Patrinos, G.P., 2010. Realities and expectations of pharmacogenomics and personalized medicine: impact of translating genetic knowledge into clinical practice. Pharmacogenomics 11, 1149–1167.

Vayena, E., Prainsack, B., 2013. Regulating genomics: time for a broader vision. Sci. Transl. Med. 5.

Vozikis, A., Cooper, D.N., Mitropoulou, C., Kambouris, M.E., Brand, A., Dolzan, V., Fortina, P., Innocenti, F., Lee, M.T., Leyens, L., Macek, M., Al-Mulla, F., Prainsack, B., Squassina, A., Taruscio, D., van Schaik, R.H., Vayena, E., Williams, M.S., Patrinos, G.P., 2016. Test pricing and reimbursement in genomic medicine: towards a general strategy. Public Health Genomics 19, 352–363.

Wang, X., Liotta, L., 2011. Clinical bioinformatics: a new emerging science. J. Clin. Bioinformatics. 1, 1.

CHAPTER 9

Minimum Information Required for Pharmacogenomics Experiments

J. Kumuthini*, L. Zass*, Emile R. Chimusa[†], Melek Chaouch[‡], Collen Masimiremwa[§]

*Centre for Proteomic and Genomic Research, Cape Town, South Africa, [†]University of Cape Town, Cape Town, South Africa, [‡]Institute of Pasteur of Tunisia, Tunis, Tunisia, [§]African Institute of Biomedical Science and Technology, Harare, Zimbabwe

9.1 PHARMACOGENOMICS

Pharmacogenomics is an integral part of pharmacology, defined as the study of the impact of genetic variation on drug response, both in terms of efficacy and toxicity (Ma and Lu, 2011). Considering the health burden currently posed by adverse drug reactions (ADRs), with prevalence ranging from 2% to 21% worldwide (Chan et al., 2016), the pharmacogenomics discipline is a crucial stepping stone toward improved healthcare and medical treatment, reduced healthcare costs and ADR occurrence, as well as the future of personalized medicine (Daly, 2013; Fakruddin and Chowdhury, 2013). When genetic variation is associated with drug response, it can influence clinical decisions, such as treatment strategies, dosage, and (or) drug alterations (Fakruddin and Chowdhury, 2013).

Modern pharmacogenomics studies range from single- and multigene analysis to whole-genome single nucleotide polymorphism (SNP) investigations (Ma and Lu, 2011). Drugs undergo both pharmacokinetics (PK) and pharmacodynamics (PD) actions when consumed (Ratain et al., 2003a,b). PK encompasses four processes: absorption, distribution, metabolism, and excretion (ADME)—associated with drug processing (Ratain et al., 2003a), while PD involves the molecular action of a particular drug on its target (Ratain et al., 2003b). Researchers conducting pharmacogenomics studies often focus on and investigate genes or genetic variation associated with PK and PD, as these are most likely to result in differential drug-processing rates.

The pharmacogenomics discipline has evolved significantly since the turn of the century and more than 4000 genes, as well as copy number variants (CNVs) and SNPs within these; have been associated with variable drug response (Liu et al., 2016). Genes associated with variable drug response are

179

Human Genome Informatics. https://doi.org/10.1016/B978-0-12-809414-3.00009-7

known as very important pharmacogenes (VIPs) and include the genes encoding thiopurine methyltransferase, dihydropyrimidine dehydrogenase, cytidine deaminase, and cytochrome P450 2D6 (*CYP2D6*) (Engen et al., 2006; Santos et al., 2011; Scott et al., 2010). Pharmacogenes are often highly polymorphic and allele frequencies differ between populations (Engen et al., 2006; Hovelson et al., 2017), such as SNPs in the gene encoding the ATP Binding Cassette Subfamily B Member 1 (*ABCB1*) protein (Santos et al., 2011).

Although significant strides have been made in the pharmacogenomics discipline in the last decade, progress has still lagged behind the optimistic predictions initially made by many researchers and policymakers. A great deal of this delay is connected to the fact that individuals' responses to drugs are often multifactorial and, therefore, difficult to analyze (Roden et al., 2011). The clinical translation of findings into sensible practice remains problematic, largely due to ethical concerns, lack of clear regulatory guidance dealing with pharmacogenomics research and products, and lack of appropriate sample sizes for clinical translation (Gershon et al., 2014). In addition, technological limitations remain a significant barrier to the implementation of pharmacogenomics in practice, and bioinformatics- and big data-related limitations complicate the interpretation and evaluation of newly as well as previously generated pharmacogenomics data sets (Alyass et al., 2015; Olivier and Williams-Jones, 2011). A significant part of these analytical issues are owing to varied reporting methods used or lack thereof, varied terminologies and nomenclature employed, and varied data structures developed within the pharmacogenomics field, especially given the prominent technological boost within the genomics field in the last decade (Fan et al., 2014).

9.2   DATA STANDARDIZATION

Standardization describes the process associated with the development and (or) implementation of a broad set of guidelines with regard to a particular topic of interest, based on the consensus of end users or interest groups. Standards are developed with the aim of increasing quality and reducing cost and turnaround time (Xie et al., 2016) and have been employed in several branches of science (e.g., formal, physical, life, social, applied, and interdisciplinary sciences) and beyond (Xie et al., 2016).

In bioinformatics and genomics sciences, standards exist in several different forms, the most prominent being terminology standards or ontologies, which define consistent and agreed upon vocabulary within a given discipline or application (e.g., Pharmacogenomics Knowledgebase (PharmGKB)); data-format or structure standards, which allows consistent flow of data between independent parties or systems due to set specifications (e.g., VCF standard); and minimum reporting or document standards, which enhance interpretability and interoperability via the provision of stipulations as to the minimum information to be reported given a specific discipline or application (e.g., Minimum Information About a Proteomics Experiment (MIAPE)) (Holmes et al., 2010; Kim, 2005). If adhered to, these standards allow research investigations to be consistently useable, interoperable, comparable, and reproducible by others and prevent unintended duplication and unexplained variance within given field or application (Holmes et al., 2010; Kim, 2005).

In the context of biomedical research, and more specifically pharmacogenomics, one of the most important types of standardization is data standardization. Data standardization refers to the critical process of data integration and harmonization through the production of data formats, ontology, and reporting that is shared and consistent, thereby allowing the merger of data from diverse sources, supporting collaborative research and large-scale analytics, and promoting the sharing of sophisticated tools and methodologies among researchers and across communities and geographical locations (Chervitz et al., 2011). Considering the rapid increase in data generation in genomics and pharmacogenomics following the turn of the century, as well as the diversity of this generated data, the implementation of data, terminology, and reporting standards is essential within research fields (Chervitz et al., 2011). A large number of software, biological tools, and machinery have been developed for pharmacogenomics applications, resulting in a significant amount of heterogeneity in terms of how data is produced, managed, manipulated, stored, reported, and shared. In the context of biomedical research, lacking defined standards produces research which is not useable, interpretable, and reproducible, as is the case with many studies within the post-genomics era (Chervitz et al., 2011).

In an effort to address the aforementioned concerns, the FAIRsharing initiative (https://fairsharing.org/) was recently established, encapsulating the previously established standards and guidelines from biosharing (https://biosharing.org) (Wilkinson et al., 2016). FAIRsharing functions as a curated, informative, and educational resource on data and metadata standards, interrelated to databases and data policies, and aims to promote FAIR (Findable, Accessible, Interoperable, and Reproducible) principals with the biomedical research field (Wilkinson et al., 2016). As previously mentioned, the FAIRsharing initiative was born from the Minimum Information for Biological and Biological Investigations (MIBBI) foundry, which housed reporting standards within the field (Taylor et al., 2008). In 2009, MIBBI was reconstructed into BioSharing, culminating in the launch of the BioSharing Portal in 2011 (McQuilton et al., 2016). Thereafter, the BioSharing portal became the ELIXIR Registry of Standards, functioning as part of the ELIXIR Interoperability Platform, and then, in response to user feedback and to reflect the broadened scope of FAIR principals, was redesigned into FAIRsharing in 2017.

Currently, FAIRsharing functions as a tool for "omics" standards, including genomics, proteomics, metabolomics, transcriptomics, and glycomics, as well as resource discovery and interoperable bioscience (Sansone et al., 2012). FAIR-sharing functions as a tool for "omics" also contains several pharmacogenomics standards, including data-format and reporting standards, such as the Toxicology Data Markup Language (ToxML) (https://fairsharing.org/bsg-s000539) and the CDISC Laboratory Data Model (CDISCLAB) (https://fairsharing.org/bsg-s000165), as well as pharmacogenomics ontologies, such as the PharmGKB Ontology (PharmGKB-owl) (https://fairsharing.org/bsg-s002745) and the Pharmacogenomic Relationships Ontology (PHARE) (https://fairsharing.org/bsg-s002697). The rest of the chapter will primarily focus on one particular pharmacogenomics reporting standard; Minimum Information required for a Drug Metabolising Enzymes and Transporters (DMET) Experiment (MIDE) (https://fairsharing.org/bsg-s000628) (Kumuthini et al., 2016).

9.3 MINIMUM INFORMATION REQUIRED FOR A DMET EXPERIMENT

9.3.1 Background

As previously mentioned, several pharmacogenomics markers have been identified over the years, and ongoing efforts serve to identify novel markers and address discrepancies previously identified; continued research is crucial in order to expand on our current knowledge of pharmacogenomics and pharmacogenomics markers, but also to guide future drug development strategies and dosage guidelines implementations (Johnson, 2013). Several genetic variants are currently recommended for pharmacogenomics testing prior to drug prescription.

In the current biomedical research climate, traditional methods of genotyping take up a significant amount of time and generate limited data. Alternatively, as they generate more data and are less time-consuming, microarrays and targeted sequencing have become standard. Thus, a number of high-throughput and specialized microarrays have been and are being developed (Peters and McLeod, 2008). Next-generation sequencing remains one of the most popular genotyping methods employed; however, array-based genotyping remains among the most reasonable and cost-effective methods of conducting a semi-targeted investigation and screening large sample sizes (Jung and Young, 2012).

In order to enhance pharmacogenomics research, Affymetrix has developed microarrays which probe recognized genetic variations within genes associated with both PK and PD, including those genes validated by the United States Food and Drug Administration (FDA) agency (Peters and McLeod, 2008). The Affymetrix DMET (Drug Metabolizing Enzymes and Transporters) Plus Premier Pack is one such array; a mid-scale genotyping platform, which enables highly multiplexed genotyping of over 1900 variants, covering over 200 genes, identifying both common and rare variants, including SNPs, copy number variants (CNV), and indels (Burmester et al., 2010).

Data generated from DMET Plus arrays may provide crucial knowledge for pharmacogenomics interpretation, as well as clinical applications (Burmester et al., 2010). The DMET technology has progressed over the years, increasing the demand for novel software and analysis tools, algorithms, web applications, and specific statistical techniques to aid pharmacogenomics research and interpretation (Sissung et al., 2010). Aiming toward personalized medicine, a large number of studies, investigating pharmacogenomics markers and using DMET technology, have been conducted globally. Although these platforms are constantly improving, African variation is still notably lacking.

In an effort to harmonize and standardize the aforementioned studies for widespread interpretability, interoperability, reproducibility, and usability and to address concerns involving the management and effective use of the information generated, MIDE was developed and submitted to FAIRsharing (then MIBBI). MIDE assesses and addresses the issues concerning the data and metadata associated with DMET experiments (Kumuthini et al., 2016). The assessment focuses on the relevant information required for reporting to public databases as well as publishing houses and provides guidelines that would facilitate effective DMET data interpretation, sharing, reporting, and usage across the scientific community in the future (Kumuthini et al., 2016). Although MIDE was specifically designed for DMET reporting, it can be extended for application with regard to other pharmacogenomics microarrays as well, for example, PharmacoScan Solution, developed by Applied Biosciences, and AmpliChip CYP450, developed by Roche.

9.3.2 DMET Console Software Analysis

Following sample processing and hybridization, the microarray is scanned and probe intensities are converted to numeric value (Affymetrix, 2011). These results are stored in an image file, which provides the base for computer-based data analysis, using the DMET console software. DMET data analysis can be divided into three phases; primary, secondary, and tertiary analysis (Fig. 9.1) (Affymetrix, 2011).

Primary analysis consists of preprocessing in order to normalize initial probe intensity values and summarize them into allele-specific values (Affymetrix, 2011). The primary work package, required for DMET data analysis, is provided by Affymetrix. Alternatively, data processing may be conducted on command. The DMET Console graphical user interface provides tools for quality control; however, it does not provide tools for statistical analysis of an experiment.

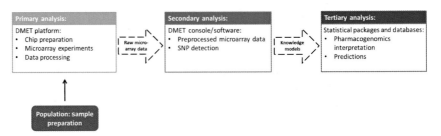

FIG. 9.1

Graphical representation of the DMET Plus analysis pipeline.

Additionally, the DMET Console software facilitates tertiary analysis, such as phenotype prediction.

Several additional files are required for DMET analysis; including sample-related data, stored as intensity, genotyping, and sample information files; reference files, DMET Plus Genotyping Files (for primary analysis) and DMET Plus Marker Annotation and Allele Translation Files, downloadable from the Affymetrix website (www.affymetrix.com); and Metabolizer Bin Files, required in some versions if the workflow includes phenotype prediction. User modifications to the Metabolizer Bin File should always be included when submitted to data repository. DMET Console is regularly updated and enables several genotyping methods. The algorithms employed for genotype calling also differ between versions; therefore, the genotyping method used should always be provided when submitting data.

During secondary analysis, using DMET Console or similar software, genotype calling for each marker is conducted (Affymetrix, 2011). Genotyping with DMET Console can be performed using one of two algorithms; the default method, fixed boundary analysis, which compares signals generated by the sample to predefined clustering models in order to make the genotype calls, and the alternate method, dynamic boundary analysis, which adapts the cluster models according to the data being analyzed.

Additional work packages, which facilitate data incorporation into databases such as dbSNP and the Pharmacogenomics Knowledge Base (PharmGKB, http://www.pharmgkb.org), are required to conduct tertiary analysis (Affymetrix, 2011). PharmGKB stores pharmacogenomics data and allows users to submit various data formats, including genotype and phenotype data. The database also accepts raw microarray data.

9.3.3 MIDE

The essential information regarding a DMET investigation involves information about the experiment itself (Kumuthini et al., 2016). In keeping with good laboratory practice, this information includes, but is not limited to, all information about the experimental design, samples used, extracts prepared, labeling, hybridization procedures and measurement specifications need to be captured, quality control information, and data extraction information. Deviations from the DMET protocol should also be reported in a publications and data submission (Kumuthini et al., 2016). Standardizing the information captured and reported, as well as the manner in which information is captured and reported, in comparable and consistent fashion, is essential to increase the readers' understanding of data and reports, the usability, and interpretability of reported data. Therefore the "Minimum Information required for a DMET Experiment (MIDE)" guideline was developed (https://fairsharing.org/bsg-s000628) (Kumuthini et al., 2016).

MIDE's format was based upon the guidelines for microarrays, as developed by Brazma et al. (2001). The format ventured to adhere to the criteria introduced by the developers of the The Minimum Information About a Proteomics Experiment (MIAPE) and Minimum Information About Peptide Array Experiment (MIAPepAE), namely sufficiency and practicability (Botha and Kumuthini, 2012; Taylor et al., 2007). Certain elements, included in the guideline, represent routine genotyping report information, such as, description of assay platform or methodology; sample statistics, including covariates and expected allele frequencies; and alleles measured and their correlation with drug-processing properties or encoded protein activity.

As illustrated in Tables 9.1–9.3, MIDE subdivides the minimum information required for a DMET experiment under study-specific, sample-specific, and experiment-specific information. Subsequently, these minimum information elements are also defined as required (R), information that needs to be reported with any given DMET experiment, or optional (O), information that enhances interpretability and interoperability if given. These tables also include example entries, which indicate how MIDE may be adhered to or implemented within a given case study. As illustrated in Table 9.1, study-specific information involves information about the aim and setup of the study.

9.3.4 Discussion

Standardizing DMET experiment reporting will not only aid research and research quality, but also facilitate collaboration between institutions (Medicine (US), 2013). The reporting standard is of particular relevance to authors, reviewers, data managers and curators, analysis software developers, and journal editors. Additionally, adopting common DMET data-reporting standards allows aggregation and integration of data from different platforms and secondary data (e.g., Q-PCR data, clinical and epidemiological data, drug interaction data, etc.), respectively, ultimately aiding the extraction of useful information in terms of a research and clinical perspective (Kumuthini et al., 2016). Several key concepts were considered in the development of MIDE:

Table 9.1 Minimum Study-Specific Information Required for a DMET Experiment

Element	Importance	Description	Example
Aim of the study	O	Aim(s) and ID of the study	Profile various SNPs between the Xhosa, Caucasian, and mixed-ancestry populations
Link to publication(s)	R	(PMID or DOI) or electronic record	All files containing results reported from analysis are located at XXX
Leading institution or source	O	Name and address of the institute where study/PI is based. Main PI's name/researcher ID (or other unique identifier)	(J. Kumuthini), (CPGR, Upper level, St Peter's Mall, Cnr Anzio and Main Road, Observatory, 7925)
Date(s)	O	Date study begun and completed	(Started: 10/11/2013) (Completed: 14/09/2014)
Sample collection, storage, shipping	O	How sample was collected, stored, and shipped	DNA samples submitted by collaborator with information about collection, storage, and shipping
Sample processing, labeling, hybridization	O	Sample processing details	Samples were processed in accordance to the Affymetrix DMET Plus protocol
Scanning	R	Instrument, software, parameters	Instrument: GeneChip Scanner 3000 7G Software: AGCC
Probe signal normalization	R	Software, parameters including reference data versions	DMET Console V1.0
Protocol(s)	O	Name/ID	Affymetrix DMET plus array protocol
Marker level summarization, software, parameters	R	Technical analysis workflow used: name of software including versions and parameters used	Marker level summarization, software, (DMET Console V1.0), (fixed genotype boundary 2), (maximum confidence score parameters‡ threshold 0.001, and minimum prior observations)
Genotyping method, software, parameters	R	Technical analysis workflow used: name of software including versions and parameters used	(DMET Console V1.0), (use of the standardized star allele nomenclature under DMET-analyzer tool)
Tertiary data analysis	O	Technical analysis workflow used: name of software including versions, parameters used and database versions and URL of the date of access	DMET Console_1_3_64 bit was used to perform genotype analysis and allele translation on the Xhosa and mixed-ancestry DMET data. Output submitted to DMET analyzer1.0.1 to identify significantly discriminative SNPs. Annotation Analysis using dbSNP and PharmGKB
Funding information	O	Funding details	NIH, grant number 001
Ethical approval	O	Ethics details	NA

O, optional; R, required.
As illustrated in Table 9.1, study-specific information involves information regarding the study as a whole.

Table 9.2 Minimum Sample-Specific Information Required for a DMET Experiment

Element	Importance	Description	Example
Sample name	R	Specimen annotation	See attached XXX
Biomaterial	O	Species or cell line name	Homo sapiens
Covariates	R	Clinical observations: ethnic group, gender, age, weight, height, region, survival (yes/no), other; disease state, treatment or normal of each sample	Ethnicity: mixed, Caucasian, Xhosa groups
Quality control steps taken	R	Type of QC checks carried out for the study	RNA concentration and quality—nanodrop spectrophotometer DNA integrity—gel electrophoresis QC to confirm amplification—Gel electrophoresis
Replicates	R	Identify, if any, replicates are technical/biological by differentiating sample IDs	No technical replicates
Cell line used	O	Tissue part, source provider, distributor, company, catalogue no., conditions of storage, contact details of laboratory	NA
Cell culture conditions	O	Conditions and characteristics of cell culture	NA
Genotype call rate	R	For each sample	95% average call rate across all samples

O, optional; R, required.
As illustrated in Table 9.2, samples-specific information involves information regarding the sample material, quality, and preprocessing relevant to the study.

Table 9.3 Minimum Experiment-Specific Information Required for a DMET Experiment

Element	Importance	Description	Example
Experimental aim	R	Description of experimental aim associated with a study (ID), for example, identify association between SNPs and genotype variations to a disease(s), use of genetic markers to predict response to medicine	Establish simple analysis pipeline to achieve the study's main objective
Summary of results	R	What was the outcome of the experiment/study?	Preliminary SNP profiles generated between Xhosa and mixed-ancestry populations for all samples with 95% call rate. Huge discrepancy between observed SNPs. 13 SNPs with a P-value $\leq .05$ indicated a significant differential distribution between the two population groups

Continued

Table 9.3 Minimum Experiment-Specific Information Required for a DMET Experiment—cont'd

Element	Importance	Description	Example
Experimental design	R	For example, compound treatment design, dose response design, stimulus or stress design, injury design, and other	Evaluation of presence of ADME variants in different populations
Data files	O	Table showing sample/raw data file/ processed data file associations containing following information. Naming convention or nomenclature chosen for the study	See attached file XXX
File name	O	Full names of files (raw data and analyzed data) and locations	See attached file XXX
File format	R	Type of formats it is available in, for example, ARR, CEL, CHP, among others	ARR, CEL, CHP
Explanation of missing data	O	Why file(s) or sample(s) were missing. Missing sample data which needs to be added before the software will analyze the data	No missing data

O, optional; R, required.
As illustrated in Table 9.3, experiment-specific information involves information regarding the experimental aim, design, and required information for re-analysis or reproduction.

9.3.4.1   Reproducibility

The MIDE guidelines enable DMET experiments to be more reproducible and verifiable following publication because the information required for is reported in systematic fashion (Chervitz et al., 2011). It also enhances the interpretability of DMET experiments due to clear and cohesive guidelines. Finally, such consistent reporting will also effectively enable and enhance the metaanalysis of similar data and publications

9.3.4.2   Comparability and Reusability

DMET experiments can be compared effectively if data-reporting standards, such as MIDE, are implemented (Chervitz et al., 2011). This extends to data extraction as well. Quality of data supersedes quantity; therefore, employing consistent data-reporting and extraction methods in DMET investigations greatly increases the ability to resolve biologically meaningful information.

9.3.4.3   Interoperability

Pharmacogenomics is continuously evolving to accommodate the integration of data from various resources (Xie et al., 2016). Implementing data-reporting standards for each resource aids the development of pipelines which consolidate the different platforms and the standards by which they follow.

9.3.4.4 *Implementation and Development of MIDE*

The reporting guideline, in its current stage, can be adapted to additional phar-macogenomics or microarray applications (Kumuthini et al., 2016). High-quality data are the foundation for deriving reliable biological conclusions; however, too often large differences in data quality have been observed between datasets generated from the same platforms (Shi et al., 2004, 2005). Standardized reporting, such as MIDE, challenges these inconsistencies. In some cases, poor quality may not be due to the inherent platform concerns, but due to lack of technical proficiency of the laboratory personnel who per-form analysis. Therefore, training and monitoring scientists in the purpose, use, and benefits of reporting guidelines and standards are crucial.

In order to facilitate the use of MIDE and automate the generation of MIDE reports, the MIDE guidelines are also provided in an XML schema, MIDE ver-sion 1.0, in line with the requirements, which were previously established by the MIBBI foundry. The MIDE standard is hosted at http://bioweb.cpgr.org.za/mide/mide.xsd. The model provides the ability to define cardinality of XML elements and define type to the XML element attributes. No well-defined DMET ontology currently exists; therefore, an ontology, which would ensure the use of consistent terminology to describe elements in the standard, was not integrated into MIDE.

9.4 PHARMACOGENOMICS STANDARDIZATION: CHALLENGES

Several challenges hinder the broad implementation and adoption of standard-ization in pharmacogenomics. The lack of education and training opportunities, especially at tertiary level, encapsulating the concepts of standardization within science, presents a significant barrier in regard to the aforementioned (Sansone and Rocca-Serra, 2012). Standard-developing bodies also need to be particularly conscientious regarding the applicable ethical policies within the specific field or discipline in which the standard is being developed or implemented. The percep-tions that standardization is only beneficial for future science endeavors and that standardization is too difficult to employ, and resistance to change, also present significant barriers to widespread standardization. Enhancing the education and advocacy of standardization efforts and FAIR standards could resolve many of the aforementioned concerns and, in this regard, publishing and funding bodies play a significant role (Sansone and Rocca-Serra, 2012).

However, standardization implementation in research and service laboratories may yet remain complicated and could be associated with significant financial investment in some cases. Tools and resources are required to address these con-cerns, simultaneously enhancing the accessibility and usability of complex pharmacogenomics standards and aiding the evaluation of developed and in-development standards (Richesson and Krischer, 2007). This is particularly important because significant overlap may exist between standards and several gaps remain as well. The importance of standard maintenance and the reduction of standard redundancy cannot be overstated (Richesson and Krischer, 2007).

9.5   CONCLUSIONS

Standardization has been a hot topic and experienced a significant surge in "omics" science since the turn of the 21st century, following the production of the first human genome (Xie et al., 2016). The process plays a significant role in the biomedical research communities' endeavors to progress to a future where personalized genomics and personal medicine can be fully realized. To that end, recently produced pharmacogenomics data-reporting standards which adhere to FAIRsharing principles, such as MIDE (for DMET and (or) drug toxicology array experiment and data reporting), are crucial. The MIDE guidelines benefit the scientific community currently conducting and planning to conduct DMET, or similar experiments, with tools that ease and automate the generation of standardized reports using the Standardized MIDE XML schema. These tools, in turn, will facilitate the sharing, dissemination, and reanalysis of pharmacogenomics data sets through the production of accessible, findable, and interoperable pharmacogenomics data.

Advocating the development, use, and implementation of standards is crucial in order to further achieve the aforementioned goals and benefit the biomedical research community. Along with standardization researchers and standard developers, funding bodies, publishers, and data repositories play a crucial role in enhancing standard use. Publications, such as BioMed Central, are increasingly endorsing the FAIRsharing initiative. Making data findable and accessible in a comprehensive and consistent manner is crucial for driving the pharmacogenomics science and personalized medicine forward. FAIRsharing drives these efforts, encouraging data harmonization, sharing, findability, accessibility, interoperability, and reproducibility. Several challenges, such as education and maintenance, remain. Nonetheless, concerted efforts by international bodies have gone a long way in encouraging standard integration and endorsing use in the pharmacogenomics discipline and beyond, such that standardization has become an essential component of the science.

If implemented successfully, pharmacogenomics standardization can significantly facilitate challenges related to the analysis of pharmacogenomics data sets, as well as alleviate data futility. Most importantly, such standardization can promote the discovery of novel information within the discipline, ultimately aiming to be effectively integrated into clinical practice.

References

Affymetrix. White Paper, DMET Plus Algorithm (2011), https://assets.thermofisher.com/TFS-Assets/LSG/brochures/dmet_plus_algorithm_whitepaperv1.pdf (Accessed 10 January 2018).

Alyass, A., Turcotte, M., Meyre, D., 2015. From big data analysis to personalized medicine for all: challenges and opportunities. BMC Med. Genomics 8, 33. https://doi.org/10.1186/s12920-015-0108-y.

Botha, G., Kumuthini, J., 2012. Minimum information about a peptide array experiment (MIAPepAE). EMBnet.journal 18, 14–21. https://doi.org/10.14806/ej.18.1.250.

Brazma, A., Hingamp, P., Quackenbush, J., Sherlock, G., Spellman, P., Stoeckert, C., Aach, J., Ansorge, W., Ball, C.A., Causton, H.C., Gaasterland, T., Glenisson, P., Holstege, F.C., Kim, I.F., Markowitz, V., Matese, J.C., Parkinson, H., Robinson, A., Sarkans, U., Schulze-Kremer, S., Stewart, J., Taylor, R., Vilo, J., Vingron, M., 2001. Minimum information about a microarray experiment (MIAME)-toward standards for microarray data. Nat. Genet. 29, 365–371. https://doi.org/10.1038/ng1201-365.

Burmester, J.K., Sedova, M., Shapero, M.H., Mansfield, E., 2010. DMET microarray technology for pharmacogenomics-based personalized medicine. Methods Mol. Biol Clifton NJ 632, 99–124. https://doi.org/10.1007/978-1-60761-663-4_7.

Chan, S.L., Ang, X., Sani, L.L., Ng, H.Y., Winther, M.D., Liu, J.J., Brunham, L.R., Chan, A., 2016. Prevalence and characteristics of adverse drug reactions at admission to hospital: a prospective observational study. Br. J. Clin. Pharmacol. 82, 1636–1646. https://doi.org/10.1111/bcp.13081.

Chervitz, S.A., Deutsch, E.W., Field, D., Parkinson, H., Quackenbush, J., Rocca-Serra, P., Sansone, S.-A., Stoeckert, C.J., Taylor, C.F., Taylor, R., Ball, C.A., 2011. Data standards for omics data: the basis of data sharing and reuse. Methods Mol. Biol Clifton NJ 719, 31–69. https://doi.org/10.1007/978-1-61779-027-0_2.

Daly, A.K., 2013. Pharmacogenomics of adverse drug reactions. Genome Med. 5, 5. https://doi.org/10.1186/gm409.

Engen, R.M., Marsh, S., Van Booven, D.J., McLeod, H.L., 2006. Ethnic differences in pharmacogenetically relevant genes. Curr. Drug Targets 7, 1641–1648.

Fakruddin, M., Chowdhury, A., 2013. Pharmacogenomics—the promise of personalized medicine. Bangladesh J. Med. Sci. 12, 346–356. https://doi.org/10.3329/bjms.v12i4.11041.

Fan, J., Han, F., Liu, H., 2014. Challenges of big data analysis. Natl. Sci. Rev. 1, 293–314. https://doi.org/10.1093/nsr/nwt032.

Gershon, E.S., Alliey-Rodriguez, N., Grennan, K., 2014. Ethical and public policy challenges for pharmacogenomics. Dialogues Clin. Neurosci. 16, 567–574.

Holmes, C., McDonald, F., Jones, M., Ozdemir, V., Graham, J.E., 2010. Standardization and omics science: technical and social dimensions are inseparable and demand symmetrical study. Omics J. Integr. Biol. 14, 327–332. https://doi.org/10.1089/omi.2010.0022.

Hovelson, D.H., Xue, Z., Zawistowski, M., Ehm, M.G., Harris, E.C., Stocker, S.L., Gross, A.S., Jang, I.-J., Ieiri, I., Lee, J.-E., Cardon, L.R., Chissoe, S.L., Abecasis, G., Nelson, M.R., 2017. Characterization of ADME gene variation in 21 populations by exome sequencing. Pharmacogenet. Genomics 27, 89–100. https://doi.org/10.1097/FPC.0000000000000260.

Institute of Medicine (US), 2013. Standardization to Enhance Data Sharing. National Academies Press, Washington, DC.

Johnson, J.A., 2013. Pharmacogenetics in clinical practice: how far have we come and where are we going? Pharmacogenomics 14, 835–843. https://doi.org/10.2217/pgs.13.52.

Jung, S.-H., Young, S.S., 2012. Power and sample size calculation for microarray studies. J. Biopharm. Stat. 22, 30–42. https://doi.org/10.1080/10543406.2010.500066.

Kim, K., 2005. Clinical Data Standards in Health Care: Five Case Studies. California Health Care Foundation.http://www.chcf.org/. [(Accessed 8 August 2017)].

Kumuthini, J., Mbiyavanga, M., Chimusa, E.R., Pathak, J., Somervuo, P., Van Schaik, R.H., Dolzan, V., Mizzi, C., Kalideen, K., Ramesar, R.S., Macek, M., Patrinos, G.P., Squassina, A., 2016. Minimum information required for a DMET experiment reporting. Pharmacogenomics 17, 1533–1545. https://doi.org/10.2217/pgs-2016-0015.

Liu, X., Yang, J., Zhang, Y., Fang, Y., Wang, F., Wang, J., Zheng, X., Yang, J., 2016. A systematic study on drug-response associated genes using baseline gene expressions of the cancer cell line encyclopedia. Sci. Rep. 6 22811. https://doi.org/10.1038/srep22811.

Ma, Q., Lu, A.Y.H., 2011. Pharmacogenetics, pharmacogenomics, and individualized medicine. Pharmacol. Rev. 63, 437–459. https://doi.org/10.1124/pr.110.003533.

McQuilton, P., Gonzalez-Beltran, A., Rocca-Serra, P., Thurston, M., Lister, A., Maguire, E., Sansone, S.-A., 2016. BioSharing: curated and crowd-sourced metadata standards, databases and data policies in the life sciences. Database (Oxford). https://doi.org/10.1093/database/baw075.

Olivier, C., Williams-Jones, B., 2011. Pharmacogenomic technologies: a necessary "luxury" for better global public health? Glob. Health 7, 30. https://doi.org/10.1186/1744-8603-7-30.

Peters, E.J., McLeod, H.L., 2008. Ability of whole-genome SNP arrays to capture "must have" pharmacogenomic variants. Pharmacogenomics 9, 1573–1577. https://doi.org/10.2217/14622416.9.11.1573.

Ratain, M.J., William, K., Plunkett, J., 2003a. Principles of Pharmacokinetics. In: Holland-Frei Cancer Medicine. sixth ed. BC Decker, Hamilton.

Ratain, M.J., William, K., Plunkett, J., 2003b. Principles of Pharmacodynamics. In: Holland-Frei Cancer Medicine. sixth ed. BC Decker, Hamilton.

Richesson, R.L., Krischer, J., 2007. Data standards in clinical research: gaps, overlaps, challenges and future directions. J. Am. Med. Inform. Assoc. 14, 687–696. https://doi.org/10.1197/jamia.M2470.

Roden, D.M., Wilke, R.A., Kroemer, H.K., Stein, C.M., 2011. Pharmacogenomics: the genetics of variable drug responses. Circulation 123, 1661–1670. https://doi.org/10.1161/CIRCULATIONAHA.109.914820.

Sansone, S.-A., Rocca-Serra, P., 2012. On the evolving portfolio of community-standards and data sharing policies: turning challenges into new opportunities. GigaScience 1, 10. https://doi.org/10.1186/2047-217X-1-10.

Sansone, S.-A., Rocca-Serra, P., Field, D., Maguire, E., Taylor, C., Hofmann, O., Fang, H., Neumann, S., Tong, W., Amaral-Zettler, L., Begley, K., Booth, T., Bougueleret, L., Burns, G., Chapman, B., Clark, T., Coleman, L.-A., Copeland, J., Das, S., de Daruvar, A., de Matos, P., Dix, I., Edmunds, S., Evelo, C.T., Forster, M.J., Gaudet, P., Gilbert, J., Goble, C., Griffin, J.L., Jacob, D., Kleinjans, J., Harland, L., Haug, K., Hermjakob, H., Sui, S.J.H., Laederach, A., Liang, S., Marshall, S., McGrath, A., Merrill, E., Reilly, D., Roux, M., Shamu, C.E., Shang, C.A., Steinbeck, C., Trefethen, A., Williams-Jones, B., Wolstencroft, K., Xenarios, I., Hide, W., 2012. Toward interoperable bioscience data. Nat. Genet. 44 (2), 121–126. https://doi.org/10.1038/ng.1054.

Santos, P.C., Soares, R.A., Santos, D.B., Nascimento, R.M., Coelho, G.L., Nicolau, J.C., Mill, J.G., Krieger, J.E., Pereira, A.C., 2011. CYP2C19 and ABCB1gene polymorphisms are differently distributed according to ethnicity in the Brazilian general population. BMC Med. Genet. 12, 13. https://doi.org/10.1186/1471-2350-12-13.

Scott, S.A., Khasawneh, R., Peter, I., Kornreich, R., Desnick, R.J., 2010. Combined CYP2C9, VKORC1 and CYP4F2 frequencies among racial and ethnic groups. Pharmacogenomics 11, 781–791. https://doi.org/10.2217/pgs.10.49.

Shi, L., Tong, W., Goodsaid, F., Frueh, F.W., Fang, H., Han, T., Fuscoe, J.C., Casciano, D.A., 2004. QA/QC: challenges and pitfalls facing the microarray community and regulatory agencies. Expert. Rev. Mol. Diagn. 4, 761–777.

Shi, L., Tong, W., Fang, H., Scherf, U., Han, J., Puri, R.K., Frueh, F.W., Goodsaid, F.M., Guo, L., Su, Z., Han, T., Fuscoe, J.C., Xu, Z.A., Patterson, T.A., Hong, H., Xie, Q., Perkins, R.G., Chen, J.J., Casciano, D.A., 2005. Cross-platform comparability of microarray technology: intra-platform consistency and appropriate data analysis procedures are essential. BMC Bioinformatics 6 (Suppl 2), S12. https://doi.org/10.1186/1471-2105-6-S2-S12.

Sissung, T.M., English, B.C., Venzon, D., Figg, W.D., Deeken, J.F., 2010. Clinical pharmacology and pharmacogenetics in a genomics era: the DMET platform. Pharmacogenomics 11, 89–103. https://doi.org/10.2217/pgs.09.154.

Taylor, C.F., Paton, N.W., Lilley, K.S., Binz, P.-A., Julian, R.K., Jones, A.R., Zhu, W., Apweiler, R., Aebersold, R., Deutsch, E.W., Dunn, M.J., Heck, A.J.R., Leitner, A., Macht, M., Mann, M., Martens, L., Neubert, T.A., Patterson, S.D., Ping, P., Seymour, S.L., Souda, P., Tsugita, A., Vandekerckhove, J., Vondriska, T.M., Whitelegge, J.P., Wilkins, M.R., Xenarios, I., Yates, J.R., Hermjakob, H., 2007. The minimum information about a proteomics experiment (MIAPE). Nat. Biotechnol. 25, 887–893. https://doi.org/10.1038/nbt1329.

Taylor, C.F., Field, D., Sansone, S.-A., Aerts, J., Apweiler, R., Ashburner, M., Ball, C.A., Binz, P.-A., Bogue, M., Booth, T., Brazma, A., Brinkman, R.R., Michael Clark, A., Deutsch, E.W., Fiehn, O., Fostel, J., Ghazal, P., Gibson, F., Gray, T., Grimes, G., Hancock, J.M., Hardy, N.W., Hermjakob, H., Julian, R.K., Kane, M., Kettner, C., Kinsinger, C., Kolker, E., Kuiper, M., Le Novère, N., Leebens-Mack, J., Lewis, S.E., Lord, P., Mallon, A.-M., Marthandan, N., Masuya, H., McNally, R., Mehrle, A., Morrison, N., Orchard, S., Quackenbush, J., Reecy, J.M., Robertson, D.G., Rocca-Serra, P., Rodriguez, H., Rosenfelder, H., Santoyo-Lopez, J., Scheuermann, R.H., Schober, D., Smith, B., Snape, J., Stoeckert, C.J., Tipton, K., Sterk, P., Untergasser, A., Vandesompele, J., Wiemann, S., 2008. Promoting coherent minimum reporting guidelines for biological and biomedical investigations: the MIBBI project. Nat. Biotechnol. 26, 889–896. https://doi.org/10.1038/nbt.1411.

Wilkinson, M.D., Dumontier, M., Aalbersberg, I.J., Appleton, G., Axton, M., Baak, A., Blomberg, N., Boiten, J.-W., da Silva Santos, L.B., Bourne, P.E., Bouwman, J., Brookes, A.J., Clark, T., Crosas, M., Dillo, I., Dumon, O., Edmunds, S., Evelo, C.T., Finkers, R., Gonzalez-Beltran, A., Gray, A.J.G., Groth, P., Goble, C., Grethe, J.S., Heringa, J., t'Hoen, P.A.C., Hooft, R., Kuhn, T., Kok, R., Kok, J., Lusher, S.J., Martone, M.E., Mons, A., Packer, A.L., Persson, B., Rocca-Serra, P., Roos, M., van Schaik, R., Sansone, S.-A., Schultes, E., Sengstag, T., Slater, T., Strawn, G., Swertz, M.A., Thompson, M., van der Lei, J., van Mulligen, E., Velterop, J., Waagmeester, A., Wittenburg, P., Wolstencroft, K., Zhao, J., Mons, B., 2016. The FAIR guiding principles for scientific data management and stewardship. Sci. Data. 3, 160018. sdata201618. https://doi.org/10.1038/sdata.2016.18.

Xie, Z., Hall, J., McCarthy, I.P., Skitmore, M., Shen, L., 2016. Standardization efforts: the relationship between knowledge dimensions, search processes and innovation outcomes. Technovation, Innov. Stand. 48–49, 69–78. https://doi.org/10.1016/j.technovation.2015.12.002.

Further Reading

de Leon, J., Susce, M.T., Johnson, M., Hardin, M., Maw, L., Shao, A., Allen, A.C.P., Chiafari, F.A., Hillman, G., Nikoloff, D.M., 2009. DNA microarray technology in the clinical environment: the AmpliChip CYP450 test for CYP2D6 and CYP2C19 genotyping. CNS Spectr. 14, 19–34.

CHAPTER 10

Human Genomic Databases in Translational Medicine

Theodora Katsila*, Emmanouil Viennas†, Marina Bartsakoulia*, Aggeliki Komianou*, Konstantinos Sarris*, Giannis Tzimas‡, George P. Patrinos*,§,¶

*Department of Pharmacy, School of Health Sciences, University of Patras, Patras, Greece, †University of Patras, Faculty of Engineering, Department of Computer Engineering and Informatics, Patras, Greece, ‡Department of Computer and Informatics Engineering, Technological Educational Institute of Western Greece, Patras, Greece, §Department of Pathology—Bioinformatics Unit, Faculty of Medicine and Health Sciences, Erasmus University Medical Center, Rotterdam, The Netherlands, ¶Department of Pathology, College of medicine and Health Sciences, United Arab Emirates University, Al-Ain, United Arab Emirates

10.1 INTRODUCTION

Research into the genetic basis of human disorders has advanced in scale and sophistication, leading to very high rates of data production in many laboratories, while electronic healthcare records become increasingly common features of modern medical practice. Therefore, it should be possible to integrate all of this information in order to establish a detailed understanding of how variants in the human genome sequence affect human health. In the last decade, major advances have been made in the characterization of genes that are involved in human diseases and advances in technology have led to the identification of numerous genomic variants in these genes. It rapidly has become clear that the knowledge and organization of these alterations in structured repositories will be of great importance not only for diagnosis, but also for clinicians and researchers. Genomic databases are referred to as online repositories of genomic variants, described for a single (*locus-specific*) or more (*general*) genes or specifically for a population or ethnic group (*national/ethnic*).

In this chapter, the key features of the main types of genomic databases that are frequently used in translational medicine will be summarized. In particular, emphasis will be given to specific examples from the previously mentioned database types in order to: (i) describe the existing and emerging database types in this domain, (ii) emphasize their potential applications in translational medicine, and (iii) comment upon the key elements that are still missing and holding back the field.

Human Genome Informatics. https://doi.org/10.1016/B978-0-12-809414-3.00010-3

10.2   HISTORICAL OVERVIEW OF GENOMIC DATABASES

Victor McKusick was the pioneer in this field in 1966, when the first serious effort towards summarizing DNA variations and their clinical consequences, the Mendelian Inheritance in Man (MIM), was published, as a paper compendium of information on genetic disorders and genes (McKusick, 1966). This is now distributed electronically (Online Mendelian Inheritance in Man (OMIM), http://www.omim.org) and regularly updated (Amberger et al., 2015). Also, the first "locus-specific database" collecting genomic variants from a single gene has been published in 1976, including 200 genomic variants from the globin gene in a book format, at that time, and has led to the HbVar database for hemoglobin variants and thalassemia mutations (Hardison et al., 2002; Patrinos et al., 2004; Giardine et al., 2007b, 2014).

A decade later, David Cooper began listing variants in genes to determine which one was the most common (Cooper et al., 1998). In the mid-1990s, the Human Genome Organization-Mutation Database Initiative (HUGO-MDI) was established by another visionary researcher, the late Richard Cotton, in order to organize "Mutation Analysis"; this new, at that time, domain of genetics (Cotton et al., 1998) then evolved into the Human Genome Variation Society (HGVS: http://www.hgvs.org). Today, the stated objective of the HGVS is "...to foster discovery and characterization of genomic variations, including population distribution and phenotypic associations."

Nowadays, this field is rapidly expanding and there are hundreds of databases of genomic variants available on the Internet. However, not all genomic databases fulfill quality requirements, while others have been built by researchers "on the side" for their own use.

10.3   GENOMIC DATABASE TYPES

The various depositories that fall under the banner of "genomic databases" can be categorized into three main types: *General* (or *central*) *variation* databases (GVDs), *locus-specific* databases (LSDBs), and *national/ethnic genomic* databases (NEGDBs).

GVDs attempt to capture all described variants in all genes, but with each being represented in only limited detail. The included phenotype descriptions are generally quite cursory, making GVDs of little value for those wishing to go deeper into genotype-phenotype correlations and phenotypic variability. GVDs tend to include only genomic variants leading to profound clinical effects that result in Mendelian patterns of inheritance, while sequence variations associated with no, minor, or uncertain clinical consequences are rarely catalogued. Thus, GVDs provide a good overview of patterns of clinically relevant genomic variations, but almost no fine detail to aid proper understanding. The best current example of a GVD would be the Human Gene Mutation Database (http://www.hgmd.org; Stenson et al., 2014), which, by June 2018, contained 157,131 different records in 6480 different genes in the public release and 224,642 different records in 8784 different genes in the HGMD Professional Release (V2018.1). In this type of databases, there are no specific field experts to maintain them, but they rather include published data for causative genomic variants, their distribution, and references, mostly using automated data (text) mining routines. Each causative genomic variant is entered only once in order to avoid confusion between recurrent and identical-by-descent lesions and the phenotypic description associated to the mutation is very limited, preventing any study on phenotypic variability. These databases frequently are referred as "mile wide and inch deep," as they included variants from many genes, but with a limited description. However, such databases may become increasingly useful in the postgenomic era with the advent of next-generation sequencing since they can contribute towards the understanding of novel tentatively pathogenic variants (Karageorgos et al., 2015). ClinVar (http://www.ncbi.nlm.nih.gov/clinvar) is another such example of GMDs, focusing exclusively on pathogenic variants leading to inherited diseases (Landrum et al., 2014, 2016). A detailed survey on the GVDs currently available has been previously compiled by George et al. (2008).

In contrast to GVDs, LSDBs focus on just one or a few specific genes (Claustres et al., 2002; Mitropoulou et al., 2010), usually related to a single disease entity. They aim to be highly curated repositories of published and unpublished mutations within those genes and, as such, provide a much-needed complement to the core databases. Data quality and completeness is typically high, with roughly half of the stored records pertaining to otherwise unpublished variants. The data are also very rich and informative and the annotation of each variant allele includes a full molecular and phenotypic description. Therefore, contrary to GVDs, these databases are referred to as "inch wide and mile deep." For example, LSDBs will typically present each of the multiple discoveries of recurrent mutational events, thereby allowing mutational hot-spots to be identified: and when these variants occur upon different chromosomal backgrounds (linked to other variants) such that they result in several, or different, disease features, these correlations are also recorded. A good example of an LSDB would be HbVar database (http://globin.bx.psu.edu/hbvar), initially reported by Hardison et al. (2002) and subsequently updated on a frequent basis with new features and data updates (Patrinos et al., 2004; Giardine et al., 2007b, 2014); a relational database of variants leading to structural hemoglobin variants and thalassemia and providing information on pathology, hematology, clinical presentation, and laboratory findings for a wealth of DNA alterations. Gene/protein variants are annotated with respect to biochemical data, analytical techniques, structure, stability, function, literature references, and qualitative and quantitative distribution in ethnic groups and geographic locations (Patrinos et al., 2004; Giardine et al., 2007b). As is common in LSDBs, entries can be accessed through summary listings or user-generated queries, which can be highly specific. A comprehensive listing of the currently available LSDBs can be found at http://www.hgvs.org and in the literature (Cotton et al., 2007).

Finally, NEGDBs, also known as National/Ethnic Mutation Databases or NEMDBs, are repositories documenting the genetic composition of an ethnic group and/or population, the genetic defects, leading to various inherited disorders, and their frequencies calculated on a population-specific basis. The emergence of the NEGDBs is justified from the fact that the spectrum of genomic variations observed for any gene or disease will often differ between population groups worldwide, and also between distinct ethnic groups within a geographical region. Not only do NEGDBs help elaborate the demographic history of human population groups, they are also a prerequisite to the optimization of national DNA diagnostic services. In other words, they will provide essential reference information for use in the design of targeted allele detection efforts for clinical use and may also serve to enhance awareness among healthcare professionals, bioscientists, patients, and the general public about the range of most common genetic disorders (and their environmental correlates) suffered by particular population groups.

Beyond the aforementioned main database types, DNA variation is also recorded in various databases, such as those provided at the national Center of Biotechnology Information, namely dbSNP and dbGAP (http://www.ncbi.nlm.nih.gov/projects/SNP; Agarwala et al., 2016) and the HAPMAP Data Coordination Center (DCC: http://www.hapmap.org; The International HapMap Consortium, 2003). These resources are important in helping to complete the picture for any gene or region of interest, by summarizing all the variants that are typically not included in GVDs, LSDBs, and NEGDBs. In brief, GMDs, LSDBs, and NEGDBs share the same primary purpose of representing DNA variations that have definitive or likely phenotypic effect; in other words, clinical interest. They tackle this goal from very different perspectives, and there is clearly a need for these three types of resource in the various disciplines of human genetics and genomics, particularly genetic testing.

In the following paragraphs, the basic aspects of a representative resource from GVDs, LSDBs, and NEGDBs will be presented in detail, in relation to their applications in translational medicine, since these resources are more closely related to molecular genetic testing than any other genomic database type.

10.4 MODELS FOR DATABASE MANAGEMENT

In its strict definition, a database is a collection of records, each of which contains one or more fields (i.e., pieces of data) about a certain entity (e.g., DNA sequences, alleles) that has a regular structure and that is organized in such a way that the desired information can easily be queried and retrieved. Development of databases is relied on the model that the curator, that is, the person or group of persons who are responsible for developing, updating, and ultimately maintaining a genetic database, will chose for setting up one. There are mainly three types of database management models, from low to higher degrees of complexities, each with its own advantages and drawbacks.

Nowadays, relational databases are the most frequently used ones, since they are very efficient for dealing with large volumes of information than any other database types. A relational database is based on data organization in a series of interrelated tables. Also, information can be retrieved in an extremely flexible manner by using structured data queries. Although interest in this model was initially confined to academia, subsequently, relational databases became the dominant type for high-performance applications because of their efficiency, ease of use, and ability to perform a variety of useful tasks that had not been originally envisioned. The dominant query language for relational databases is the semistandardized structured query language (SQL), for which different SQL variants exist, the main ones being Microsoft SQL (http://www.microsoft.com/sql), MySQL (http://www.mysql.com), and postgreSQL (http://www.postgresql.org). Although critics claim that SQL is not consistent with the relational model, it works extremely well in practice and no replacement is on the horizon. The requirement of specialized software for developing a relational database can potentially be a disadvantage, since significant computer proficiency is required.

In the very first genomic databases, information on genomic variants was provided in plain text websites; in other words, the simplest "database" format. Even as this structure cannot be considered as a database in a strict sense, it was often used in the past for several database projects. The advantages for such a model were the development and maintenance simplicity, since no specific software was required. However, there were no true data querying options, apart from the standard searching tool provided by the respective Internet browser, while the database was very difficult to maintain in case of expanded datasets. In addition, flat-file databases were the simplest database types, particularly for small-scale datasets and simple applications. These databases had modest querying capacity and could accommodate small to moderately big datasets, while their development required average computing skills, even though they were based on simple software. The first version of the *ETHNOS* software (Patrinos et al., 2005a; see below) was developed, using the flat-file database model.

10.5 GENERAL VARIATION DATABASES: DOCUMENTATION OF VARIANTS OF CLINICAL SIGNIFICANCE

10.5.1 ClinVar Database

ClinVar (http://www.ncbi.nlm.nih.gov/clinvar; Landrum et al., 2016) is a freely available public resource of genotype-phenotype relationships with supporting evidence of clinical utility. As such, ClinVar reports relationships asserted between human genomic variation and observed health status and the history of that interpretation. ClinVar records include submissions reporting variants found in patient samples, assertions made regarding their clinical significance, information about the submitter, and other supporting data. The alleles described in submissions are mapped to reference sequences and reported according to the HGVS nomenclature. ClinVar was originally launched in 2012, with the first full public release in April 2013 (Landrum et al., 2014). The initial dataset included variations from OMIM, GeneReviews, some locus-specific databases (LSDB), contributing testing laboratories, and others.

At present, ClinVar has 211,118 records submitted in total, from which 120,667 records have assertion criteria and 190,933 records an interpretation, representing 160,119 unique genomic variants from 27,261 genes from 579 submitters. From these, 23 genomic variants are accompanied with practice guidelines, while 11,979 variants are accompanied with assertion criteria and are submitted by multiple submitters and have no conflicts (database last accessed 17 September 2016).

ClinVar supports data submission of differing levels of complexity. In other words, the submission may be as simple as a representation of an allele and its interpretation (sometimes termed a variant-level submission), or as detailed as providing multiple types of structured observational (case-level) or experimental and/or functional evidence about the effect of the certain genomic variant on phenotype. A unique feature of ClinVar is that it archives previous versions of submissions, meaning that when submitters update their records, the previous version is retained for review. http://www.ncbi.nlm.nih.gov/clinvar/docs/submit. The level of confidence in the accuracy of variant calls and assertions of clinical significance mainly depends on the supporting evidence, so this information, when available, is collected and made visible to users. Because the availability of supporting evidence may vary, particularly as far as retrospective data aggregated from the published literature are concerned, the archive accepts submissions from multiple groups, and aggregates related information, to reflect in a transparent manner both consensus and conflicting assertions of clinical significance.

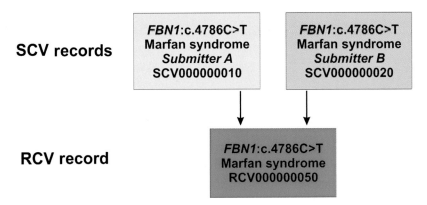

FIG. 10.1

ClinVar records are aggregates by genomic variant and phenotype. Two independent submitters
A and B contribute to ClinVar the same variant in the *FBN1* gene, leading to Marfan syndrome. Each record
has a different SCV ID, but are both parts of a unique reference ClinVar record with the same RCV ID.

ClinVar accessions follow the format SCV000000000.0 and are assigned to
each submission. In cases where there are multiple submissions about the same
variant/condition relationship, these are aggregated within ClinVar's data flow
and reported as a reference accession of the format RCV000000000.0. As a
result, one variant allele may be included in multiple RCV accessions whenever
different phenotypes may be reported for that variant allele (Fig 10.1).

ClinVar currently includes clinical assertions for variants identified through:
(a) clinical testing, where clinical significance is reported as part of the genetic
testing process in CLIA certified or ISO 1589 accredited laboratories,
(b) research, for variations identified in human samples as part of a research
project, and (c) extraction from the literature, where reporting of the phenotype
is extracted directly from the literature without modification of authors' statements. At present, ClinVar does not include uncurated sets of data from
genome-wide association studies (GWAS), although variants that were identified through GWAS and have been individually curated to provide an interpretation of clinical significance are in scope.

ClinVar includes the following representations:

- *Medical conditions*: ClinVar aggregates the names of medical
  conditions with a genetic basis from resources such as SNOMED CT,
  GeneReviews, Genetic Home Reference, Office of Rare Diseases, MeSH,
  and OMIM, while it also aggregates descriptions of associated traits
  from Human Phenotype Ontology (HPO), OMIM, and other sources.
  Every source of information is adequately tracked and can be used
  in queries.

- *Genomic variants*: Human variations are reported to the user as sequence changes relative to an mRNA, genomic, and protein reference sequence (if appropriate), according to the Human Genome Variation Society (HGVS; www.hgvs.org) standard. Genomic sequences are represented in RefSeqGene/LRG coordinates, as well as locations on chromosomes (as versioned accessions and per assembly name, such as NCBI36/hg18 and GRCh37/hg19). Novel variations are accessioned in NCBI's variation databases (dbSNP and dbVar).
- *Relationships among various phenotypes and variants*: ClinVar is designed to support the evolution of the current knowledge about the relationship between genotypes and medically important phenotypes. By aggregating information about variants observed in individuals with or without a clinical phenotype, ClinVar supports the establishment of the clinical validity of human variation.

Each ClinVar record contains the following parameters:

- *ClinVar Accession and Version*: Accession number/version number separated by a decimal (SCV000000000.0) assigned to each record and reference accession number/version separated by a decimal (RCV000000000.0) assigned to submitted sets of assertions about the same variation/phenotype.
- *Identifiers for each variant allele or allele set*: These include HGVS expressions, published allele names, and database identifiers, where applicable.
- *Attributes of each phenotype*: These include the name of the phenotype and/or clinical condition, description(s) and defining features, its prevalence in the population, and again database identifiers, where applicable.
- *Description of the genotype-phenotype relationship*: This includes a review status of the asserted relationship, the submitter of the assertion, and its clinical significance. It also includes the summary of the evidence for clinical significance, the total number of the observations of a particular genotype/allele in individuals with a given phenotype, the total number of observations of genotype/allele in those individuals without the phenotype, the various family studies and study design, and a description of the population sampled. The description also includes in vitro and in silico studies and animal models, where available, and also the mode of inheritance, and the relevant citations.
- *Submission information*: Includes submitter description, submission and update dates, and data added by central NCBI computation.

The information aggregated in ClinVar is reported in the viewer in the most accessible presentation possible. Linking within ClinVar and links out are minimized where possible to make sure that the greatest amount of information is visible with the fewest possible number of uninformative reference numbers. Any conflict or uncertainty about a variant/phenotype association will be reported explicitly, as ClinVar only reports conclusions from external data submitters. Clinical laboratories can integrate the information available from ClinVar into their workflow, both submitting variants and associated assertions of clinical significance and using the available information to identify the clinical significance of already documented variants.

10.5.2 Data Sharing in ClinVar

The information documented in ClinVar is freely available to users and organizations to ensure the broadest utility to the medical genetics community. Emphasis is given so that data structures are designed to facilitate data exchange so that data can be shared bilaterally. Also, attribution is important to identify the source of variants and assertions, to facilitate communication and to give due credit to data submitters (see also Chapter 6). Each submitter is explicitly acknowledged, with pointers to more detailed submitter contact information to facilitate communication and collaboration within the genetics community.

10.6 LOCUS-SPECIFIC DATABASES IN TRANSLATIONAL MEDICINE

Detection of DNA sequence variation can be very efficiently performed using a plethora of molecular diagnostics techniques, both low- and high-throughput. LSDBs can facilitate molecular diagnosis of inherited diseases in various ways. For example, LSBDs can assist in ascertaining whether a DNA variation is indeed causative, leading to a genetic disease, or benign. Similarly, some high-quality LSDBs provide detailed phenotypic information that is related to disease-causing genomic variants. Ultimately, LSDBs can assist in the selection of the optimal mutation detection strategy.

Soussi and coworkers were among the first to present LSDBs' challenges and opportunities on the basis of the p53 paradigm (Soussi et al., 2006). Just recently, laboratory data on *BRCA1* and *BRCA2* genetic variants were compared to those provided by BIC, ClinVar, HGMD Professional, LOVD, and UMD databases (Vail et al., 2015). Findings reported substantial disparity of variant classifications within and among the databases considered, suggesting that LSDBs' use in clinical practice is still challenging.

10.6.1 Comparison Among Various LSDBs

Several hundred of LSDBs are available today on the Internet, which sometimes makes it difficult to choose the "best" LSDB. In addition, there are often more than one LSDB per gene locus, from which it is hard to determine the "reference" and best curated LSDB. This fact generates confusion to potential users as to which LSDB to choose. This can be particularly worrying, since not all databases conform to the proposed quality guidelines, or are curated or updated frequently, while they are rather diverse in terms of content and structure.

In 2010, Mitropoulou et al. performed a thorough domain analysis of the 1188, at that time, existing LSDBs in an effort to comprehensively map data models and ontology options, on which the existing LSDBs are based, in order to provide insight into ways the field should further develop and to produce recommendations towards implementation of LSDBs for use in a clinical and genetic laboratory setting. This effort came as a follow-up of the comparative analysis of Claustres et al. (2002), dictated by the rapid growth of LSDBs and the vast data-content heterogeneity that characterizes the field. These LSDBs were assessed for a total of 44 content criteria pertaining to general presentation, locus-specific information, database structure, data collection, variant information table, and database querying.

A key observation that derived from this analysis is the fact that more and more LSDBs are generated using an, often downloadable, LSDB management system, hence adequately tackling the issue of data-content heterogeneity existed in the dawn of LSDB era in the early 2000. Overall, several elements have helped to advance the field and reduce data heterogeneity, such as the development of specialized database management systems and the creation of improved data querying tools, while a number of deficiencies were identified, namely the lack of detailed disease and phenotypic descriptions for each genetic variant and links to relevant patient organizations, which, if addressed, would allow LSDBs to better serve the clinical genetics community. Based on these findings, Mitropoulou et al. (2010) proposed an LSDB-based structure, which would contribute to a federated genetic variation browser and also allow the seamless maintenance of genomic variation data. To this end, this effort constituted a formal "requirements analysis" that was undertaken by the GEN2PHEN project (http://www.gen2phen.org) aiming to (a) contribute guidelines upon which the LSDB field can be further evolved, (b) formalize the data models and the nomenclature systems being utilized by the entire LSDB community, and (c) maximize synergy with groups involved in the LSDB field.

10.6.2   Identification of Causative Genomic Variants

Usually, in diagnostic laboratories, if a missense variant is detected, additional experiments need to be conducted prior to concluding that the variant in question is in fact causative in the family. This is particularly important in case of whole genome sequencing where a plethora of variants are reported. In the absence of a functional test, the segregation of the variant in the affected family members, the absence of this variation in a panel of at least 100 control samples, the prediction of the biochemical nature of the substitution, the region where the variation is located, and the degree of conservation among species are some of the arguments for the causative nature of the variant. As this approach is often time-consuming, the use of a LSDB can provide researchers with valuable information to help in such a decision process. If the variant has been reported as a causative one, its full description and the corresponding literature is provided in the LSDB. Furthermore, a comprehensive LSDB does not only include the reference sequence of the gene, but also the description of structural domains and data about interspecies conservation for each protein residue.

Also, as many recent publications include large data sets, it is often possible to observe errors or use of wrong variant nomenclature, due to reference to an old sequence (up to 10% of errors for some publications). The use of LSDBs can be helpful to this end. Several LSDBs include an automatic nomenclature system, based on a reference sequence. In other words, variants' entry is done such that the variant nucleotide is automatically checked against the reference sequence at the respective variant position and named based on the official (HGVS) nomenclature (den Dunnen and Antonarakis, 2001). Interestingly, Mutalyzer, a dedicated module to automatically produce any sequence variation nomenclature (http://www.lovd.nl/mutalyzer; Wildeman et al., 2008), enables unambiguous and correct sequence variant descriptions to avoid mistakes and uncertainties that may lead to undesired errors in clinical diagnosis. Mutalyzer handles most variation types, that is, substitutions, deletions, duplications, etc., and follows current HGVS recommendations.

Finally, a handful of LSDBs include data presentation tools to visualize their content in a graphical display. VariVis is a generic visualization toolkit that can be employed by LSBDs to generate graphical models of gene sequence with corresponding variations and their consequences (Smith and Cotton, 2008). The VariVis software package can run on any web server capable of executing Perl CGI scripts and can interface with numerous database management systems and even flat-file databases. The toolkit was first tested in A_1ATVar, a LSDB for *SERPINA1* gene variants, leading to α_1-antitrypsin deficiency (Zaimidou et al., 2009) and can be integrated into generic database management systems used for LSDBs development (see also Section 10.8).

10.6.3 Linking Genotype Information With Phenotypic Patterns

LSDBs are far more than just inert repositories, as they include analyzing tools, which exploit computing power to answer complex queries, such as phenotypic heterogeneity and genotype-phenotype correlations. The vast majority of LSDBs, especially all LOVD-based LSDBs (see Section 10.8), provide phenotypic descriptions in abstract format. Other custom-built databases, such as HbVar, phenotypic descriptions, are significantly more detailed, for example, providing information on the clinical presentation of thalassemia carriers and patients together with their hematological indices for every hemoglobin and thalassemia variants (Giardine et al., 2007a). All LSDBs attempt to enforce controlled vocabulary to facilitate straightforward data querying. Therefore, phenotype data in the vast majority of LSDBs are presented in a very basic way, such as in the form of free text entries and/or with very little detail. There is a definite need for this situation to be improved and data content to be harmonized and a general wish for the comprehensive analysis of phenotypes to occur, a goal termed "phenomics" (Gerlai, 2002; Hall, 2003; Scriver, 2004), supported by the necessary informatics solutions.

Also, it is possible that patients with the same causative variant will have a completely different phenotype, resulting from a number of modifier genes. This valuable information could be useful for predictive medicine. A good example is given by the distribution of mutations of the *FBN1* gene (MIM# 134797) that are associated with Marfan syndrome (MFS) and a spectrum of conditions phenotypically related to MFS, including dominantly inherited ectopia lentis, severe neonatal MFS, and isolated typical features of MFS. MFS, the founding member of heritable disorders of connective tissue, is a dominantly inherited condition characterized by tall stature and skeletal deformities, dislocation of the ocular lens, and propensity to aortic dissection (Collod-Beroud and Boileau, 2002). The syndrome is characterized by considerable variation in the clinical phenotype between families and also within the same family. Severe neonatal MFS has features of the MFS and of congenital contractural arachnodactyly present at birth, along with unique features such as loose, redundant skin, cardiac malformations, and pulmonary emphysema (Collod-Beroud and Boileau, 2002). A specific pattern of causative variants is observed in exons 24–26 in association with the neonatal MFS. In fact, 73.1% of variants are located in this region in the neonatal form of the disease, but only 4.8% of variants associated with a classical MFS are located in these exons (FBN1 database: http://www.umd.be; Collod et al., 1996; Collod-Beroud et al., 2003).

An interesting project attempted to interrelate human phenotype and clinical data in various LSDBs with data on genome sequences, evolutionary history, and function from the ENCODE project and other resources in genome browsers. PhenCode (Phenotypes for ENCODE; http://www.bx.psu.edu/ phencode; Giardine et al., 2007b) is a collaborative, exploratory project to help understand phenotypes of human mutations in the context of sequence and functional data from genome projects. The project initially focused on a few selected LSDBs, namely HbVar (*HBA2*, *HBA1*, and *HBB* genes), PAHdb (*PAH* gene; Scriver et al., 2000), etc. Variants found in a genome browser can be tracked by following links back to the LSDBs for more detailed information. Alternatively, users can start with queries on mutations or phenotypes at an LSDB and then display the results at the genome browser to view complementary information such as functional data (e.g., chromatin modifications and protein binding from the ENCODE consortium), evolutionary constraint, regulatory potential, and/or any other tracks they choose. PhenCode provides a seamless, bidirectional connection between LSDBs and ENCODE data at genome browsers, which allows users to easily explore phenotypes associated with functional elements and look for genomic data that could explain clinical phenotypes. Therefore, PhenCode not only is helpful to clinicians for diagnostics, it also serves biomedical researchers by integrating multiple types of information and facilitating the generation of testable hypotheses to improve our understanding of both the functions of genomic DNA and the mechanisms by which it achieves those functions. These and other types of data provide new opportunities to better explain phenotypes.

10.6.4 Selection of the Optimal Variant Allele Detection Strategy

As LSDBs collect all published and unpublished genomic variants, they are very useful to define an optimal genetic screening strategy, especially when targeted resequencing is needed. Therefore, an overview of the distribution of variants at the exonic level can help to focus on specific exons, where most of the variants are located. The best example is given by the study of the TP53 gene involved in up to 50% of human cancers (Soussi, 2000). This gene is composed of 11 exons from which 10 are transcribed in a 393 amino acids protein. The distribution of over 25,000 variants reported in the TP53 database available either in a UMD (http://www.umd.be) or LOVD format (http://www.lovd.nl; see Section 10.8) shows that approximately 95% of pathogenic variants are located in 4 out of the 11 exons of the gene (exons 5–8; Beroud and Soussi, 2003). This observation has led 39% of research groups to search for mutations only in these exons, whereas 13% performed a complete scanning of the gene (Soussi and Beroud, 2001). Although this strategy is cost-effective, one needs to be careful in case of a negative result and should perform a complete scanning in order to avoid bias, as described in Soussi and Beroud (2001).

Similarly, summary listings of variations documented in LSDBs (such as those provided in LOVD-based LSDBs) will help to choose the best experimental approach. For example, if most variants are nonsense, the protein truncation test could be one of the best approaches, whereas sequencing is considered the golden standard for variant detection. Certain LSBDs may contain additional information about primers and technical conditions to help new research groups or diagnostic laboratories to establish their own diagnostic procedures, such as those for the various genes involved in muscular dystrophies (http://www.dmd.nl) and thalassemias (http://www.goldenhelix.org/xprbase; a companion database to HbVar; Giardine et al., 2007b).

Finally, in few LSDBs, explicit information is provided regarding the variant allelic pattern associated with a population/ethnic group and/or geographical region. This information can be extremely helpful in stratifying variant allele detection strategies. In other words, in ethnic groups with a more or less homogeneous mutation pattern, mutation screening efforts can be either targeted to those genomic regions that the majority of mutations have been reported, or to a specific variant allele detection technique, that is, ARMS or restriction enzyme analysis, thus saving time and resources. Coupling LSDBs with NEGDBs resources (see also Section 10.7), if available, would further facilitate these efforts. However, as previously explained, extreme caution should be taken in case of a negative result that would require complete mutation scanning in order to avoid bias.

10.7   NATIONAL/ETHNIC GENOMIC DATABASES: ARCHIVING THE GENOMIC BASIS OF HUMAN DISORDERS ON A POPULATION-SPECIFIC BASIS

NEGDBs are genomic variant repositories, recording extensive information over the described genetic heterogeneity of an ethnic group or population. These resources have recently emerged, mostly driven by the need to document the varying mutation spectrum observed for any gene (or multiple genes) associated with a genetic disorder, among different population and ethnic groups (Patrinos, 2006).

In general, the NEGDBs available to date can be divided into two subcategories:

- The "National Genetic" (or Disease Mutation) databases, the first ones that appeared online, record the extant genetic composition of a population or ethnic group, but with limited or no description of mutation frequencies. The first NEGDB to come online was the Finnish database (http://www.findis.org; Sipila and Aula, 2002), which, although rich in information, provided very limited querying capacity, particularly for allelic frequencies.
- The "National Mutation Frequency" databases, providing comprehensive information only of those inherited, mostly monogenic, disorders whose disease-causing variants spectrum is well-defined. The Hellenic and Cypriot NEGDBs (Patrinos et al., 2005a; Kleanthous et al., 2006,

respectively) introduced a specialized database management system for NEGDBs that enabled both basic query formulation and restricted-access data entry so that all records are manually curated to ensure high and consistent data quality (van Baal et al., 2010).

In order to provide a simple and expandable system for worldwide population-specific allele frequency data documentation, focusing in particular on clinically revenant genomic variants, the latter NEGDB group was used as the basis for the design of Frequency of INherited Disorders database (FINDbase; http://www.findbase.org), a relational database that records frequencies of clinically relevant genomic variants worldwide (van Baal et al., 2007). Even from its very first version, FINDbase offered a user-friendly query interface, providing instant access to the list, and frequencies of the different mutations and query outputs can be either in a table or graphical format.

In 2010, FINDbase migrated to the new version of the *ETHNOS* software (see below), which included new data querying and visualization tools to further exploit the expanded FINDbase data collection (Georgitsi et al., 2011a). The data querying and visualization tools were built around Microsoft's Pivot-Viewer software (http://www.getpivot.com), based on Microsoft Silverlight technology (http://www.silverlight.net), which provides an elegant, web-based multimedia interface for population-based genetic variation data collection and retrieval. The entire system architecture is based on a three-tier client-server model (Eckerson, 1995), namely the client application, the application server, and the database server. All the FINDbase data records were converted to a set of files on a server, which were CXML and Deep Zoom-formatted (DZC) images. When the user browses the collection from a web page, the PivotViewer uses the Silverlight Control to display the files. The entire FINDbase causative mutations data collection via PivotViewer is shown in Fig. 10.2, which enables the user to interact with large datasets at once. PivotViewer enables users to smoothly and quickly arrange FINDbase data collections according to common characteristics that can be selected from the data query menu (Georgitsi et al., 2011a) and then zoom in for a closer look, by either filtering the collection to get a subset of information or clicking on a particular display item. Each display item in the form of a card (Fig. 10.3), with a chromosomal figure (derived from http://www.genecards.org) displaying the gene position, is provided for each genetic variation, along with a sidebar textbox with in-depth data concerning the particular genetic variation and population. Hyperlinks for each gene name to OMIM database and HGMD offer to the user the possibility of easily accessing additional information.

In particular, the new FINDbase data querying and visualization environment enables the user to perform simple and complex queries, visualize and sort, organize and categorize data dynamically, and discover trends across all items,

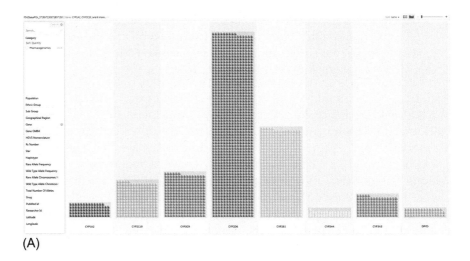

(A)

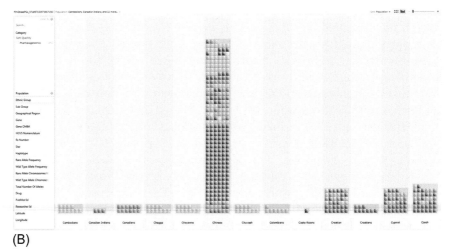

(B)

FIG. 10.2

Overview of the FINDbase pharmacogenomic biomarkers data module, based on Microsoft's PivotViewer and Silverlight technology. The querying interface is shown on the *left* and the output option can be selected at the *top-right* corner of the screen. Indicative examples include data output per gene (A) or population group (B). The different entries are shown as colored boxes, presented as display items (see Fig. 10.3). The user can zoom in for a closer look or click on a particular item to get more in-depth information.

using different views. In 2012, additional visualization tools have been implemented (Viennas et al., 2012; Papadopoulos et al., 2014), based on the Flare visualization toolkit (http://flare.prefuse.org), which provides two extra types of data query and visualization outputs, namely the Gene and Mutation Map and the Mutation Dependency Graph. The Gene and Mutation Map is

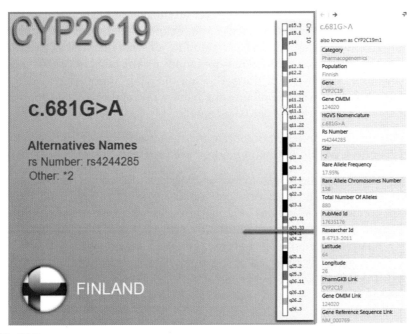

FIG. 10.3

Example of a display item provided for each pharmacogenomic biomarker (*CYP2C19* c.681 G > A), accompanied by a sidebar textbox with in-depth data concerning the particular variant and population (Finland; shown on the *right*). Each item includes the name of the allele in its official HGVS or other nomenclature systems, if available, the population for which this information is available (shown by the country's flag), and a chromosomal map, where the gene's position is indicated. Hyperlinks for each gene to PharmGKB database (http://www.pharmgkb.org) offer to the user the possibility of easily accessing additional information. Finally, each item displays the corresponding PubMed and Researcher IDs, if applicable. Similar display items are also used for the Causative Mutations and Genetic Disease Summaries modules (not shown).

based on a tree-map, which constitutes of an easy way of analyzing large data sets (http://www.cs.umd.edu/hcil/treemap). In FINDbase, the tree-map corresponds to mutation frequencies estimated for each population. Each rectangle represents a population's mutation and a specific color corresponds to each population. The area covering each node encodes the frequency of rare alleles. Each time the user clicks on a node, the occurrence of the selected mutation is shown over all populations. Similarly, the Mutation Dependency Graph visualizes the dependencies that occur among different populations on the basis of a selected genomic variant. In FINDbase, population names are placed along a circle and these populations are clustered based on the presence and/or the frequency of a certain genomic variant. A link between populations indicates that these populations have the same genomic variant in common and, by clicking on a specific population, the user can see all the relevant dependencies of that population concerning the selected genomic variant. These tools, available only for the causative mutation FINDbase module, provide the means of establishing relationships among different populations on the basis of certain genome variants and, together with the basic data querying option provided by the PivotViewer, significantly enhance the battery of the data visualization tools available to explore FINDbase data content.

Apart from the documentation of causative genomic variants, leading to inherited disorders, two additional modules have been implemented in FINDbase:

(a) *The Pharmacogenomics Biomarker module*: which comprehensively documents the incidence of pharmacogenomics biomarkers in different populations (Georgitsi et al., 2011b). This is a much-needed addition, since population- and ethnic group-specific allele frequencies of pharmacogenomic markers are poorly documented and not systematically collected in structured data repositories.

(b) *The Genetic Disease Summaries module*: which is in fact the evolution of the first flat-file *ETHNOS*-based NEMBDs (van Baal et al., 2010; Viennas et al., 2017). As with the existing FINDbase modules for causative genomic variants and pharmacogenomic biomarkers, the component services of this module were also built using the same principles and exploiting PivotViewer and Microsoft Silverlight technology. In this module, database records include the population, the genetic disease, the gene name, online Mendelian Inheritance in Man (OMIM) ID, and chromosome on which it resides which are all included in the query interface.

Given their scope, to maximize the utility of NEMDBs, the mode by which their content is provided needs to provide a seamless integration with related content in LSDBs and GMDs (Patrinos and Brookes, 2005) and with other resources and tools, such as VarioML (Viennas et al., 2017). Furthermore, as is always desirable for specialized databases, extensive links to other external information (e.g., to OMIM and to various types of genome sequence annotation) would ideally be provided as part of the necessary tying together of the growing network of genomic databases.

10.8   NEGDBs IN A MOLECULAR DIAGNOSTICS SETTING

NEMDBs can be helpful in different ways in translational medicine. First of all, they can help optimize national molecular diagnostic services by providing essential reference information for the design and implementation of regional or national-wide mutation screening efforts (see Section 10.6.4). In addition, NEGDBs can enhance awareness among clinicians, bioscientists, and the general public about the range of most common genetic disorders suffered by certain populations and/or ethnic groups. Based on populations' allele frequency spectra, customized genetic tests can be designed. Several diagnostic companies have designed kits for various genetic disorders, such as β-thalassemia, which qualitatively detect, for example, the most common Mediterranean and Asian β-thalassemia pathogenic variants (reviewed in Patrinos et al., 2005b). Most importantly, those databases can also assist in interpreting diagnostic test results in countries with heterogeneous populations, particularly where interpretation of test results in minority ethnic groups may be ambiguous or problematic (Zlotogora et al., 2007, 2009; Zlotogora and Patrinos, 2017).

10.8.1 NEGDBs and Society

Apart from their importance in a molecular diagnostic setting, NEGDBs can also contribute toward the elucidation of populations' origin and migration. History of a certain population is tightly linked with the history of its allele(s). Therefore, NEGDBs, particularly those including data from many population groups, can serve as the platform for comparative genomic studies that can reciprocally provide insights into, for example, the demographic history of human populations, patterns of their migration and admixture, gene/ mutation flow, etc.

To this end, in order for a NEGDB to maximize its accuracy and comprehensive data coverage, it is vital that certain parameters are strictly observed. In the case of recurrent mutational events (e.g., sickle cell variation), caution should be taken to precisely record the underlying genomic background on which an allelic variant has occurred. Furthermore, allelic frequencies should be calculated based on the most representative study that involves sufficient numbers of patients and controls. Estimation of absolute allelic frequencies based on multiple reports has the inherent danger of including redundant cases that can alter the calculated frequencies. Finally, the very delicate issue of anonymity should be adequately preserved, by including data only at a summary rather than individual level, so that NEGDBs' data contents consist only of number of chromosomes rather than sensitive personal details of their carriers. Several recommendations and guidelines to facilitate participation of emerging countries in genetic variation data documentation, ensuring an accurate and comprehensive worldwide data collection, have been proposed (Patrinos et al., 2011), which include: (a) creating international networks from closely related populations to enhance communication, interaction, and initiate research networks between them, (b) fostering communication, interaction, and research networks between developing and developed countries, (c) developing procedures to include clinicians and researchers in developing nations, (d) ensuring that the database management systems that are being used or developed can be utilized by those in a limited resource environment, (e) providing support to developing countries to build capacity and fully participate in the collection, analysis, and sharing of genetic variation information, (f) developing a framework to facilitate interactions between the coordinating center and national, regional, and international agencies, and (g) ensuring that all ethical, legal, religious, and social issues are thoroughly considered when NEGDBs and/or data capture projects are launched in developing countries. Based on these recommendations, a number of Country Nodes have been developed (Patrinos et al., 2012).

10.9 DATABASE MANAGEMENT SYSTEMS FOR LSDBs AND NEGDBs

Ever since the development of the first genetic databases, a number of guidelines and recommendations for mutation nomenclature (den Dunnen and Antonarakis, 2001; den Dunnen and Paalman, 2003), content, structure (Scriver et al., 1999), curation, and deployment of LSDBs (Claustres et al., 2002; Cotton et al., 2008; Mitropoulou et al., 2010) and NEGDBs (Patrinos, 2006; Patrinos et al., 2012) have been produced to encourage the harmonization of LSDB and NEGDB development and curation, respectively.

To facilitate interested parties and research groups to develop and curate their own LSDBs and NEGDBs, generic tools, known as database management systems (DBMS), have been made available for this purpose. Several off-the-shelf freely available and user-friendly software packages have been developed for LSDB development and curation, including MUTbase (Riikonen and Vihinen, 1999), Universal Mutation database (UMD; Beroud et al., 2000), Mutation Storage and Retrieval (MutStaR; Brown and McKie, 2000), and Leiden-Open (source) Variation Database (LOVD, Fokkema et al., 2005, 2011). From these, LOVD, now in V3.0 with a total of 292,327,696 (3,944,452 unique) variants in 82 LOVD installations (last accessed June 2018) and in synergy with the Human Variome Project (Kaput et al., 2009; Cotton et al., 2009; http://www.humanvariomeproject.org), the NCBI, and other online resources, is considered by far the most commonly used DBMS for LSDB development and curation. This software provides, mostly in a tabular format and in an easy-to-query manner, detailed lists of all variants documented for a specific gene locus, while a summary page allows the user to scan through graphical summaries of all genomic variants available in a specific installation (Fig. 10.4). In its newest release, LOVD extends this concept to also provide patient-centered data storage and storage of NGS data, hence allowing the deposition of variants that reside in intergenic regions with possible functional significance.

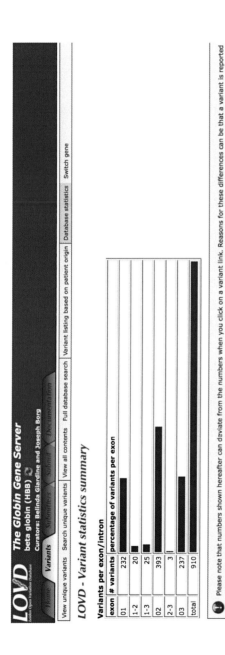

FIG. 10.4

Screen shot of part of the variant statistics summary page of a typical LOVD database for the *HBB* gene, as part of the Globin Gene Server. The different types of variants for DNA, RNA, and protein are provided in *blue* (observed variants) or *red* (predicted variants; not shown). Variants are also clustered per exons and introns, hence allowing the identification of mutational hot-spots and facilitating the design of molecular genetic screening assays.

As far as NEGDBs are concerned, ETHnic and National database Operating System (*ETHNOS*; van Baal et al., 2010) is the only DBMS currently available, on which the first-generation NEGDBs and FINDbase were developed.

These user-friendly DBMSs are designed to promote the creation of more and better LSDBs, by reducing or eliminating the requirement for substantial knowledge of computing and bioinformatics for interested parties to establish a LSDB from scratch. In addition, the use of off-the-shelf solutions positively impacts data uniformity and, contrary to the NEGDB's structure and data content that is relatively uniform since this discipline has more recently emerged (Patrinos, 2006), the use of such DBMS in the LSDB domain resulted in a significant increase of LSDB structure and data-content homogeneity (Mitropoulou et al., 2010), hence drastically resolving the issue and increasing data uniformity among these resources.

Finally, the use of a DBMS that can be run on any platform will reduce the risk of the database being "lost." If database curation for some reason, for example, lack of funding, is interrupted, data will then be transferred directly between platforms or locations and they will remain accessible to all. Potential curators will be encouraged to set up LSDBs, with the choice of using these software packages locally on their own workstations or, most importantly, having their databases hosted on a central server on the Internet.

10.10 FUTURE CHALLENGES

Notwithstanding the technical challenges, the issues of rewarding genomic data submission and sharing and the ways database research is organized and motivated are perhaps more difficult to overcome (see also Chapter 6). For example, forming consensus opinions and truly committed consortia in order to create standards is far from easy in the highly competitive world of science. This may partly explain why leading bioinformatics activities today are often conducted in large specialized centers (e.g., the European Bioinformatics Institute, and the United States National Center for Biotechnology Information) where the political influence and critical mass are such that what the resources and tools these centers produce automatically become the *de facto* standard. These groups, however, cannot build all the necessary GVDs, LSDBs, and NEGDBs that are needed, but they could help others (biological domain experts) to do it and then integrate all their efforts (Stein, 2002).

In 2006, a global initiative, called the Human Variome Project (HVP; http://www.humanvariomeproject.org), was initiated, aiming to catalogue all human genetic variation and make that information freely available to researchers, clinicians, and patients worldwide (Ring et al., 2006) and envisioning to achieve improved health outcomes by facilitating the unification of human genetic variation and its impact on human health (Horaitis et al., 2007; Kaput et al., 2009; Cotton et al., 2009).

Today, an enormous amount of content already exists in the digital universe, characterized by high rates of new information that is added, distributed, and demands attention. This big data momentum lacks applicability and societal impact at significant levels, as there are several gaps at various knowledge levels that set individuals and experts apart. Indicatively, genetic information cannot be equally understood by a researcher, a geneticist, or a clinician. For this, IT tools are anticipated to play a crucial role towards data management and decision making. In this context, DruGeVar (Dalabira et al., 2014) was developed as an online resource to triangulate drugs with genes and variants that serve as pharmacogenomic biomarkers.

Finally, the most fundamental hurdle of all that retards the field is that of limited funding. Because of this, almost all mutation databases in existence today have been built by researchers "on the side" for their own use, with a small degree of sponsorship/funding at best. To advance beyond this stage, projects need to be increased in scale, quality, and durability, and this can only happen if strategically minded funding agencies make available substantial targeted funds. The new databases that thus emerge will then need long-term support for general maintenance and further development. To solve this, the projects may ultimately need to be run as self-sustaining "businesses" that charge for data access, and/or it might be possible to develop novel forms of joint academic-corporate funding.

With great vision, the European Commission announced in 2007 the 1st call for proposals for the 7th Framework Program (FP7; thematic area HEALTH), which included a topic on "… unifying human and model organism genetic variation databases." GEN2PHEN (http://www.gen2phen.org), a 5-year (2007–13) large-scale project, with 19 participating academic and corporate entities, which was funded from this call of proposals, aimed to unify human genetic variation databases towards increasingly holistic views into genotype-to-phenotype (G2P) data and link this system into other biomedical knowledge sources via genome browser functionality. Consequently, a similar FP7 project (RD-Connect; http://www.rd-conenct.eu; Thompson et al., 2014b) is envisaged to act as a warehouse for bioinformatics solutions to facilitate research and discovery in rare diseases. Similar funding opportunities have also announced from other funding bodies, such as the National Institutes of Health in the United States, aiming to address this need.

10.11   CONCLUSIONS

It is widely accepted that LSDBs and NEGDBs are increasingly becoming valuable tools in translational medicine. Although there have been significant improvements in both qualitative (data uniformity and database quality) and quantitative terms (increase of the number of LSDBs and NEGDBs), there are still limitations in the degree of interconnection of these resources to capture all that is known and being discovered regarding pathogenic DNA variants. The main reason for this deficiency is that the modern research ethos fails to provide adequate incentives (i.e., publication options, peer recognition, funding) to encourage researchers to build new and/or curate existing databases. Apparently, the biomedical community must first appreciate the overwhelming need for further improving genetic/mutation database systems and the most adequate solution will then presumably follow.

Acknowledgments

Most of our own work has been supported by funds from the European Commission grants [GEN2-PHEN (FP7-200754), RD-Connect (FP7-305444) and U-PGx (H2020-668353)] to GPP. We are indebted to Dr. Marianthi Georgitsi and all past group members who have been involved in data curation and management for FINDbase and all other genomic databases developed by our group.

References

Agarwala, R., Barrett, T., Beck, J., Benson, D.A., Bollin, C., Bolton, E., Bourexis, D., Brister, J.R., Bryant, S.H., Canese, K., Charowhas, C., Clark, K., DiCuccio, M., Dondoshansky, I., Federhen, S., Feolo, M., Funk, K., Geer, L.Y., Gorelenkov, V., Hoeppner, M., Holmes, B., Johnson, M., Khotomlianski, V., Kimchi, A., Kimelman, M., Kitts, P., Klimke, W., Krasnov, S., Kuznetsov, A., Landrum, M.J., Landsman, D., Lee, J.M., Lipman, D.J., Lu, Z., Madden, T.L., Madej, T., Marchler-Bauer, A., Karsch-Mizrachi, I., Murphy, T., Orris, R., Ostell, J., O'Sullivan, C., Panchenko, A., Phan, L., Preuss, D., Pruitt, K.D., Rodarmer, K., Rubinstein, W., Sayers, E.W., Schneider, V., Schuler, G.D., Sherry, S.T., Sirotkin, K., Siyan, K., Slotta, D., Soboleva, A., Soussov, V., Starchenko, G., Tatusova, T.A., Todorov, K., Trawick, B.W., Vakatov, D., Wang, Y., Ward, M., Wilbur, W.J., Yaschenko, E., Zbicz, K., 2016. Database resources of the National Center for biotechnology information. Nucleic Acids Res. 44, D7–D19.

Amberger, J.S., Bocchini, C.A., Schiettecatte, F., Scott, A.F., Hamosh, A., 2015. OMIM.org: Online Mendelian Inheritance in Man (OMIM®), an online catalog of human genes and genetic disorders. Nucleic Acids Res. 43, D789–D798.

Beroud, C., Soussi, T., 2003. The UMD-p53 database: new mutations and analysis tools. Hum. Mutat. 21, 176–181.

Beroud, C., Collod-Beroud, G., Boileau, C., Soussi, T., Junien, C., 2000. UMD (Universal mutation database): a generic software to build and analyze locus-specific databases. Hum. Mutat. 15, 86–94.

Brown, A.F., McKie, M.A., 2000. MuStaR and other software for locus-specific mutation databases. Hum. Mutat. 15, 76–85.

Claustres, M., Horaitis, O., Vanevski, M., Cotton, R.G., 2002. Time for a unified system of mutation description and reporting: a review of locus-specific mutation databases. Genome Res. 12, 680–688.

Collod, G., Beroud, C., Soussi, T., Junien, C., Boileau, C., 1996. Software and database for the analysis of mutations in the human FBN1 gene. Nucleic Acids Res. 24, 137–140.

Collod-Beroud, G., Boileau, C., 2002. Marfan syndrome in the third millennium. Eur. J. Hum. Genet. 10, 673–681.

Collod-Beroud, G., Le Bourdelles, S., Ades, L., Ala-Kokko, L., Booms, P., Boxer, M., Child, A., Comeglio, P., De Paepe, A., Hyland, J.C., Holman, K., Kaitila, I., Loeys, B., Matyas, G., Nuytinck, L., Peltonen, L., Rantamaki, T., Robinson, P., Steinmann, B., Junien, C., Beroud, C., Boileau, C., 2003. Update of the UMD-FBN1 mutation database and creation of an FBN1 polymorphism database. Hum. Mutat. 22, 199–208.

Cooper, D.N., Ball, E.V., Krawczak, M., 1998. The human gene mutation database. Nucleic Acids Res. 26, 285–287.

Cotton, R.G., McKusick, V., Scriver, C.R., 1998. The HUGO mutation database initiative. Science 279, 10–11.

Cotton, R.G., Phillips, K., Horaitis, O., 2007. A survey of locus-specific database curation. Human Genome Variation Society. J. Med. Genet. 44.

Cotton, R.G., Auerbach, A.D., Beckmann, J.S., Blumenfeld, O.O., Brookes, A.J., Brown, A.F., Carrera, P., Cox, D.W., Gottlieb, B., Greenblatt, M.S., Hilbert, P., Lehvaslaiho, H., Liang, P., Marsh, S., Nebert, D.W., Povey, S., Rossetti, S., Scriver, C.R., Summar, M., Tolan, D.R., Verma, I.C., Vihinen, M., den Dunnen, J.T., 2008. Recommendations for locus-specific databases and their curation. Hum. Mutat. 29, 2–5.

Cotton, R.G., Al Aqeel, A.I., Al-Mulla, F., Carrera, P., Claustres, M., Ekong, R., Hyland, V.J., Marafie, M.J., Paalman, M.H., Patrinos, G.P., Qi, M., Ramesar, R.S., Scott, R.J., Sijmons, R.H., Sobrido, M.J., Vihinen, M., 2009. Capturing all disease-causing mutation for clinical and research use: towards an effortless system for the human variome project. Genet. Med. 11, 843–849.

Dalabira, E., Viennas, E., Daki, E., Komianou, A., Bartsakoulia, M., Poulas, K., Katsila, T., Tzimas, G., Patrinos, G.P., 2014. DruGeVar: an online resource triangulating drugs with genes and genomic biomarkers for clinical pharmacogenomics. Public Health Genomics 17 (5–6), 265–271.

den Dunnen, J.T., Antonarakis, S.E., 2001. Nomenclature for the description of human sequence variations. Hum. Genet. 109, 121–124.

den Dunnen, J.T., Paalman, M.H., 2003. Standardizing mutation nomenclature: why bother? Hum. Mutat. 22, 181–182.

Eckerson, W.W., 1995. Three tier Client/Server architecture: achieving scalability, performance, and efficiency in Client Server applications. Open Inform. Syst. 10 (3), 46–50.

Fokkema, I.F., den Dunnen, J.T., Taschner, P.E., 2005. LOVD: easy creation of a locus-specific sequence variation database using an "LSDB-in-a-box" approach. Hum. Mutat. 26, 63–68.

Fokkema, I.F., Taschner, P.E., Schaafsma, G.C., Celli, J., Laros, J.F., den Dunnen, J.T., 2011. LOVD v.2.0: the next generation in gene variant databases. Hum. Mutat. 32, 557–563.

George, R.A., Smith, T.D., Callaghan, S., Hardman, L., Pierides, C., Horaitis, O., Wouters, M.A., Cotton, R.G., 2008. General mutation databases: analysis and review. J. Med. Genet. 45, 65–70.

Georgitsi, M., Viennas, E., Gkantouna, V., van Baal, S., Petricoin, E.F., Poulas, K., Tzimas, G., Patrinos, G.P., 2011a. FINDbase: A worldwide database for genetic variation allele frequencies updated. Nucleic Acids Res. 39, D926–D932.

Georgitsi, M., Viennas, E., Gkantouna, V., Christodoulopoulou, E., Zagoriti, Z., Tafrali, C., Ntellos, F., Giannakopoulou, O., Boulakou, A., Vlahopoulou, P., Kyriacou, E., Tsaknakis, J., Tsakalidis, A., Poulas, K., Tzimas, G., Patrinos, G.P., 2011b. Population-specific documentation of pharmacogenomic markers and their allelic frequencies in FINDbase. Pharmacogenomics 12, 49–58.

Gerlai, R., 2002. Phenomics: fiction or the future? Trends Neurosci. 25, 506–509.

Giardine, B., Riemer, C., Hefferon, T., Thomas, D., Hsu, F., Zielenski, J., Sang, Y., Elnitski, L., Cutting, G., Trumbower, H., Kern, A., Kuhn, R., Patrinos, G.P., Hughes, J., Higgs, D., Chui, D., Scriver, C., Phommarinh, M., Patnaik, S.K., Blumenfeld, O., Gottlieb, B., Vihinen, M., Väliaho, J., Kent, J., Miller, W., Hardison, R.C., 2007a. PhenCode: connecting ENCODE data with mutations and phenotype. Hum. Mutat. 28, 554–562.

Giardine, B., van Baal, S., Kaimakis, P., Riemer, C., Miller, W., Samara, M., Kollia, P., Anagnou, N.P., Chui, D.H., Wajcman, H., Hardison, R.C., Patrinos, G.P., 2007b. HbVar database of human hemoglobin variants and thalassemia mutations: 2007 update. Hum. Mutat. 28, 206.

Giardine, B., Borg, J., Viennas, E., Pavlidis, C., Moradkhani, K., Joly, P., Bartsakoulia, M., Riemer, C., Miller, W., Tzimas, G., Wajcman, H., Hardison, R.C., Patrinos, G.P., 2014. Updates of the HbVar database of human hemoglobin variants and thalassemia mutations. Nucleic Acids Res. 42, D1063–D1069.

Hall, J.G., 2003. A clinician's plea. Nat. Genet. 33, 440–442.

Hardison, R.C., Chui, D.H., Giardine, B., Riemer, C., Patrinos, G.P., Anagnou, N., Miller, W., Wajcman, H., 2002. HbVar: a relational database of human hemoglobin variants and thalassemia mutations at the globin gene server. Hum. Mutat. 19, 225–233.

Horaitis, O., Talbot Jr., C.C., Phommarinh, M., Phillips, K.M., Cotton, R.G., 2007. A database of locus-specific databases. Nat. Genet. 39, 425.

International HapMap Consortium, 2003. The International HapMap project. Nature 426, 789–796.

Kaput, J., Cotton, R.G., Hardman, L., Watson, M., Al Aqeel, A.I., Al-Aama, J.Y., Al-Mulla, F., Alonso, S., Aretz, S., Auerbach, A.D., Bapat, B., Bernstein, I.T., Bhak, J., Bleoo, S.L., Blöcker, H., Brenner, S.E., Burn, J., Bustamante, M., Calzone, R., Cambon-Thomsen, A., Cargill, M., Carrera, P., Cavedon, L., Cho, Y.S., Chung, Y.J., Claustres, M., Cutting, G., Dalgleish, R., den Dunnen, J.T., Díaz, C., Dobrowolski, S., Dos Santos, M.R., Ekong, R., Flanagan, S.B., Flicek, P., Furukawa, Y., Genuardi, M., Ghang, H., Golubenko, M.V., Greenblatt, M.S., Hamosh, A., Hancock, J.M., Hardison, R., Harrison, T.M., Hoffmann, R., Horaitis, R., Howard, H.J., Barash, C.I., Izagirre, N., Jung, J., Kojima, T., Laradi, S., Lee, Y.S., Lee, J.Y., Gil-da-Silva-Lopes, V.L., Macrae, F.A., Maglott, D., Marafie, M.J., Marsh, S.G., Matsubara, Y., Messiaen, L.M., Möslein, G., Netea, M.G., Norton, M.L., Oefner, P.J., Oetting, W.S., O'Leary, J.C., de Ramirez, A.M., Paalman, M.H., Parboosingh, J., Patrinos, G.P., Perozzi, G., Phillips, I.R., Povey, S., Prasad, S., Qi, M., Quin, D.J., Ramesar, R.S., Richards, C.S., Savige, J., Scheible, D.G., Scott, R.J., Seminara, D., Shephard, E.A., Sijmons, R.H., Smith, T.D., Sobrido, M.J., Tanaka, T., Tavtigian, S.V., Taylor, G.R., Teague, J., Töpel, T., Ullman-Cullere, M., Utsunomiya, J., van Kranen, H.J., Vihinen, M., Webb, E., Weber, T.K., Yeager, M., Yeom, Y.I., Yim, S.H., Yoo, H.S., 2009. Planning the human variome project. The Spain report. Hum. Mutat. 30, 496–510.

Karageorgos, I., Giannopoulou, E., Mizzi, C., Pavlidis, C., Peters, B., Karamitri, A., Zagoriti, Z., Stenson, P., Kalofonos, H.P., Drmanac, R., Borg, J., Cooper, D.N., Katsila, T., Patrinos, G.P., 2015. Identification of cancer predisposition variants using a next generation sequencing-based family genomics approach. Hum. Genomics 9, 12.

Kleanthous, M., Patsalis, P.C., Drousiotou, A., Motazacker, M., Christodoulou, K., Cariolou, M., Baysal, E., Khrizi, K., Pourfarzad, F., Moghimi, B., van Baal, S., Deltas, C.C., Najmabadi, S., Patrinos, G.P., 2006. The Cypriot and Iranian National Mutation databases. Hum. Mutat. 27, 598–599.

Landrum, M.J., Lee, J.M., Riley, G.R., Jang, W., Rubinstein, W.S., Church, D.M., Maglott, D.R., 2014. ClinVar: public archive of relationships among sequence variation and human phenotype. Nucleic Acids Res. 42, D980–D985.

Landrum, M.J., Lee, J.M., Benson, M., Brown, G., Chao, C., Chitipiralla, S., Gu, B., Hart, J., Hoffman, D., Hoover, J., Jang, W., Katz, K., Ovetsky, M., Riley, G., Sethi, A., Tully, R., Villamarin-Salomon, R., Rubinstein, W., Maglott, D.R., 2016. ClinVar: public archive of interpretations of clinically relevant variants. Nucleic Acids Res. 44, D862–D868.

McKusick, V.A., 1966. Mendelian Inheritance in Man. A Catalog of Human Genes and Genetic Disorders, first ed. Johns Hopkins University Press, Baltimore, MD.

Mitropoulou, C., Webb, A.J., Mitropoulos, K., Brookes, A.J., Patrinos, G.P., 2010. Locus-specific databases domain and data content analysis: evolution and content maturation towards clinical use. Hum. Mutat. 31, 1109–1116.

Papadopoulos, P., Viennas, E., Gkantouna, V., Pavlidis, C., Bartsakoulia, M., Ioannou, Z.M., Ratbi, I., Sefiani, A., Tsaknakis, J., Poulas, K., Tzimas, G., Patrinos, G.P., 2014. Developments in FINDbase worldwide database for clinically relevant genomic variation allele frequencies. Nucleic Acids Res. 42, D1020–D1026.

Patrinos, G.P., 2006. National and ethnic mutation databases: recording populations' genography. Hum. Mutat. 27, 879–887.

Patrinos, G.P., Brookes, A.J., 2005. DNA, disease and databases: disastrously deficient. Trends Genet. 21, 333–338.

Patrinos, G.P., Giardine, B., Riemer, C., Miller, W., Chui, D.H., Anagnou, N.P., Wajcman, H., Hardison, R.C., 2004. Improvements in the HbVar database of human hemoglobin variants and thalassemia mutations for population and sequence variation studies. Nucleic Acids Res. 32, D537–D541.

Patrinos, G.P., van Baal, S., Petersen, M.B., Papadakis, M.N., 2005a. The hellenic national mutation database: a prototype database for mutations leading to inherited disorders in the hellenic population. Hum. Mutat. 25, 327–333.

Patrinos, G.P., Kollia, P., Papadakis, M.N., 2005b. Molecular diagnosis of inherited disorders: lessons from hemoglobinopathies. Hum. Mutat. 26, 399–412.

Patrinos, G.P., Al Aama, J., Al Aqeel, A., Al-Mulla, F., Borg, J., Devereux, A., Felice, A.E., Macrae, F., Marafie, M.J., Petersen, M.B., Qi, M., Ramesar, R.S., Zlotogora, J., Cotton, R.G., 2011. Recommendations for genetic variation data capture in emerging and developing countries to ensure a comprehensive worldwide data collection. Hum Mutat. 32, 2–9.

Patrinos, G.P., Smith, T.D., Howard, H., Al-Mulla, F., Chouchane, L., Hadjisavvas, A., Hamed, S.A., Li, X.T., Marafie, M., Ramesar, R.S., Ramos, F.J., de Ravel, T., El-Ruby, M.O., Shrestha, T., Sobrido, M.J., Tadmouri, G., Witsch-Baumgartner, M., Zilfalil, B.A., Auerbach, A.D., Carpenter, K., Cutting, G., Dung, V.C., Grody, W., Hasler, J., Jorde, L., Kaput, J., Macek, M., Matsubara, Y., Padilla, C., Robinson, H., Rojas-Martinez, A., Taylor, G.R., Vihinen, M., Weber, T., Burn, J., Qi, M., Cotton, R.G., Rimoin, D., 2012. Human variome project country nodes: documenting genetic information within a country. Hum. Mutat. 33, 1513–1519.

Riikonen, P., Vihinen, M., 1999. MUTbase: maintenance and analysis of distributed mutation databases. Bioinformatics 15, 852–859.

Ring, H.Z., Kwok, P.Y., Cotton, R.G., 2006. Human Variome Project: an international collaboration to catalogue human genetic variation. Pharmacogenomics 7, 969–972.

Scriver, C.R., 2004. After the genome—the phenome? J. Inherit. Metab. Dis. 27, 305–317.

Scriver, C.R., Nowacki, P.M., Lehvaslaiho, H., 1999. Guidelines and recommendations for content, structure, and deployment of mutation databases. Hum. Mutat. 13, 344–350.

Scriver, C.R., Waters, P.J., Sarkissian, C., Ryan, S., Prevost, L., Cote, D., Novak, J., Teebi, S., Nowacki, P.M., 2000. PAHdb: a locus-specific knowledgebase. Hum. Mutat. 15, 99–104.

Sipila, K., Aula, P., 2002. Database for the mutations of the Finnish disease heritage. Hum. Mutat. 19, 16–22.

Smith, T.D., Cotton, R.G., 2008. VariVis: a visualisation toolkit for variation databases. BMC Bioinformatics 9, 206.

Soussi, T., 2000. The p53 tumor suppressor gene: from molecular biology to clinical investigation. Ann. N. Y. Acad. Sci. 910, 121–137.

Soussi, T., Beroud, C., 2001. Assessing TP53 status in human tumours to evaluate clinical outcome. Nat. Rev. Cancer 1, 233–240.

Soussi, T., Ishioka, C., Beroud, C., 2006. Locus-specific mutation databases: pitfalls and good practice based on the p53 experience. Nat. Rev. Cancer 6, 83–90.

Stein, L., 2002. Creating a bioinformatics nation. Nature 417, 119–120.

Stenson, P.D., Mort, M., Ball, E.V., Shaw, K., Phillips, A., Cooper, D.N., 2014. The Human gene mutation database: building a comprehensive mutation repository for clinical and molecular genetics, diagnostic testing and personalized genomic medicine. Hum. Genet. 133, 1–9.

Thompson, R., Johnston, L., Taruscio, D., Monaco, L., Béroud, C., Gut, I.G., Hansson, M.G., 't Hoen, P.B., Patrinos, G.P., Dawkins, H., Ensini, M., Zatloukal, K., Koubi, D., Heslop, E., Paschall, J.E., Posada, M., Robinson, P.N., Bushby, K., Lochmüller, H., 2014b. RD-Connect: an integrated platform connecting databases, registries, biobanks and clinical bioinformatics for rare disease research. J. Gen. Intern. Med. 29, 780–787.

Vail, P.J., Morris, B., van Kan, A., Burdett, B.C., Moyes, K., Theisen, A., Kerr, I.D., Wenstrup, R.J., Eggington, J.M., 2015. Comparison of locus-specific databases for *BRCA1* and *BRCA2* variants reveals disparity in variant classification within and among databases. J. Community Genet. 6, 351–359.

van Baal, S., Kaimakis, P., Phommarinh, M., Koumbi, D., Cuppens, H., Riccardino, F., Macek Jr., M., Scriver, C.R., Patrinos, G.P., 2007. *FINDbase*: a relational database recording frequencies of genetic defects leading to inherited disorders worldwide. Nucleic Acids Res. 35, D690–D695.

van Baal, S., Zlotogora, J., Lagoumintzis, G., Gkantouna, V., Tzimas, I., Poulas, K., Tsakalidis, A., Romeo, G., Patrinos, G.P., 2010. ETHNOS: a versatile electronic tool for the development and curation of National Genetic databases. Hum. Genomics 4, 361–368.

Viennas, E., Gkantouna, V., Ioannou, M., Georgitsi, M., Rigou, M., Poulas, K., Patrinos, G.P., Tzimas, G., 2012. Population-ethnic group specific genome variation allele frequency data: a querying and visualization journey. Genomics 100, 93–101.

Viennas, E., Komianou, A., Mizzi, C., Stojiljkovic, M., Mitropoulou, C., Muilu, J., Vihinen, M., Grypioti, P., Papadaki, S., Pavlidis, C., Zukic, B., Katsila, T., van der Spek, P.J., Pavlovic, S., Tzimas, G., Patrinos, G.P., 2017. Expanded national database collection and data coverage in the FINDbase worldwide database for clinically relevant genomic variation allele frequencies. Nucleic Acids Res 45, D846–D853.

Wildeman, M., van Ophuizen, E., den Dunnen, J.T., Taschner, P.E., 2008. Improving sequence variant descriptions in mutation databases and literature using the Mutalyzer sequence variation nomenclature checker. Hum. Mutat. 29, 6–13.

Zaimidou, S., van Baal, S., Smith, T.D., Mitropoulos, K., Ljujic, M., Radojkovic, D., Cotton, R.G., Patrinos, G.P., 2009. A_1ATVar: a relational database of human *SERPINA1* gene variants leading to alpha1-antitrypsin deficiency. Hum. Mutat. 30 (3), 308–313. (in press).

Zlotogora, J., Patrinos, G.P., 2017. The Israeli National genetic database: a 10-year experience. Hum. Genomics 11, 5.

Zlotogora, J., van Baal, S., Patrinos, G.P., 2007. Documentation of inherited disorders and mutation frequencies in the different religious communities in Israel in the Israeli National Genetic Database. Hum. Mutat. 28, 944–949.

Zlotogora, J., van Baal, J., Patrinos, G.P., 2009. The Israeli national genetics database. Isr. Med. Assoc. J. 11, 373–375.

CHAPTER 11

Artificial Intelligence: The Future Landscape of Genomic Medical Diagnosis: Dataset, In Silico Artificial Intelligent Clinical Information, and Machine Learning Systems

Darrol J. Baker

The Golden Helix Foundation, London, United Kingdom

11.1 INTRODUCTION

"A knotty puzzle may hold up a scientist for a century when it may be that a colleague has the solution already and is not even aware of the puzzle that it might solve."

Prof. Amadiro, Planet Aurora. From Robots of Empire (Asimov, 1985).

It is hard to believe that artificial intelligence (AI) enters its 50th year of emergence. Throughout this book, the authors have presented innovative uses of current knowledge and published work to present a common understanding. In this chapter, the author aims to examine the role of AI in clinical genomics and the dramatic change over the next 20–50 years. Aside from the main topic, the author will look at AI's expected magnitude on the clinical environment and not venture into debates on machine or legal rights, nor the next generation of learning robots, nor the effects these rights may have on human labor markets and income inequality. We are considering the practical use of AI within academia and industry and, although we are at a very early stage, we envisage very encouraging uses and clinical outcomes even with current data standards.

Recent advances in clinical decision making and point of care (PoC) devices, including the accelerated support from clinical practice and information technology, have enabled a new breed of scientists to exploit AI, machine, and object learning: to extend where bioinformatics fails and building systems exceed current clinical thinking. Most systems and processes within an organization are seemingly semiautomatous at present. Starting with the company organization, down to pay role through to the processes of an actual job, most

Human Genome Informatics. https://doi.org/10.1016/B978-0-12-809414-3.00011-5

have standard of operation, rules of the game, or simple standard operating procedure (SOP); with AI, we look to automate these processes, learnings, and feedback to the next cycle of learning, producing results that are unexpected from what the original designer thought of. This is the power of AI to find the unthinkable and learn a new means of operation within a system, to improve the system to be more efficient at a given task, or epoch. The tradition line of "Garage in Garage our" (GiGo) does not apply in AI.

In this chapter, the author will look at many aspects of AI, albeit bias towards one's own interests; one hopes that information is taken in context of general use and indeed application for yet undiscovered pathways. The author aims to highlight the processes, tools, and programmes so that the reader can act on and activity produces better results, patient outcomes, and speed in clinical decision making. AI is an exciting time where machines are capable of independent thoughts and actions; this presents some very deep prospectives that are beyond a simple chapter and the author will point out references within this text for further reading, where possible. However, the simplest questions have difficult answers.

11.2 WHAT IS ARTIFICIAL INTELLIGENCE AND MACHINE LEARNING?

It is easy to ask, yet very hard to answer. We have little understanding of what intelligence really is and animal, indeed human, intelligence differs greatly to that of machine-learned intelligence. There are many definitions equilibriumed around *one's computer code or hardware with human capable behaviors, regarded intelligent, if humans exhibited it.*

Here lies the problem all AI systems have: how do you measure intelligence: Is it your intelligence, my intelligences, or maybe my cat Einstein's? Questions and answers are subjective and therefore quantitatively and qualitatively hard to define. Within the spectrum of clinical genomics, AIs have been compounded by voluminous other metafactors. Many AIs running currently are highly specific task-orientated machines or facilities, within the spectrum of the clinic and pharmacogenomics; we are looking at many variables and conditions in real time. Where time, real time, is a true rate-limiting step within the AI space, clinical genomics and patient diagnosis operate. Therefore, in this chapter, we are going to look at how we approach solving the problem, which is more of a determinate on whether you solve "it" or not. "It" being everything at once, of course.

11.2.1 Clinical Genomics Medical AI

Clinical AI is still in its infancy and, although we have seen some cases where AIs are antisocially used within the decision-making process, a true sentient AI solution is still many years away. The healthcare consumer has become more obsessed in their own medical treatment, from genome sequencing to building a personalized health portfolio from the data in their fitness and lifestyle trackers. Data-driven medicine has the ability to not only improve the speed and accuracy of diagnosis for genetic diseases, but also unlock the possibility of personalized medical treatments. However, these personal devices are accomplishing little or no population-wide understanding, albeit un/scrupulous companies selling metadata information from their subscriber base.

11.2.2 The Emerging Role of AI

AI dominates headlines globally, which forecast that spending on AI would rise from $8 billion in 2016 to $47 billion by 2020, according to IDC. Identified in the report as one of the sectors likely to spend the most in cognitive or AI systems, platforms such as IBM Watson Health have accomplished substantial progress using AI and cognitive computing in oncology and genomics. Lately, digital health firms offering platforms provide personalized recommendations based on insights from metadata database medical profiles and knowledge gained from millions of online consultations. AIs are now applied to large volumes of data from increasingly diverse sources. The results are guiding clinicians on everything from drug discovery, diagnosis, and treatment (High, 2012).

In this chapter, the author will look at the key countenances of artificial intelligence and machine learning for promoting health and well-being within a global setting of real-time clinical diagnosis. In particular, the emphasis will be given to specific clinical real-world examples and how this will change the shape of clinical diagnostics medicine reporting. As this will form the basis of lecture course, the authors aim to add educational points of interest for students.

11.3 ARTIFICIAL INTELLIGENCE: THE AI, THE WHOLE AI, AND NOTHING BUT THE AI

When you first learn about object relational database management systems (ORDBMS), an important quip related to the use of a primary key is: the key, the whole key, and nothing but the key so help me Codd (E. F. Codd, of IBM's San Jose Research Laboratory). This same paradigm is applicable to any complex systems rule; we need primary events, epochs, or anchors that allow one to focus on deep learning in a clinical and indeed wider world setting. Within the world of theoretical mathematics, artificial intelligence does have the luxury of repetitive error making; in a clinical setting, this is not the case. In this section, it is important to lay down the foundations of the Magrathea artificial intelligence framework and what they term "the rabbit hole."

In the United States of America, the national health expenditure is estimated at $3.4 trillion with a projected increase from 17.8% to 19.9% of the GDP between 2015 and 2025. The industry analysts estimate that the artificial intelligence in the health market is poised to reach $6.6 billion by 2021 and, by 2026, can potentially save the US healthcare economy $150 billion in annual savings. However, no sources have taken a comprehensive look at machine learning applications at America's leading hospitals (Burner et al., 1992).

Intelligent systems play a vital role within clinical medical genomics and, although we are at the start of this medical revolution, even with our early footsteps we are making important insights into patient welfare, medical research, clinical knowledge, and training (Gilks et al., 1995). This chapter explores the work by the Magrathea teams in the United Kingdom and Kuwait with the importance they place on using artificial intelligence in patient understanding and feeding information into the complex systems and in silico learning measures to guide clinician understanding and decision making in all aspects of critical patient care. This chapter will cover the processes and techniques used by Magrathea, how their teams have integrated in real-time clinical patient possibilities, and future directions for the industry on their own research.

11.3.1   Building an Artificial Intelligence Framework (AIF)

AIF is relatively easy compared to many of the challenges Magrathea have had in recent times. It is important to note, however, that the Magrathea team sees artificial intelligence as a cobweb not a steel ring: meaning the integration of people, systems, and technology has a significant chance or failure and error. Hence, within a clinical setting, great care and dedication is needed for understanding the information flow within a complex system. At the core of any complex system or child is the integration of intelligent learning or "epochs," which are initially very weak to change, adapt and widely open to many forms of failure (Yampolskiy, 2016). Although AI failure does make exciting reading or great cinematic entertainment, the weakness and openness to attack make the whole process highly subjective when we are considering using artificial intelligence and machine learning techniques within clinical genomics hospital for patient understanding where outcome and security are of vital importance. Developing epoch learning cycles within an AI framework is key for any successful AIF.

In recent years, we have seen many examples of complex system beating world leaders in chess and GO. These wins were and are inevitable as the systems are picking up on not just game play, but, as the years go by, on past games, biometric data, and player "tells" and even predictive moves, which are then used in analysis, although some see this being creepy in some way. The application of these epochs is done for disease understanding, predication, and control: now and into the future. Magrathea employ the same tools only in a clinical setting, feeding this information back into the learning cycle, discussed later.

Interestingly, AlphaGO (Tian, 2016) made its name defeating the high-profile GO player Lee Sedol, the world's best player of GO, 4-1. However, one would expect this; AlphaGO (Fig. 11.1) is an automaton and is learning a simple task or game that we call epoch. What is most impressive is that Lee Sedol actually won one game; not that AlphaGO was going to win. Again, AI is a cobweb and subject to error. GO is a game of around 150 moves that can involve 10,360 possible configurations, "more than there are atoms in the Universe" (Koch, 2016). Combining epochs for real-world understanding, the Magrathea team power used complex systems and deep learning linked to patient diagnostic treatment where patients are seeing benefits of early diagnosis of diseases.

Artificial intelligence within the clinical genomics plays a vital role in galvanizing tools within a clinical setting. As of yet, little research covers the "linking" of

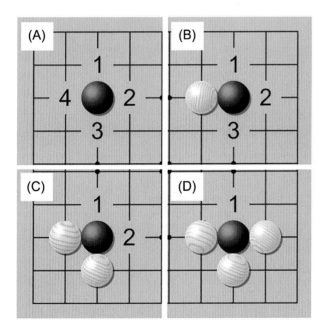

FIG. 11.1

(A–D) AlphaGo: Google's artificial Intelligence AlphaGo made the front cover of Nature. A new approach used to create AlphaGo Demis Hassabis, an artificial intelligence researcher, neuroscientist, computer game designer, and world-class gamer, who works at Google DeepMind, wrote: "So when we set out to crack Go, we took a different approach. We built a system, AlphaGo, which combines an advanced tree search with deep neural networks." *From: pbs.twimg.com.*

the clinical decision-making steps together. At Magrathea, they have at the core a state-of-the-art clinical practices and traditional patient framework run by the Genatak teams in Kuwait and Nottingham, England. At the outset of the human genome project (IHGSC, 2004) and further next-generation sequencing (NGS) of human genomes for clinical understanding, a new system was needed to deal with the huge volumes of data and clinically relevant interpretations. The original core project was a very small 200 whole-human genome-sequencing (WHGS) project, which later expanded to a further 750 WHGS; at the outset, clinical validation of the whole data and data understanding was extreme difficult in the fall of 2010.

Multivariant AI requires some technical preknowledge and good mathematical and algorithmic thinking abilities to perform the related research, for the applications to work well and properly (Forbus, 1984). The primary learning techniques employed within the current framework are namely:

- Fuzzy logic
- Monte Carlo tree search
- Artificial neural networks (ANNs)
- One-time pad decryptions
- Evolutionary algorithms.
- Nature-based algorithms: particle swarm, ant colony, bee colony, and bird murmuration (internal research) optimized algorithms
- Hybrid support algorithms
- Poly-dimensional topology
- Irrational pattern functions.
- Random sequence extrapolations

In this framework, clinical parallel intelligence can be used in three operational modes:

- *Primary*: Clinical epoch learning and training: the parallel intelligence is used for establishing the virtual clinical artificial system. In this mode, the clinical artificial system is indeed very different from the real healthcare system, and not much interaction is required (Mac Parthaláin and Jensen, 2015).
- *Secondary*: Clinical experiment and evaluation: clinical parallel intelligence is used for generating and conducting computational experiments in this clinical epoch mode, for testing and evaluating clinical diagnostic patient scenarios and solutions. For given clinical disease cases, teams are using combinations; if not all, algorithmic code to form the bases of the answer to our initial diagnostic postulates (Wang et al., 2016a,b).
- *Tertiary*: Evolved artificial intelligence system: using epoch machine learning that is feed back into any downstream questioning. Epochs are

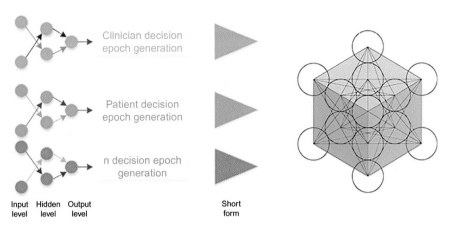

FIG. 11.2

Clinical parallel intelligence for decision-making: an architecture of parallel control and management for complex systems, the execution framework of the parallel systems with the clinical pipeline. Clinical input/outputs follow a multivariant hexagonical (poly-dimensional) AI questions.

repeatedly staged with machine learning events and come in a variety of input types, volumes, and velocities (Fig. 11.2). This is a staged framework of "incorrect/correct" or "known/unknown" understandings; only after the programmes have evolved to a "learned" status does the valid programme go live for clinical use. That is the "learned" or "cleared" status, which is clinical validated to the given certified approval board. They also have fast-track systems for "dirty status" or nonboard approval or research-only projects, reducing anomalies within the clinical decision-making process. In a virtual sandbox scenario testing for disease outbreaks, all that play in this serious testing environment resort to "dirty runs" since traditional roots always fail scenario goals.

11.3.2 Security and Quantum Computing (QComp)

The "rabbit hole" is infinitely deep and will take the clinician way out of the realms of normal medicine to the world of mathematical, a world few clinicians seldom tread, hence, the difficult nature and learnings needed for even basic correspondence with disparate disciplines. Put simply, QComp is important for speeding up the processing of information within the AI framework. The anomaly detecting security, Independent verification and validation (IV&V), and Monte Carlo simulations are deep part of data science, but what about clinical deep learning. Can QComp be repurposed to dramatically speed up convolutional and recurrent neural nets, adversarial, and reinforcement learning with their multitudes of hidden layers that usually just slows everything down? As it turns out, they found it could. Moreover, the results are quite astounding.

The standard description of how QComp can be used falls in these three categories: simulation, optimization, and sampling. This might throw you off the track a bit until you realize that any clinical diagnostic problem can be seen as either an optimization or a sampling problem. The most important work has been building the deep learning framework for how QComp will disrupt the way we approach clinical diagnostic understanding.

When communication and computing technologies are employed especially within clinical patient understanding, it is important to ensure high levels of security. More so now with the trend for global social-communication, keeping processes at stable optimum levels in order to maintain accurate clinical reporting processes and all in a timely manner are vital. Security is, however, an ever-moving target (Graham, 2017); the three ways the team are ensuring security are:

- *Data security*: Ensuring security for data kept in local or private cloud, variety of evolving encryption techniques, and personnel access rights to ensure data flow.
- *Security for malicious activities*: Grooming techniques (Mitnick and Simon, 2011), Hacking, ratting, onsite theft, backdoor, and bug fixes in software/hardware space.
- *Security for environmental factors*: Back-up plans for data storage, recovery with little downtime as possible, and interrupting data flow. An off-grid historic data storage.

11.3.3   Complex Knowledge Management Systems I

Clinical genomic data interpretation minus a knowledge management system routinely creates a chaotic range of viewpoints. Even with traditional pipeline approaches, we see an anarchic choice of reporting; "Top 10, 20 etc.…." lists of variant findings, where neither the healthcare worker nor patient has little understanding of reporting. Within the teams framework, these unique forms of information or Epochs can be stored and used in a variety of ways: from information centers to retrieval systems to support the information system through AI learning initiatives (de Vasconcelos, et al., 2017).

Controlling this new technology is vitally important to direct information, resources, actions, and long-term patient planning, alongside other activities within the healthcare system. The team found within a clinical setting that the introduction of complex system management greatly aided efficient patient diagnosis, treatment, and recovery rate (Zhao et al., 2017). Moreover, they are now in a beneficial state to prevent illness before an economic impact emerges within the healthcare system longitudinally.

Clinical genomic knowledge management systems (CG-KMS) are at an early stage of innovation, with implementation still nascent in many countries. CG-KMS holds the promise of becoming key technologies for knowledge creation and management in the future (Hameurlain et al., 2017; Hu, et al., 2017).

Epochs were originally nicknamed "shoplifters" since they were designed to take data from a variety of sources. Usually, without the users knowing and, in some, epoch runs from "clinicians" that didn't know their SOP, protocols, and understanding was lifted. Only the AI-epoch took information from the appropriate GC-KMS; in most cases, the AI did not "know" what they were taking or why; this was pure epoch-learned events and reporting (Fig. 11.3A). The research team openly found it very "strange" at the beginning of the processing. This is the strange world of epochs and epoch primes. Epoch primes are similar to the primary key systems within a database management system (DBMS); only with epoch primes, we are priming to a key artificial intelligence (Kai) event or *Kai Event* (Fig. 11.3B). At this point, once we have the primed locked, loaded, and verified, we can evolve the epoch-cycle development within enterprise level systems, such as: MathWorks Matlab, Wolfram Mathematica, and your in-house bespoke Hadoop environments.

Compared to traditional healthcare systems, Magrathea work in a very different way. Within the AI framework, they have a simple process of all information; this includes the painful task of standardizing all inputs to an "internal" standard, allowing the end users Epoch Developer Kit access to process the information, in essence to:

- Select, acquire, preserve, organize, and manage all digital collections.
- Design the technical architecture for clinical decision making.
- Describe the content and attributes of all items within the unknown and known universe (all user input, metadata, ERC, biopsy, X-ray, etc.).
- Plan, implement, and support digital Epoch learning.
- Consultation and transmit services.
- Develop very friendly graphical user interface (GUI) over the clinical network (clinic and patient side).
- Relative standards and policies for clinical data access and use.
- Design, maintain, and transmit pharmacogenomic and other product information.
- Protect digital intellectual property in network environment.
- Complete information security.

At each stage of this process, Epoch learning is a key for real-time artificial intelligence understanding and learned feedback to users within the framework of Kai events and the epoch cycle (Fig. 11.4).

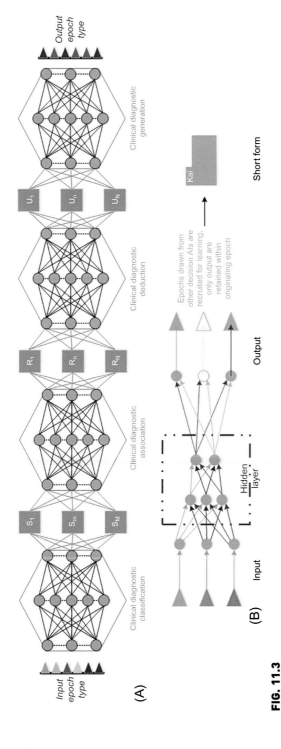

FIG. 11.3

(A) Clinical genomic epoch generation, this instance, we are seeing the simplified network structure and interaction of one multivariant hexagonical AI questions between the clinician and patients AI systems. (B) Groups responses, called "Kai events," are key to epoch learning and anchoring the epochs to multi-AI algorithms. Allowing for completed AI learning or "clinically verified" pathways, learning or traditionally known data anchored as "Kai" events for building poly-dimensional diagnostic understanding.

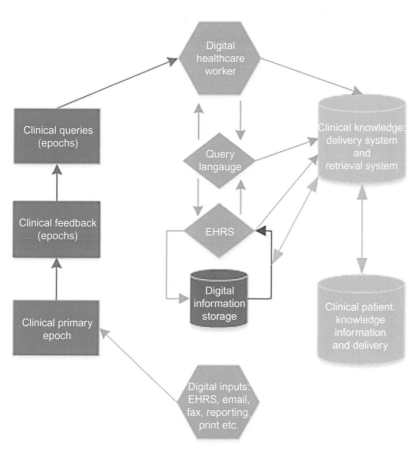

FIG. 11.4
Digital information access and retrieval system within a clinical genomic healthcare system.

Key Epoch stages of clinical complex knowledge acquisition are present to take the conclusion by configuring information by the intelligent system. Within the context of this study, the system was programmed in line with the following processes:

- Clinical acquisition of all datatypes.
- Representation/diagnostic reporting at all levels using a variety of GUIs.
- Clinical validation.
- Inferencing of applications to physician, healthcare worker, and the patient.
- Valid clinical (CAP/CLIA and EN ISO 14155) explanation and acceptance.
- Continued learning, development, and teaching for a variety of educational levels.

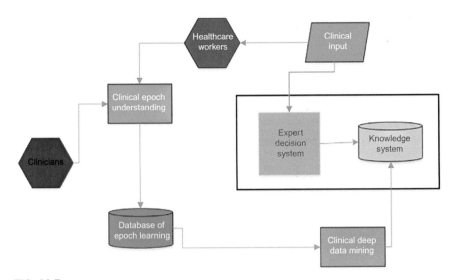

FIG. 11.5

Basic clinical decision-making process within the hidden process region: a simplified version of the inputs
we have within a clinical genomics healthcare system.

The clinical systems represented in pure logic based on clinically validated
results, clinical premises, and logical facts (Fig. 11.5).

All clinical functions are designed with these inputs based on the healthcare
workers input and experience over many years of clinical decision making.
Moreover, the outputs are produced in two ways: traditional clinician understanding and in-depth knowledge of a given case and the clinical inferences
and results, all of which are presented to the healthcare worker via GUI. The
logic we employ overall falls into the following categories:

- *Propositional logic*: Allowing logical results to be either true or false (Gao
  et al., 2017). After what it is "known," a Kai Events triggered and becomes
  an input used to derive new epoch events. Current clinical guidelines from
  the International Disease classification (ICD) are used to determine
  whether new suggestions are true (T) or false (F). From this, further logical
  operators can be employed, ICD links to patient reporting, and feedback
  to clinical staff.
- *Lists*: Decision lists and trees are employed for all kinds of representation
  of information found in a hierarchical structure (Santofimia, et al., 2017).
  They are a vast array of examples including: Laboratory protocols, disease
  profiles, list of expert "people in the field," list of drugs open to the
  particular condition, and next steps/what to do lists. They are used
  when objects are grouped, categorized, and/or assessed based on
  the relationship to the Kai Event or relationship to a given epoch.

When the relationship between epochs groups are demonstrated with clear logical connections, an epoch tree structure is constructed.

- *Decision tree*: Similar to clinical semantic net that are connected with hard and learned rules. The clinical rule-based decision trees are obtained after expert knowledge representation; clinical decision process is started to deduce the condition (Stamate, et al., 2017).
- *Case-based reasoning*: Clinical, although it plays a significant sideliner role within the framework of epoch and any other learning event (He and Tian, 2017). It is the founding part where human input into the system is a key and indeed one of the areas with the highest security and definitive representation of "go live" understanding by estimating solving of new problems by input solutions from older similar problems as base. An input can only be made at this point to an existing epoch from a previous period of reconsidering. It is important to note that this is also an area for "Grooming" hacking and, consequently, an area for the highest security SOP level.
- *Markov decision processes*: Magrathea clinically model, plan, and control the decision epochs within the mathematical frame under the partial controls of a random or decision plan. This is a vast area; we model many things that are limited to the power of our compute farm and available data. AI methods are used greatly to link models and understanding (Yu et al., 2017; Špačková and Straub, 2017).
- *Neuro-Fuzzy processes*: Magrathea have been routinely assimilating all clinical human (and indeed other data) neural network to learning and communication network. Magrathea performs numerous parameters and interpretation is complicated as we draw on all healthcare data for epoch learning; once the "learning curve" is at a clinical epoch grade, this moves to a rule-based approach.
- *Digital blackboard system*: It is another important sideliner artificial intelligence system interpretation based on the results of brainstorming within healthcare professionals. The DBS is used onsite for cooperative studies. We have linked digital characters reading, speech recognition, and motion capture artificial intelligence to feed this into the framework to speed epoch understanding, learning, and learned capabilities (Giese et al., 2017).
- *Rule-based systems*: Magrathea's artificial intelligence models are sets of rules to determine a given decision within epoch learning or a Kai event. Rules do need to be agreed and this can be an area for colossal human ego and conflict par-excellence. The difficulties of determination of proper rules for various situations and the absence of degrees of accuracy are the disadvantages of these systems; however, the epoch learning systems help rewrite these conditions and we have had some limited success in self-determination of rules not discussed in this work.

The idea behind this use of logic is to allow for a variety of "hits" from several sources; indeed, the logic allows for further development into natural writing, voice, image, and motion studies and yet undiscovered tools. One such, we believe, to be of interest is Xbox Kinect (codenamed Project Natal during development) as a line of motion sensing input devices for Xbox video game consoles and, recently within motion capture, virtual reality-based rehabilitation (Chanpimol et al., 2017). Their "Halo" team has used sensing for a whole range of data and image capture exercises: security triggers, motion capture for breathing, and predictive motion (epilepsy, stroke, and myocardial infarction). The Halo team is in conjunction with other devices looking to produce and include biometric patient data into R&D and supplementary epoch testing within the wider Magrathea framework.

Such additional devices include PoC devices such as the wearable watches and embedded chips (Ra et al., 2017) for additional inputs into the system. The Magrathea team is looking into working with PoC companies for future research and development.

Although these will, at present, contribute almost anecdotal information as most healthcare systems are closed loops, we do see this changing with subsequent generations moving to a digital lifestyle platform (Fig. 11.6). These expert systems provide diagnostic or given therapeutic information based on clinical patient-specified EHRS data, so the inclusion of these devices, although important, is diagnostically limited due to the lack of clinical certification. These framework epochs provide more information to the EHRS and produce sophisticated logic. These systems are perfect decision support systems for clinical diagnostic understanding and decision making. These decision-making tools aid with patient care, transitions, and orders for good patient outcome. These workflows help in the deduction of many types of errors within clinical diagnostics and the removal of inefficiency with the given healthcare system.

11.3.4   Epochs Intelligence for Clinical Diagnostics Understanding

Kai events are anchor points within intelligent systems framework allowing for a hybrid fusion integration event, effecting organic learning from very different data components. Once we have anchors in place to a given postulate, be that the traditional variant genomic searching from databases and metadata source (Hu et al., 2017; Ziegler, 2017) or from the clinical side patient record management (Vuokko et al., 2017), electrocardiograms (Vu et al., 2017), or hospital laboratory results (Ialongo et al., 2017), all can be "linked" via the epoch intelligence. In this section, we examine how epoch intelligence is used within real-time clinical decision making.

Convolution neural networks (CNNs) are key tools for deep learning and are especially suited for image recognition. You can construct a CNNs architecture,

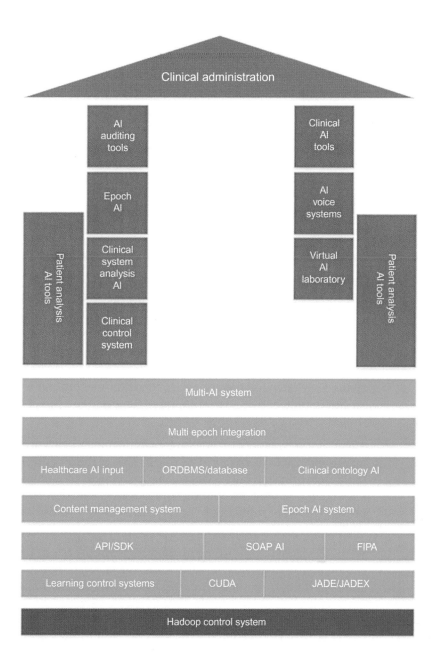

FIG. 11.6

The "House" framework, architectural development for clinical decision-making AI. Working with healthcare and patient input for a continued understanding and development of epoch sophistication. The healthcare worker interaction is key to epoch erudition for critical intelligence addition to AI frameworks. The key with any poly-dynamic AI systems is complete systems inclusion, updating and learning within any epoch paradigm.

train a network, and use the trained network to predict class labels or numeric responses. You can also extract features from a pretrained network and use these features to train a linear classifier. Neural network framework Epochs also enable you to perform transfer learning; that is, retrain the last fully connected layer of an existing CNN on new data.

In this instance, we can combine epoch learning from image analysis (Fig. 11.7), where the primary image from a given test (i.e., genomic data, written information, slide, biopsy, and X-ray) can be compared with a given set of known and unknown datasets. From this, we can develop epochs that are learning over many generations until the decision making is to our acceptable clinical standard. With the image, we train the epochs to learn individual approaches to preparation by a multistage pooling process followed by a process of convulsion and connections to know dataset and understanding within a Monty Carlo AI framework system. At this stage, input is granular and the child-epoch via a standard convolutional neural network (Simard et al., 2003; Chen et al., 2017) is at a very early stage of staging.

The interesting thing about this approach is we can detect who did what image based on known image samples within an initially supervised framework. The more image assigned to a given operatory, the greater the accuracy for prediction. As an aside, this approach is used to determine good "end user or A's' method" (Waller and Fawcett, 2013) decision making and is a powerful tool for finding the correct person to answer a given epoch, clinical decision, or clinical referral.

After training routinely 10–50,000 epochs, the first models achieve best performance with an accuracy rate of 98.6%. In each of the experiments, the top layer of the network has a fully connected layer followed by a nonlinearity that predicts the probability distribution over classes given the input sentence. These predicted pooled inputs link to the ICD10 disease classification system for end users' report generation.

We have adapted this learning method for many aspects of understanding for real-time predictive recognition systems; this is rather interesting in its adaptability, increasing our understanding by evolving epochs, to delve into clinical image understanding which has been sadly lacking within the medical field. Our epochs into skin, breast, and colorectal cancers are unpublished research due to patient confidentiality; however, we aim to present what data we can within this work.

The medical world is truly challenging artificial intelligence and machine learning limits. The Magrathea group has pioneered the use of artificial intelligence epoch anchoring. Similar to the traditional primary keys common in ORDBMS, but have anchors that are key to not only binding artificial intelligence flow

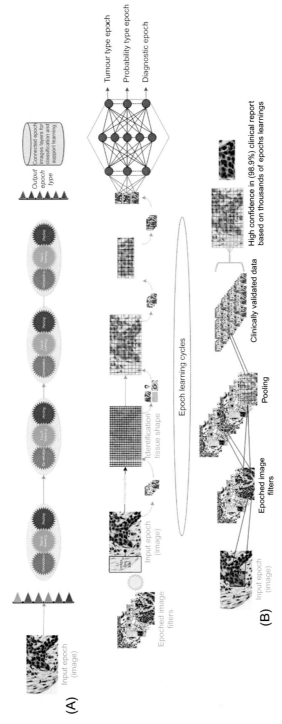

FIG. 11.7

(A) Epoch trained in blocks of 10–50,000 rounds of learning before a clinical model is acceptable for patient diagnosis. One practical application of epoch learning is in the analysis of EMR data, images, or biochemical test. (B) Learning cancer sample is one example where accuracy of very early patient diagnosis presenting with a given set of seemingly unrelated symptoms. This process is having profound results within the spectrum of WHGS and pharmacogenomics.

mechanisms, but also are point holders to link data and inputs from external sources, such as GUI, healthcare workers, ERCs, PoC, or other clinical R&D devices. Our goal is the linking of healthcare coordinated multifaceted artificial intelligence approaches to key clinical findings from traditional genomic pipelines (bloods to CAP/CLIA called clinical variant and interpretation). Moreover, pulling this information in real time to the physician and, where applicable, the patient and patient-wearable devices.

11.3.5   Clinical Patient Understanding

Although this industry is in its very infancy and, in reality, is still only accessible to a small number of wealthy patients, the founding principles are to make this process economically accessible to all globally. The foundation and basis for scaling is set within the artificial intelligence and the author on this work would welcome all and any input to further the understanding, training, and education development.

Important to epoch processing is having not only a robust clinical framework for phrase questionings and process managers, but to also having a clinical team skilled in this process. We have placed great effort in the training and educational aspects of this continued undertaking at all levels. This part of the undertaking continues to be the most challenging and rewarding aspect of the work. In areas where there is little or no understanding, we had to teach, learn, or relearn techniques.

From the outset, this task was and is highly challenging indeed. We do hear billionaire celebrity hype globally, be it Mr. Gates (Rawlinson, 2015), Mr. Musk, or Professor Hawkins (Victor, 2015). However, although this may shine a light on their ego, its shows a complete lack of understanding within the field of how fragile these systems really are both on the coding and real-world level. As we have seen in previous chapters, sharing data across boarders still shares the same time old issues of transparency. Moreover, with the advent of clinical genomics, this can be deeply personal and subject to wider governmental control and patient consent issues.

The cyber-attacks on the 12th of May 2017 against the NHS in the United Kingdom that affected many machines, systems, and servers (Collier, 2017) serves as a warning to vulnerabilities of complex systems and the ease by which they can falter (Fig. 11.8). Artificial intelligence systems, far from being strong and impenetrable as depicted in the movies, are in fact a lattice of interconnect AIs that are easily destroyed, a cobweb not a ring of steel. The importance of feedback, learning, storage, and rollback does evolve an understanding of protecting one's source code, database, and algorithms with the highest level of protection, but like most IT and biological system their fundamental core is fragile.

FIG. 11.8
Ransomware: Cyber-attack on UK hospitals causes delayed surgeries and cancelled appointments.

11.3.6 Epoch Deep Belief Neural Networks

The framework for epoch learning relies resolutely on deep belief networks (DBNs); these networks are layered forms with undirected connections between its top layers and downward directed connections between all its lower layers. We employ this throughout the learning process and cross-reference with other epoch learning cycles routinely to promote greater clinical understanding (Fig. 11.9A).

Indeed, with deep epoch input, the DBNs can have extensive learning or data construction epochs, if a DBN is to infer posterior probabilities and make generative predictions. Restricted Boltzmann Machines formed by adjacent layers in DBN are pretrained from bottom-up layer by layer; then the whole network is fine-tuned with back propagation to serve discriminative purposes (Tieleman, 2008) or up down algorithms for generative purposes learning within these systems, leading to the reconstruction of the neural net as epoch's cycle through learning and teaching events (Fig. 11.9B).

Datasets have very strict processing (Mocanu et al., 2017); it is of vital importance to the framework that a prediction process performed by the proposed neural network is clearly known and adhered to by the organization. In addition, it should form the fundamental basis of all AI educational, training, and SOPs (Fig. 11.10).

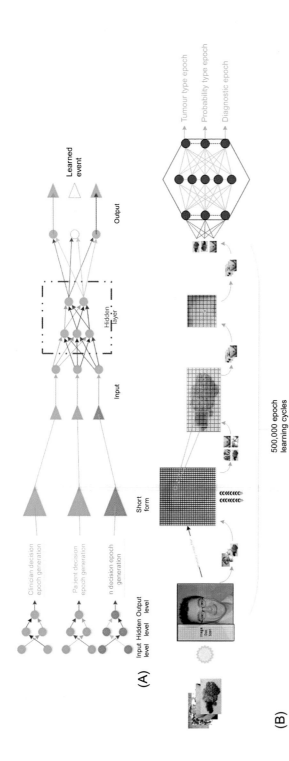

FIG. 11.9

The deep belief network and epoch learning: (A) Illustrates the connection between two inputs, in this case clinician and patient resulting in a simple output. The input epochs can come with a variety of flexibility, persevering changing projections of Gaussian randomized epoch learning between the projected signal and output vectors. (B) Convolution neural networks are essential tools for deep learning and are especially suited for clinical image recognition and epoch learning. The patient image is "loaded" in the framework and the neural network examines all known images for comparison of the given sample (tumor normal pairs, X-ray or any other input images, handwriting, voice, or many other CNN inputs). Aspects of the image are compared and epochs are learned against known data from a variety of users, sources, and inputs until a clinical threshold is met and decision is made; this feeds into the framework for further analysis and further epoch training. In this instance, an upper torso skin cancer analysis for this patient: we are using epoch learned from image, facial, and clinical learnings. Images analyses are key epoch with topographical multivariant AI used, looking into a variety of input data types: color, size, shape, 3D topology, and other metadata inputs. *From http://emedicine. medscape.com/article/1100753-overview.*

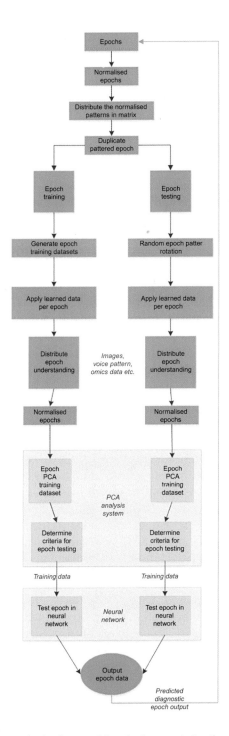

FIG. 11.10

The Epoch learning framework. This is an example of a given epoch learning framework at resting state. Learning cycles, or epochs, is a key for the building of any clinical AI framework.

The basis of these frameworks differs for all input types; in this chapter, we have been focusing on image prediction for data extracted and fed to the layered neural-net architecture within a given network. In this instance, the input for the object recognition module is the RGB tiff image, moving picture experts group video file or motion capture software for the objects: the region of interest is automatically extracted through a point cloud-based table top segmentation. Objects are represented as compact feature vectors and fed into a self-organizing map and principal component analysis neural networks. The layers work to build an epoch picture of clinical diagnostics and decision making. The frameworks are highly adaptable including an almost limitless number of epoch layers and layering of layers. The only thing we make as an analogy is the layering done in traditional cartoon making. The building of a picture *"layers, layered, and layering"* moves and hence produces the cartoon. Our framing is in real-time decision making linked to AI anchor points. The anchoring of the epochs is a key here and this chapter's scope limits further particularization. Consequently, layers and building of layers can have disparate novel inputs, favoring screen of many disease types including skin cancer, biopsies, and anywhere visual confirmation needed for absolute target identification.

11.3.7   Educational Aside

The Gulf is famous for oil exports to the world (Kelly et al., 2017). Indeed, on visiting an oil field and refinery, you would not expect to gain insight into your process, think not of oil but DNA. This is far from the case, the Gulf countries deal in single, multiple, and remote pipelines; going to/from different diverse locations, with storage sites, geopolitical issues, policy, law, and pipeline breakage (point mutation, translocations, copy number, and SVs are all seen in many ways). We see astonishing security issues, theft (in many innovative ways), and all of this controlled within a real-time environment and a means of remuneration. The base mathematical algorithms that run these systems are seemingly identical to tools we employ. Hence, epoch learning and inputs can come from a variety of lateral thinking (De Bono and Zimbalist, 2010).

Our basic types of learning undertaken within neural networks are supervised and unsupervised (Soheily-Khah et al., 2017). Within the framework of biological systems, both are used and combined with learning from previously discussed anchors. The anchors are key links to the "real world" and are vital for points of user interaction—the common concepts to all approaches, including nonneural network, pattern recognition, genetic algorithms, Bayesian learning, and discriminant analysis as discussed previously. These relationships do not tell us anything about cause and effect. However, the discovery of association between ranges of values and outcomes can lend clues for further investigation (Ni et al., 2017). In networks with all nodes interconnected, it is difficult to interpret these weights. With some algorithms, the weight interpretation is much more straightforward. In either approach, two phases are involved in neural network development: the *training phase* and the *testing phase*. Only after the completion of these two steps can we use the model for classification (Appice et al., 2017). Unsupervised learning can be used when little is known about the data set. It only requires a set of input vectors, with lack of information one must take care in interpreting the output in a clinical setting. Primary concern is the number of unknown categories; some means of deciding the appropriate number of clusters must be defined, and the author is happy to join forces on any data concerns.

11.3.8 Extracting Information From the Medical Record

This scheme for Magrathea epoch learning is by far the most challenging of all undertakings and the Gulf has taken a longitudinal view. The opportunity for setting leads to pooling current and best practice from leading research establishments around the world; indeed, sending out masters and doctoral student to gather these insights is a global norm—to build a complex framework that is able to gather information, pool legacy, and capture all other data into the system. Magrathea are utilizing many aspects of pattern recognition to digitize this information and epoch learning cycles include digital pens analysis from Magrathea clinicians. Within the many healthcare systems, Magrathea have multiple languages and learnings to consider each needing a separate epoch and team of trained experts.

Although many hospital information systems have some portions of the medical records in digital form, the handwritten medical record exists in most hospitals and clinics. The handwritten format introduces a number of problems. The development of medical decision-making aids, through the use of neural network, is of vital importance because of the decision's impact on the health of individuals. Hence, it is more important in the medical field than in most other fields to test the model as thoroughly as possible. In clinical practice, it is nearly impossible to achieve 100% accuracy, a goal that physicians would like to achieve. Indeed, Magrathea employ epoch binary swarming as proposed by Kennedy and Eberhart (1995) for pattern recognition and understanding. Magrathea variation of a particle swarm optimization is inspired by the behavior within the clinical space of a given group such as clinical behavior of clinicians and the clinical behavior of nurses in an emergency setting. The practical application is only now becoming known and will have a ground-breaking impact on clinical medicine. All examples of swarming are in a very early stage of development within the healthcare system. The best that we can hope for is a careful analysis of possible sources of error, epoch learning, and real-world comparison. As of yet, Magrathea are researching epoch swarming, but are limited on compute power to only 7000 cores. The author must note AI-epoch swarming, although of uttermost global significance is an immeasurable compute undertaking! In addition, as discussed early, Magrathea are looking at alternatives in QComp for more affordable options.

Epochs referenced to both ICD9 or 10 reference databases funded by the World Health Organisation and, to a lesser extent, the Electronic Medical Records and Genomics (eMERGE) Network and National Institutes of Health organized and funded consortium of US medical research institutions. The rules are being expanded and used with the artificial intelligence framework of all anchors linking object relational data, metadata, and metadata headers from "clean" and "known" metadata sources, which total 1200 metadata sources and some 10,000,000 data types within the AI framework. In the wider world of metagenomics, pan-omics-profiling, and omics-data, there are vast resources going into the many billions. Within the confinement of the healthcare systems, Magrathea are somewhat protected from this deluge of information, but this does beg another question of how does one feed clean data into a system, and how does one know this is relevant to patient understanding. Magrathea, at present, are making these calls based on biases and epoch-learned understanding. In the end, it comes down to how information is collected and compiled within, and one suspect this will resolve the issue of information sharing and walled gardening of data by each data-powerful country.

11.3.9 Eliciting Information From Experts

The idea of eliciting information from experts is not as straightforward as it seems. First, the Epoch Analyst Trainer (EAT aka. "*Eaters*") is a person who collects the data and must know the information that is required (Li et al., 2017). One cannot simple go to a clinician and say: "*Tell me about colorectal cancer.*" The questioning must be focused. Magrathea's initial approach is a variation of: "*How do you make decisions regarding the presence of colorectal cancer?*" Usually, more specific questionings needed for a particular condition and symptoms to be sought and Magrathea draw on thousands of years of knowledge gathered since Galen (Harris, 1973). Caution is needed for the Epoch Analyst indigence, if EATs ask questions in a prespecified format; the expert will fit their answers into this format and may exclude some of the actual decision-making process. Consequently, some of the unsolved problems that remain need better representation of high-level knowledge; on the other hand, a more adequate understanding of the human reasoning process and enhanced understanding of physiological concepts—concepts that may produce systems that utilize deep or causal reasoning strategies (Moravčík et al., 2017). However, they are looking into a swarmed evolved epoch methodology, discussed previously, and involve both character and voice recognition AI (Fig. 11.11).

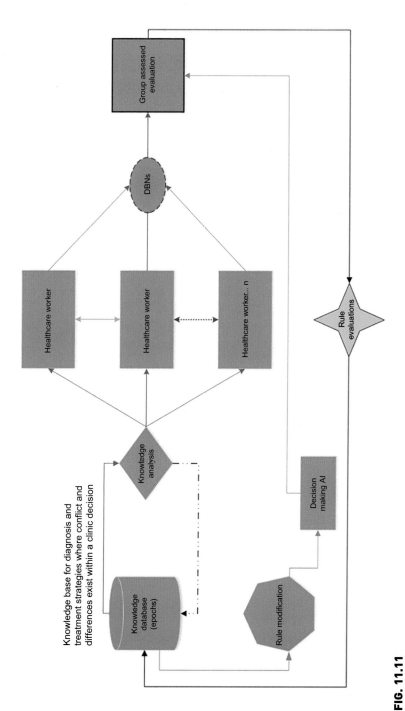

FIG. 11.11

The prediction process, performed by the proposed neural network within the DBN framework. Eliciting information from experts for expert systems.

11.3.10   The Next Steps in Clinical AI

The Magrathea framework is exceptionally robust and clinically battle-tested for over 5 years of epoch learnings. Magrathea are evolving epochs and improving the framework as and when new "internet of things" come online. The goal is total-real-time analysis and, with this, comes a strong need for metadata security: data, personnel, employees, epoch, algorithms, computer code, and this list goes on. Magrathea see the natural "life"/"language" processing and visual world of AI to be the Magrathea's next big step in developing epoch learning; allowing AI to investigate disease in the wider population with great efficiency and, indeed, accuracy of clinical diagnosis, discovery, understanding, and outcome optimization.

The future directions are based on voice- and image-activated artificial intelligence and deep knowledge understanding. This Magrathea believe will improve epoch training and understanding and the artificial intelligence can adapt to voice/image and thus can act not only as a means of input, but for security setting in conjunctions with other standard security protocols. As securing the environment is now key for data protection. Magrathea are basing this understanding on a concept system of deep recognition of speech commands in a natural language using patterns and antipatterns of commands.

11.4   CONNECTING ARTIFICIAL INTELLIGENCE AND MACHINE LEARNING METADATA ANALYSIS FOR CLINICAL DIAGNOSTIC DISCOVERY

Deep learning and complex artificial intelligence systems within a clinical genomic setting are daunting undertaking, and one is not to be involved half-heartedly. The traditional dogma of finding "true" clinical bioinformatists covers these areas: biology, computing, and healthcare. One must separate out biology and healthcare, for simple reasons. Healthcare covers all clinical aspects from the application of biology—the fixing part if you will, and parts covering the real world. Biology is cold hard facts based on current scientific knowledge. Way back in the early days of "*bioinformatics*," not even bioinformatics existed. It was more a composite of modules from departments that frankly were geographically and philosophically desperate. Finding people who cross these chasms is still difficult and practically impossible in the "*ye olde*" days; more so, for countries outside of the European-North American bubble. The rest of the globe sent their daughters and sons to these countries for learning and introduction of best practices and educational development programmes to their countries. These, to this day, are slowly emerging networks (Zayed, 2016a,b). Funding for early projects and courses was very difficult due to the massive costs associated with genomic medical sequencing. However, in the advent of the Human Genome Project (Collins et al., 1998), this all changed. Now most academic institutes run a degree programme with significant modules in bioinformatics, computational biology, and/or statistical genetics. Better institutes have full degree, masters, and doctorial programmes covering a plethora of informatics research undertakings. However, next to none run deep learning in clinical genomics. The patient's need for clinical understanding is fuelling the demand for training in this field, more so in global genetic counselling services field (Modell, 2003). Modern lifestyles are pushing for the integration of clinical, biological, healthcare, and personal "omics" data—in real time for instant understanding, the so-called Omics profiling (Li-Pook-Than and Snyder, 2013) or multiple omics deep learnings—Pan Omics Profiling (POPs).

The AI framework started out of pure frustration with the inability of the computation world "to simply link" bioinformatics findings with clinical patient understanding. A list of "hits" was seen as some major success, yet had little or no relation to the patient condition or diagnosis; this was due to the dogma discussed previously. One was happy linking databases to arbitrary findings, but had little or no regard of the patient—simply a sample to data push to the clinician, at best: "The genome had a trait for such and such therefore it must be so." The team came at this solution from the patient and physician side, after spending many years involved in traditional bloods to clinically called variant analysis. The samples got processed in a clinically validated laboratory following strict adherence to Centres for Disease Control and Prevention (CDC) (Wallace et al., 2014), Clinical Laboratory Improvement Amendments (CLIA) (CMS GOV, 2016), College of American Pathologists (CAP), American College of Medical Genetics and Genomics (Richards et al., 2015), and Clinical and Laboratory Standards Institute (Wayne, 2005) certification guidelines and an environment with strict adherence to Health Insurance Portability and Accountability Act regulations. It is an ongoing topic for healthcare professionals throughout the country and RDNs can benefit from participating in those discussions. Alas, the team came to another digital dead end. Although the results were good, automated pipelines for wet, dry, and moist lab settings, the same issues arose like from the early days of Sanger-sequencing file assemble: What does the data mean? This relied on what Magrathea called a PoC person, someone within the system who adds the key input that made the whole endeavor happen, consequently, making the endeavor into building the framework almost pointless, with the rate-limiting step always being the individuals' ability and time to confirm a finding. At the time, the team were working on acute cases involving fast growing cells and extremely sick children. The turnaround times (TATs) are the devil in this story and the reduction of TAT "for everything" is at the heart of the AI. This is inescapable, at present, within the realms of machine learning, artificial intelligence systems, and something that the Magrathea team works on. The issues to variant calling pipelines were robust, automated, and needed little tweaking from the team, so Magrathea concentrated their effort on the patient physician artificial intelligent system, integration of machine learning within complex systems, and the ability for long-term training programmes. Ironically, the deep learning network architectures were later used, to improve in-house base callers, an issue seen with pore-channel development having to "slow" readings rates for base callings.

There are many companies offering NGS services, variant analysis, and some clinical interpretation using a variety of databases, algorithms, applications, and bespoke programming; these companies are great if you have the luxury of time or indeed wealthy patients, with respect to a vanity exercise and do nothing for population level events. In a clinical setting, time to diagnostic decision making is critical and the primary motivation is clinical TAT; TATs for every step within the clinical decision-making process. The Magrathea framework was born out by a simple lack of genomic interpretation within a complex clinical and healthcare setting. The Magrathea framework combines standard NGS pipelines alongside links to front-line medicine for patients and healthcare workers. The Magrathea framework also uses state-of-the-art artificial intelligence and machine learning algorithms to feedback into the decision making for real-time patient diagnosis. From its outset, the Magrathea framework is an ultra-high end data integration of NGS, variant, companion diagnostics, research and development, clinical electronic healthcare record (EHR), hospital records, technology, and physician-based artificial intelligence all in real time (Fig. 11.12). Indeed, this allows the flow of information into the system when it comes online or is learned from Kai or epoch events. Within the framework, a delay triggers secondary and tertiary responses; the deep learning networks draw upon deep knowledge sources from epoch data that are already available helping projects. For example, if clinicians are waiting on WHGS interpretation data and have other data to handle, the AI can use data from a variety of inputs: panels, arrays, targeted sequencing, exome, and many others linked to patient understanding EHR via the ICD10 AI for disease classification. This data is stored, and reporting is a highly priced resource for the given healthcare unit, therefore robust security is in place and further trigged when WHGS or "omics-data" progresses to the data stream.

When one considers that a country's clinical genomic data is its gold mine for clinical genome understanding and potential durability targets for manipulation, this will have a massive economic impact on a nation's wealth prospects. One can only agree with Genomics England's point of view to allow access to use and no download access for any of their data. After all, Governments do not usually have the habit of giving something for nothing. Moreover, given the ability to spend hundreds of millions of pounds within a healthcare community, one would want to see some meaningful results for one's voters by setting

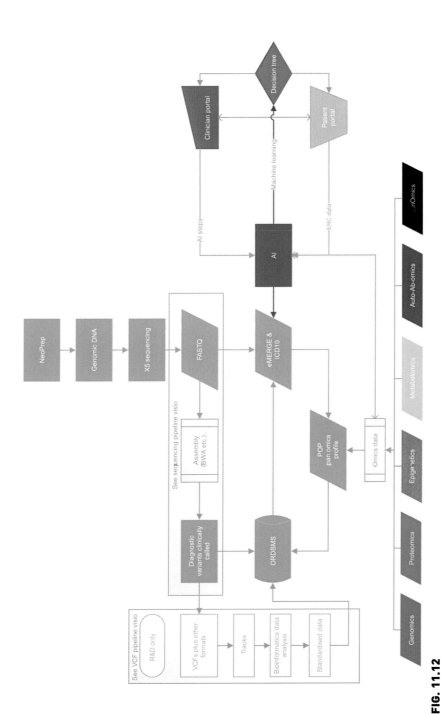

FIG. 11.12

Magrathea NGS sample handling pipeline. This is a complete overview of the Magrathea process. Clinical reporting in Magrathea: The patient and physician GUI for the Magrathea patient for clinical reporting. Although in an early phase of adoption feedbacks from all members in pilot studies have welcomed the ease of access and use.

strict goals and rule to work within the payer's framework, databases, and software. The integration of these tools into the NHS, research communities, and national biobanks is of vital importance to any country's future growth within the genome space. This allows for the total control of the clinical identification and findings process: genomes, software, and third-party applications. The results give an all-round saving in time for prognosis, decision, and medicine purchasers on a country-wide basis, R&D, and many more unknown areas at this time. In a very similar way, the framework automates this process with the use of real-time processing to deal with complex issues and feedback into the decision-making process discussed at length earlier in this chapter.

Most genomic pipelines suffer from their inability to feed meaningful results back into the healthcare system, physician, or patient where results can be acted upon. It is clear that the current crop of large data files is not practical for integrating into a healthcare system on a larger scale. The raw data and even the BAM to VCF (Goyal et al., 2017) generation pipelines are clearly not efficient for real-time analysis or clinical diagnosis. We have seen from Genomics England the promotion of App specific analysis to utilize the data locally for local NHS hospitals and indeed promotion of companies that are adopting this framework for clear result understanding and an end to the "Top list of hits" of results, leading to further delays and lack of understanding. Moreover, the use of "internal" social media to share experience and understanding with the collective framework works well for efficient diagnosis of conditions and is the basis of one aspect of the clinical framework. The disconnection between the variants being called to patient understanding is a real problem within global healthcare systems; time is lost in many aspects of patient care and treatment. The majority of service offerings are at best a simple reporting of findings, a link to clinician-signed report, and this understanding needs to be fed back into the electronic healthcare system for further understanding and detailing of correct pathways to diagnosis, help, and best known practice for the given condition.

The limiting step at present in the clinical diagnosis of patients is the turnaround time for sequencing in the decision pipeline. This is dramatically being reduced as new technologies come into the market: from high-end-sequencing companies through to the PoC, and use once technologies from companies, such as Oxford Nanopore with new kits providing 5–10 Gb per MinION flow cell, raw accuracy 92% (1D), 97% (2D), supporting very high consensus accuracies, 5 min library prep—combining with real-time data streaming for very fast access to data. The opening of the PromethION, the high-throughput, high-sample number, on-demand sequencing device, is based on the same Nanopore sensing technology as the MinION and GridION (Fig. 11.13). New enrichment techniques, progress of the mobile phone-compatible SmidgION device, and much more are happening and one can only see Oxford Nanopore and indeed other such similar companies globally pushing the call

FIG. 11.13

GridION X5: Benchtop DNA/RNA sequencer. A fast and impressively quick way to sequence and analyze genomic data. A very quick way to include new epoch learning into a multivariant AI framework.

and coverage rate to reduce false positive and false negative calling needed for clinical accurate decision making.

The AI framework follows a similar pipeline to Genomics England: they outsource all genome-sequencing projects, until a local Illumina HiSeq facility was built and the utilization of a governmental-run local site and virtual private clouds for genome understanding started. Unlike Genomic England, the team used internal programming for downstream NGS and metadata analysis. The Magrathea framework analysis is based on true artificial intelligence and machine learning methodologies for real-time understanding of genomic data integration into a whole genomic and epigenetic clinical gene analysis, as with Genomics England taking longitudinal approaches to patient records, medication, and understanding. The team used a new class of technologies such as nonsequential programming (Cobb, 1978), symbolic processing (Watt, 2006; Davenport et al., 1988), knowledge engineering (Schreiber et al., 2000), and uncertain reasoning (Cohen, 1984; Russell et al., 2003). Coupled to traditional artificial intelligence approaches in expert systems (Giarratano and Riley, 1998), neural networks (Hornik et al., 1989), natural language processing (Chowdhury, 2003; Joshi, 1991), fuzzy logic (Klir and Yuan, 1995; Mamdani and Assilian, 1975; Zadeh, 1997), and semantic processing (Poldrack et al., 1999) are all within the scope of a mathematical framework for metagenomics studies. Allowing decision making on the physician/patient side to be dynamically linked to the traditional NGS clinically called variants on the sequencing side. The artificial intelligence/machine learning methodologies are also employed at the variant calling stage, although these are still in an early research phase of epoch learning cycles. The team currently focuses on the development of Apps algorithms (Bartsch et al., 2016; Sevel et al., 2016), using ANNs for medical diagnostics and statistics understanding including: Cancer systems biology and bioinformatics, identification of prognostic, diagnostic, and predictive biomarkers for disease and therapy and development of prognostic indices for cancer, characterization of tumors from immunohistochemical markers in breast cancer, and the molecular diagnostic modeling of cancer disease characteristics from mass spectrometry (MS) data of tissue and serum. They also work with teams on strain and species characterization of bacterial pathogens from MS and sequence data in an effort to develop an artificial intelligence model immune response in colorectal cancer (Al-Mulla et al., 2016; Al-Temaimi et al., 2016).

Genome interpretation is very hard, reading at a rate of six words per second would take 35 years to read your genome (Pavlidis et al., 2016). The systematic development of software for the long term and not a postdoc/seat of the pants solutions is difficult and expensive to build. Challenges are born out of real-world clinical patient informatics, not bred in the lab; clinicians need to be able to easily gather and learn information from the wider world and that takes time to deal with massive data and building good public reference datasets that are fluid and future proof. Magrathea need integrative ways to incorporate RNA data compared with genome data; in addition, asking many difficult questions of clinical findings for epoch learnings: Why do we see two different transcripts expressed at different levels? Why we see two alleles expressed? Then use clinical findings and tools to trace this back to the genetics. Curating data on the functional annotation side of a given gene and what it does comes from in vitro studies from model organisms (Goldstein and King, 2016). Clinicians need good public reference genomes for a variety of ethnic groups for detailed understanding. It is for this reason the AI framework was set up to answer the difficult questions, pushing further understanding needed for artificial intelligence, and machine learning tools to complement. For example, the understanding of knockout genes in a mouse and how to observe how protein-protein interactions are, and how interaction is important going forward to aid clinical understanding of the nested phenotypes in a given disorder region. Underpinning all of these is the need to have good terminology, the standardization of metadata approaches, and going further downstream for artificial intelligence and machine learning neural networks with the Magrathea framework and epoch learning system.

At present, the machine learning aspect of the framework is a rather basic workflow: whereby data training sets is data derived from the traditional NGS pipelines that Magrathea see from many companies; in the frameworks case, a Dragen Bio-IT processor system by Edico Genome (Dutton, 2016). Here, the genome approach is to move quickly from sequence file input to VCF caller and the called data is then moved to an object relationship database management systems (Daraio et al., 2016). Then, the artificial intelligence workflow takes over, producing a final database that can be manipulated in very simple form and indeed with very little compute or space needed. Magrathea have a machine teaching "learning room" that classifies the epochs in three ways: new, supervised, or unsupervised data. The latter is data good enough to be used for any feature or information exactions and deemed "Gold" by the conditions called on the system. Supervised data is data that Magrathea internally call "naughty child (NC) data," which is data that in some way is not conforming to the specified inputs or learning epoch cycles; again, derived from the conditions called by the system. Finally, "new data" is fresh data into the system that is used to learn and push the NC data to an unsupervised state and more so to learn from unsupervised data how not to become NC data. The system is to work on very simplistic operations (App Bots) that work on each specific variant and not on the whole analysis. Machine learning algorithms employ a group of data scrubbers, findings, and results in some common characteristics of the project. Note, this is fluid and open, but open to interpretation in many ways is a gross understatement. The framework AI also allows for predictive modeling of the data epochs and, if needs be, for genomic or GUI purposes and the annotation of data. Due to the findings and output relational databases being small, the findings have been parsed and indeed held on very small devices indeed; from simple app-based approaches for mobile phones, tablets, and watches to Microbit (Ball et al., 2016; Schmidt, 2016) processors looking into small aspects of a given disease condition or indeed PoC devices for commercial scaling of findings and indeed reduced cost analysis in areas of the world that have little or no infrastructure.

The aim is not only to compare whole-human genome understanding, but also to enhance this with genomes from alternative areas or indeed with respect to the human digestive system genomes that are native in the human body (Koren and Ley, 2016; Zmora et al., 2016). As one can imagine the compute, power is vast and yet the underlying finding principles of the Magrathea framework is that data should run on a standard laptop. Of course, the initial back-end processing is highly compute-intensive; however, in the front end, the GUI needs to be of such simplistic taste that one should be able to use it with little or no training. Alas, this later endeavor is still a work in progress and subject to repeated changes and as the Apps evolve to suit the needs and indeed knowledge of the variant studied for the healthcare workers (Fig. 11.14).

The resulting epoch structure allows the system to "learn" from input and each generation of analysis and, interpretation happens in the system "learns and

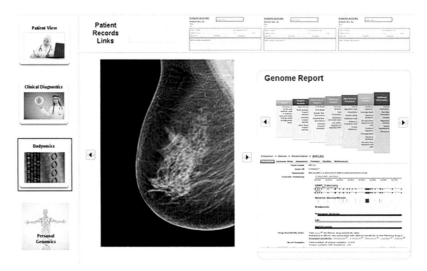

FIG. 11.14

Physician reporting, given access to the data and results, can stream real-time results in the decision of Artificial Intelligence and present when needed. This include NGS findings and latest information from clinical laboratory and specialist. In this instance, we are seeing results from a breast cancer scan and supported reporting GUI; clinical reports not shown here.

feeds back" and with the help of user direction adaption, be it from NGS, array panels, metadata inputs, "best known practices" for omics epochs, or projects. In this instance, the "learning" is directed toward the community of physicians. Like Genomics England, the approach to use an internal social media GUI to understand works well. Magrathea's metaanalysis draws on data from commonly used social media tools, where this social data is agreed to or can be accessed. This can also include medical records or indeed tribal information from where this is utilized by Professor Fahd Al-Mulla in Kuwait (Al-Mulla et al., 2013), for private patient care and understanding.

Within the Magrathea framework, each omics study starts life as an epoch. An epoch is the starting processing position for a given condition: epoch for Alzheimer's, Parkinson's, diabetes, various cancer types, etc. Then, later epochs are specifically designed for the given issue of the variants or indeed for metadata informatics of many kinds of learning. The use of analysis within the Magrathea framework of metadata and data compression tools is aimed not only for the comparison of an individual's genome, but for analysis of pairs, trios, and larger cohort studies from around the world. With more metadata being curated, this can be incorporated with other epochs generated either within the Magrathea framework or within the Magrathea Genomics software developers kit or application programming interface for any third-party use.

Variant-specific epoch learning is linked to pharmacogenomics, epigenetics, and many other data sources under the general term of omics epochs can and are processed within the "learning" phase of the processing.

The GUI reporting system within the Magrathea framework is a simple patient physician input. The clinical side of the analysis is aimed to create a quick reply for standard disease types as a score within the International Statistical Classification of Diseases and Related Health Problems, usually called by the short-form name International Classification of Diseases (ICD); the version Magrathea use is ICD9, 10, 10-CM, PCS, and 10-CA. Magrathea are also incorporating the eMERGE and Genomics England network protocols to the artificial intelligence. Magrathea use the same artificial Intelligence and machine learning approach for ICD10, and all aspects of EHR data, including images, biochemistry, and findings, where possible, to draw upon (Zweigenbaum and Lavergne, 2016). The Magrathea framework takes a given protocols and indeed even in silico designed protocols that are drawn upon when one is investigating a given epoch condition. Like Genomics England, there is a collaborative aspect to this tool, allowing for users to share findings within a secure platform. The only difference is that the artificial intelligence scales the response to a risk score and profile. At this stage, certain procedures have a gold pathway, whereby a condition treatment is given and agreed upon by the given healthcare system and/or insurance grouping in a way to create a patient treatment programme; this "processed logged artificial intelligent system" allows for a "*Best Practice*" approach to the treatment and care of patients. This leads to a "*Hot Hit*," a list of known pathways that are automatically okayed treatment pathways for individuals to select and utilize. For some, these are great ways to share concept medical care or indeed for making aware doctors about key diagnostic findings pertinent to the patient and aid patient healing or legal implications. In recent years, the team has concentrated efforts on pharmacogenomics analysis and genotyping data, to aid miss diagnosis of drugs and the reduction of adverse drug reactions. The end output is a "Patient Diagnostic," a simple report which can actually be sent to the patient; one idea is to have a patient card with known adverse drug reaction variant present. The process from initial diagnosis through to outcome is completely process-logged, that is, all key strokes, notes, findings, and steps are fed into the decision-making artificial intelligence to aid future analysis in this area. The datasets like Genomics England's datasets are securely held and only open to key-carded and finger print-read individuals (Fig. 11.15).

The current direction of the Magrathea team is the manipulation of the whole artificial intelligence framework to look at approaches to downstream alignments, reporting variants and what a given genome looks like. More important to the team is the idea of throwing out the concept of a "Healthy" or "Sick" person or genome. The idea is born out that all clinicians to date have patients

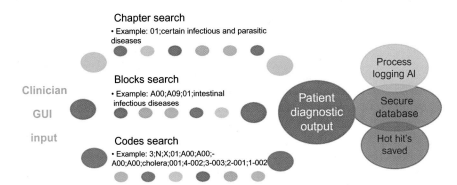

FIG. 11.15

Clinical ICD10 GUI overview. The AI framework owes a lot to the ICD classification system. A list of clinical facts and understanding the AI can link clinical decision making to the NGS AI decision framework. Within the social GUI, clinicians and researchers can build on epoch learning; building a "hit list" or "gold standards" for patient care (drug availability, insurance cover, and treatment), SOP, or presentation at initial consultation. Biometric security is used to access this system.

presenting to them with illnesses and, hence, the healthcare system is reacting to the patient, reacting to the condition, and usually at the very late stage in the fight. The Magrathea team wants the fight against a given disease state to be from a position where they attack the disease, before the disease state becomes an issue or to look into an underlining family condition from a very early stage and aid the individual or family. The Magrathea team believes this will profoundly change the way healthcare will be managed in a clinical longitudinal manner.

The Magrathea team is also using the artificial intelligence framework internally for long-term population disease understanding. This is indeed a tricky problem to understand, the how and why we get sick, which sounds like a simple question. However, when you consider actually how do genotypes and phenotype get sick in a population? This will aid clinical approaches in combating many of the leading causes of healthcare disease in the world and not just be limited to the chosen few individuals with healthy bank balances. The Magrathea team needed to reconstruct the last common ancestors of all people who carry at each position that person's particular haploid genome: an ancestral genome (Bourque et al., 2006; Baba et al., 2009) or ancestral reconstruction (Wang et al., 2016a,b; Muffato et al., 2010; Vakirlis et al., 2016). That reference genome served as a more objective less problematic standard than the cobbled together version we have now, the so-called "Standard Reference Genome" or the ancestral genome verses a chimeric composite reference genome. The Magrathea team is looking for a reference genome that is healthy: What is a "healthy" variant? When you consider that the variant that contributes to cancer risk (Margadant and Sonnenberg, 2010) may also be involved in wound healing (Scappaticci et al., 2005), we need to understand more about the concept of how genotype types get sick and the epoch learning cycles are producing artificial intelligent clinical population understanding.

We believe the work we do will aid understanding in reference genomes for a given population, group, or tribe to combat disease. To look into the concept of genomic changes over time to the given genome of interest and with the advent of the artificial intelligence framework and metadata, artificial intelligence means to determine how and why the mutation not only was, but how it started (Kurokawa et al., 2007). Moreover, when we have insertion or deletions next to a cytosine is there more insertions or deletions? This is an empirical question that the team believe you cannot answer unless you know the change in the ancestral genome where the insertions or deletion evolved from. The artificial intelligence and machine learning epoch workflows help also to infer summary patterns in single nucleotide variants, substitutions, and evolution to the reference genome, allowing us to determine what happens to the mutation after it arose. It is early stages yet, but the finding will aid druggable targets for many years to come. Indeed, the artificial intelligence framework allows your physician to run off the latest information on that genome, exome, targeted genome, and so on and so forth using ICD10 protocols linked via the artificial intelligence framework to the latest and best version of rare variant burden tests (Wu et al., 2011) and the team employs many more tools. This is very difficult indeed and statistically demanding; Magrathea compound this issue further with the addition of modern companion diagnostic tools. Tools to look at diseased genotype-phenotype associations in so-called "big data" sets may trace to variants that are too rare individually but together cluster to become part of the genome that is significant for a disease. Or indeed diseased genotype versus phenotype associations on big data sets or diseased variants are too rare to be determined individually, but together as a cluster in a given part of the genome can be determined, characterized, and complied on a wider scale. This is the power of an integrated artificial intelligence approach to genome medicine. The artificial intelligence framework is also easy to use and on a command front and was born with the ability for adaption legacy and sustainability.

Magrathea have three main areas of user need:

1. Urgent diagnostics (Hillman et al., 2011): This falls into two main groups: sick children and cancer patients. The Magrathea team is highly confident in artificial intelligence framework with this fight; however, they openly accept that the technology is still playing catch up to these diseases and have the tools to win the majority of the battles; it is time and, moreover, TATs that are defeating the artificial intelligence at

present. These TATs are vastly reducing and, to the entire field, it is not fast enough, but every year better people, systems, and technology are being deployed to aid the fight. At present, speed comes at a huge cost and only available to the chosen few, but like TATs are dramatically reducing.

2. *Urgent prognosis* (Tejpar et al., 2010) is vital for prenatal use and couples screening (Whole genome couples screening). Magrathea believe, although there are many questions, the key points are if, in the long run, whole genomes will help or can help people find, marry, and have healthy children with anybody.

3. *Prognostic* (Elston and Ellis, 1991) clinical diagnostic use for the wider population that is fed into the healthcare system, like at Genomics England that aids the long-term understanding of disease. The so-called birth of old age understanding of disease risk, as Magrathea Genomics put it, enables us predict and act upon a given disease states that can be tracked over time to understand how people with particular genotypes or phenotypes get sick in very particular ways (Harris et al., 2007).

Magrathea's long-term aims are geared towards not short-read-based analysis, but more too long-read sequencing or even chromosomal length sequencing in the near future. The artificial intelligence is set to infer haplotype whose variants are found together on the same copy of a given chromosome, allowing them to know how these particular variants interact with each other or other variants. The Magrathea artificial intelligence framework is not there, yet does have enough learning cycles to have a good understanding of in silico artificial intelligence clinical data insights that are being deployed right now. Behind the artificial intelligence epoch framework is a very simple workflow and decision make tree; the main processes are concerned with understanding if the variant is real, who does the variant belongs to, is the genome sick or healthy, what is the biology of the gene variant, what does this gene variant do in the body, and what does the process have to do with the disease. The Magrathea artificial intelligence determines what is real and which genome the variant belongs to. This allows the physicians and research groups to be able to determine results and to figure out which genomes share a given gene variant at the right zygosity to be of interest. In one internal example, the artificial intelligence assumes it reoccurs in a group of children and was looking for a sharing of homozygous variants, a homozygous variant that is not in the control group or within the family or cohort group; the artificial intelligence looks into the wider population for who and how common is the variant within this population, a simple example that draws upon all aspects of the artificial intelligence framework and machine learning epochs that are being learned into the system. These are artificial intelligence questions that help plan and answer clinical diagnostics in a faster way and will be a main stay of future diagnostics reporting in medicine.

Although the Magrathea team laughs when people say "Big Data," it does not matter what the datasets are or indeed the size. The process management of simple questioning is at the basis of the Artificial Intelligence and machine learning techniques and there is a dynamic foundation that evolves with the scale of the questioning. For large scale data analysis, Magrathea need to use rare testing or case control cohorts or for functional annotations the Magrathea artificial intelligence short listing of variants often provides questions, indeed many questions: a short list of what variant is real, who the variant is in and, in this, the artificial intelligence has a list of very ready questions that have been determined by the artificial intelligence epoch framework, but the Magrathea team has a Pop (PoC person) point, to use or pose within the machine learning workflow, and which of these variant short listings are worth following up on—combining what is known or guessable based on the artificial intelligent and machine learning Magrathea framework under the backend processing on the clinical side using the ICD platform. In this way, Magrathea can utilize what we know about a given gene, from traditional models and with the artificial intelligence become more involved with the prioritising within the machine learning framework is effective for diagnostic treatment. This allows the clinical decision-making artificial intelligence predictive modeling of applications to determine more functionally relevant information to what is actually going on with the patient rather than traditional "top hit list" with clinician calling. Of course, an important part of any artificial intelligence is not to reinvent the wheel, and therefore, the artificial intelligent system does draw upon user inputs such as the Pop points and so-called screen to brain understanding (McCarthy, 1960). Here, the healthcare workers' input simply is quick and better than technology, in the early stage of epoch learning cycles: the artificial intelligence framework is after integrating all of the people, system, and technology we have available to aid the fight against disease in its many fascists (Leventhal et al., 2016).

11.5 CONCLUSIONS

It is very clear that this century will be defined by artificial intelligent systems and the integration complex systems within the medical field. Everyday life is littered with sporadic yet huge influences on our daily lives. However, artificial intelligent systems are yet to impact global life for the better. In terms of health, well-being, food production, or safety, it is slowly happening and is needed in a world full of interpretable information needing guidance. This AI framework is one such early systems that makes inroads into the clinical genomic world. There are many more that will slowly integrate as and when compute, storage, and servers cope with the speed that this world needs. Turnaround times and time itself is the burglar in this instance and AI is waiting for the people, systems, and technology to catch up.

The process of sample to understanding and treatment will be the long-lasting legacy of this process, when genome analysis moves ever closer to real-time analysis. It is where important steps made by global AI teams in artificial intelligence and others in the field cannot link data from people, systems, and technology together, by accumulating meaning to these findings for systems, processes, and pathways to patient pioneered by Genomics England. The variant-specific learned epochs will be the new Google, Twitter, or Facebook of the 22nd century or maybe a massive-controller App AI such as in the mentioned social media companies, where these genomics/medical Apps naturally connect to when the society moves progressively to a pure digital information society.

Like all things, they have phases and cycles and genomics is no different. We are in a phase of data gathering "Beagle exercise" over analysis and one can make obvious comparisons to Darwin. Regrettably, the large scale does push aside that devoted analyst, craft-person, and artistry within the medical world. The people who can make changes and a difference are being silenced by the larger projects. One does hope the effort Genomics England and Magrathea have made goes some way for allowing freedom of expression and questioning of the medical status-quo: centralized pooled resources to do the day-by-day things, yes, but retaining the artistic license to inspire a new age of clinical genomic entrepreneurship.

Acknowledgments

The author would like to thank Professor Fahd Al-Mulla, Sebnem Daniels Baker, Ayla Baker, Hyram Baker, and Yvonne Hisemen for their valued and continued support of this new scientific idea and research. Darrol Baker is also the founder of the Magrathea Group.

References

Al-Mulla, F., Bitar, M.S., Taqi, Z., Yeung, K.C., 2013. RKIP: much more than Raf kinase inhibitory protein. J. Cell. Physiol. 228 (8), 1688–1702.

Al-mulla, F., Thiery, J.P. and Bitar, M.S., 2016. Method for Determining Risk of Metastatic Relapse in a Patient Diagnosed With Colorectal Cancer. U.S. Patent 20,160,068,912, Kuwait University.

Al-Temaimi, R., Kapila, K., Al-Mulla, F.R., Francis, I.M., Al-Waheeb, S., Al-Ayadhy, B., 2016. Epidermal growth factor receptor mutations in nonsmall cell lung carcinoma patients in Kuwait. J. Cytol. 33 (1). p.1.

Appice, A., Guccione, P., Malerba, D., 2017. A novel spectral-spatial co-training algorithm for the transductive classification of hyperspectral imagery data. Pattern Recogn. 63, 229–245.

Asimov, I., 1985. Robots and Empire. Grafton Books, London.

Baba, T., Kuwahara-Arai, K., Uchiyama, I., Takeuchi, F., Ito, T., Hiramatsu, K., 2009. Complete genome sequence of Macrococcus caseolyticus strain JSCS5402, reflecting the ancestral genome of the human-pathogenic staphylococci. J. Bacteriol. 191 (4), 1180–1190.

Ball, T., Protzenko, J., Bishop, J., Moskal, M., de Halleux, J., Braun, M., Hodges, S., Riley, C., 2016. Microsoft touch develop and the BBC micro: bit. Proceedings of the 38th International Con ference on Software Engineering Companion. ACM, pp. 637–640.

Bartsch, G., Mitra, A.P., Mitra, S.A., Almal, A.A., Steven, K.E., Skinner, D.G., Fry, D.W., Lenehan, P.F., Worzel, W.P., Cote, R.J., 2016. Use of artificial intelligence and machine learning algorithms with gene expression profiling to predict recurrent nonmuscle invasive urothelial carcinoma of the bladder. J. Urol. 195 (2), 493–498.

Bourque, G., Tesler, G., Pevzner, P.A., 2006. The convergence of cytogenetics and rearrangement-based models for ancestral genome reconstruction. Genome Res. 16 (3), 311–313.

Burner, S.T., Waldo, D.R., McKusick, D.R., 1992. National health expenditures projections through 2030. Health Care Financ. Rev. 14 (1), 1–29.

Chanpimol, S., et al., 2017. Using Xbox Kinect motion capture technology to improve clinical rehabilitation outcomes for balance and cardiovascular health in an individual with chronic TBI. Arch. Physiother. 7 (1), 6.

Chen, Y.-H., et al., 2017. Eyeriss: an energy-efficient reconfigurable accelerator for deep convolutional neural networks. IEEE J. Solid State Circuits 52 (1), 127–138.

Chowdhury, G.G., 2003. Natural language processing. Annu. Rev. Inf. Sci. Technol. 37 (1), 51–89.

CMS GOV, 2016. How to Apply for a CLIA Certificate, Including International Laboratories. Available from: https://www.cms.gov/Regulations-and-Guidance/Legislation/CLIA/How_to_Apply_for_a_CLIA_Certificate_International_Laboratories.html. [(Accessed 8 October 2016)].

Cobb, G.W., 1978. A measurement of structure for unstructured programming languages. ACM SIGSOFT Softw. Eng. Notes 0163-5948. 3 (5), 140–147.

Cohen, P.R., 1984. Heuristic Reasoning About Uncertainty: An Artificial Intelligence Approach. Stanford University, Stanford, CA.

Collier, R., 2017. NHS ransomware attack spreads worldwide. CMAJ 189 (22), E786–E787.

Collins, F.S., et al., 1998. New goals for the US human genome project: 1998–2003. Science 282 (5389), 682–689.

Daraio, C., Lenzerini, M., Leporelli, C., Naggar, P., Bonaccorsi, A., Bartolucci, A., 2016. The advantages of an ontology-based data management approach: openness, interoperability and data quality. Scientometrics 108 (1), 441–455.

Davenport, J.H., Siret, Y., Tournier, E., 1988. Computer Algebra: Systems and Algorithms for Algebraic Computation. Academic Press, San Diego, CA, ISBN: 978-0-12-204230-0.

De Bono, E., Zimbalist, E., 2010. Lateral Thinking. Viking.

de Vasconcelos, J.B., et al., 2017. The application of knowledge management to software evolution. Int. J. Inf. Manag. 37 (1), 1499–1506.

Dutton, G., 2016. From DNA to diagnosis without delay. Clinical OMICs 3 (3), 28–29.

Elston, C.W., Ellis, I.O., 1991. Pathological prognostic factors in breast cancer. I. The value of histological grade in breast cancer: experience from a large study with long-term follow-up. Histopathology 19 (5), 403–410.

Forbus, K.D., 1984. Qualitative process theory. Artif. Intell. 24 (1), 85–168.

Gao, X.-L., Hui, X.-J., Zhu, N.-D., 2017. Relative Divergence Degree and Relative Consistency Degree of Theories in a Kind of Goguen Propositional Logic System. Quantitative Logic and Soft Computing 2016. Springer International Publishing, pp. 89–100.

Giarratano, J.C., Riley, G., 1998. Expert Systems. PWS Publishing Company, Boston, MA.

Giese, H., et al., 2017. State of the Art in Architectures for Self-Aware Computing Systems. Self-Aware Computing Systems. Springer International Publishing, pp. 237–275.

Gilks, W.R., Richardson, S., Spiegelhalter, D. (Eds.), 1995. Markov Chain Monte Carlo in Practice. CRC press, Boca Raton, FL.

Goldstein, B., King, N., 2016. The future of cell biology: emerging model organisms. Trends Cell Biol. 26 (11), 818–824.

Goyal, A., et al., 2017. Ultra-fast next generation human genome sequencing data processing using DRAGEN TM Bio-IT processor for precision medicine. Open J. Genet. 7 (01), 9.

Graham, C., 2017, NHS Cyber Attack: Everything You Need to Know About Biggest Ransomware Offensive in History, http://www.telegraph.co.uk/news/2017/05/13/nhs-cyber-attack-everything-need-know-biggest-ransomware-offensive.

Hameurlain, A. et al., (Eds.), 2017. Transactions on Large-Scale Data-and Knowledge-Centered Systems XXXI: Special Issue on Data and Security Engineering. In: vol. 10140. Springer, Heidelberg.

Harris, C.R.S., 1973. The Heart and the Vascular System in Ancient Greek Medicine, From Alcmaeon to Galen. Oxford University Press, Cambridge, MA.

Harris, T.B., Launer, L.J., Eiriksdottir, G., Kjartansson, O., Jonsson, P.V., Sigurdsson, G., Thorgeirsson, G., Aspelund, T., Garcia, M.E., Cotch, M.F., Hoffman, H.J., 2007. Age, gene/environment susceptibility–reykjavik study: multidisciplinary applied phenomics. Am. J. Epidemiol. 165 (9), 1076–1087.

He, W., Tian, X., 2017. A longitudinal study of user queries and browsing requests in a case-based reasoning retrieval system. J. Assoc. Inf. Sci. Technol. 68 (5), 1124–1136.

High, R., 2012. The Era of Cognitive Systems: An Inside Look at IBM Watson and How It Works. IBM Corporation, Redbooks.

Hillman, S.C., Pretlove, S., Coomarasamy, A., McMullan, D.J., Davison, E.V., Maher, E.R., Kilby, M.D., 2011. Additional information from array comparative genomic hybridization technology over conventional karyotyping in prenatal diagnosis: a systematic review and meta-analysis. Ultrasound Obstet. Gynecol. 37 (1), 6–14.

Hornik, K., Stinchcombe, M., White, H., 1989. Multilayer feedforward networks are universal approximators. Neural Netw. 2 (5), 359–366.

Hu, M., et al., 2017. Decision tree-based maneuver prediction for driver rear-end risk-avoidance behaviors in cut-in scenarios. J. Adv. Transp. 2017.

Ialongo, C., Pieri, M., Bernardini, S., 2017. Artificial neural network for total laboratory automation to improve the management of sample dilution: smart automation for clinical laboratory timeliness. J. Assoc. Lab. Autom. 22 (1), 44–49.

IHGSC, 2004. Finishing the euchromatic sequence of the human genome. Nature 431, 931–945.

Joshi, A.K., 1991. Natural language processing. Science 253 (5025), 1242–1249.

Kelly, J.D., et al., 2017. Crude-Oil Blend Scheduling Optimization of An Industrial-Sized Refinery: A Discrete-Time Benchmark. Foundations of Computer Aided Process Operations, Tucson, AZ.

Kennedy, J., Eberhart, R., 1995. Particle swarm optimization. IEEE International Conference on Neural Networks.

Klir, G., Yuan, B., 1995. Fuzzy Sets and Fuzzy Logic. vol. 4. Prentice hall, New Jersey.

Koch C. How the computer beat the Go master. Sci. Am. http://www.scientificamerican.com/article/how-the-computer-beat-the-go-master/. Published March 19, 2016. (Accessed 8 August 2016).

Koren, O., Ley, R.E., 2016. The human intestinal microbiota and microbiome. In: Yamada's Textbook of Gastroenterology, pp. 617–625.

Kurokawa, K., Itoh, T., Kuwahara, T., Oshima, K., Toh, H., Toyoda, A., Takami, H., Morita, H., Sharma, V.K., Srivastava, T.P., Taylor, T.D., 2007. Comparative metagenomics revealed commonly enriched gene sets in human gut microbiomes. DNA Res. 14 (4), 169–181.

Leventhal, H., Phillips, L.A., Burns, E., 2016. The common-sense model of self-regulation (CSM): a dynamic framework for understanding illness self-management. J. Behav. Med. 39 (6), 935–946.

Li, C.-C., et al., 2017. Personalized individual semantics in computing with words for supporting linguistic group decision making. An application on consensus reaching. Inform. Fusion 33, 29–40.

Li-Pook-Than, J., Snyder, M., 2013. iPOP goes the world: integrated personalized Omics profiling and the road toward improved health care. Chem. Biol. 20 (5), 660–666.

Mac Partháláin, N., Jensen, R., 2015. Fuzzy-rough feature selection using flock of starlings optimisation. Fuzzy Systems (FUZZ-IEEE), 2015 IEEE International Conference on. IEEE.

Mamdani, E.H., Assilian, S., 1975. An experiment in linguistic synthesis with a fuzzy logic controller. Int. J. Man Mach. Stud. 7 (1), 1–13.

Margadant, C., Sonnenberg, A., 2010. Integrin–TGF-β crosstalk in fibrosis, cancer and wound healing. EMBO Rep. 11 (2), 97–105.

McCarthy, J., 1960. Programs with common sense. In: RLE and MIT Computation Centre, pp. 300–307.

Mitnick, K., Simon, W.L., 2011. Ghost in the wires: my adventures as the world's most wanted Hacker. Little, Brown and Company, Boston, MA, ISBN: 978-0-316-03770-9.

Mocanu, D.C., et al., 2017. Estimating 3D trajectories from 2d projections via disjunctive factored four-way conditional restricted boltzmann machines. Pattern Recogn. 69, 325–335.

Modell, B., 2003. Recommendations for introducing genetics services in developing countries. Nat. Rev. Genet. 4 (1), 61–68.

Moravčík, M., et al., 2017. DeepStack: Expert-Level Artificial Intelligence in No-Limit Poker. arXiv preprint arXiv:1701.01724.

Muffato, M., Louis, A., Poisnel, C.E., Crollius, H.R., 2010. Genomicus: a database and a browser to study gene synteny in modern and ancestral genomes. Bioinformatics 26 (8), 1119–1121.

Ni, Z., et al., 2017. Research on fault diagnosis method based on rule base neural network. J. Control Sci. Eng. 2017.

Pavlidis, C., Nebel, J.C., Katsila, T., Patrinos, G.P., 2016. Nutrigenomics 2.0: the need for ongoing and independent evaluation and synthesis of commercial nutrigenomics tests' scientific knowledge base for responsible innovation. OMICS 20 (2), 65–68.

Poldrack, R.A., Wagner, A.D., Prull, M.W., Desmond, J.E., Glover, G.H., Gabrieli, J.D., 1999. Functional specialization for semantic and phonological processing in the left inferior prefrontal cortex. Neuroimage 10 (1), 15–35.

Ra, H.-K., et al., 2017. I am a smart watch, smart enough to know the accuracy of my own heart rate sensor. Proceedings of the 18th International Workshop on Mobile Computing Systems and Applications. ACM.

Rawlinson, K., 2015. Microsoft's Bill Gates Insists AI is a Threat. BBC News 29 January 2015.

S. Richards, N. Aziz, S. Bale, D. Bick, S. Das, J. Gastier-Foster, W. W. Grody, M. Hegde, E. Lyon, E. Spector, K. Voelkerding, and H. L. Rehm; on behalf of the ACMG Laboratory Quality Assurance Committee. Standards and guidelines for the interpretation of sequence variants: A joint consensus recommendation of the American College of Medical Genetics and Genomics and the Association for Molecular Pathology. Gend. Med., Volume 17. Number 5. May 2015, 405-24.

Russell, S.J., Norvig, P., Canny, J.F., Malik, J.M., Edwards, D.D., 2003. Artificial Intelligence: A Modern Approach. vol. 2. Prentice Hall, Upper Saddle River.

Santofimia, M.J., et al., 2017. Hierarchical task network planning with common-sense reasoning for multiple-people behaviour analysis. Expert Syst. Appl. 69, 118–134.

Scappaticci, F.A., Fehrenbacher, L., Cartwright, T., Hainsworth, J.D., Heim, W., Berlin, J., Kabbinavar, F., Novotny, W., Sarkar, S., Hurwitz, H., 2005. Surgical wound healing complications in metastatic colorectal cancer patients treated with bevacizumab. J. Surg. Oncol. 91 (3), 173–180.

Schmidt, A., 2016. Increasing computer literacy with the BBC micro: bit. IEEE Pervasive Comput. 15 (2), 5–7.

Schreiber, A.T., Akkermans, H., Anjewierden, A., Dehoog, R., Shadbolt, N., Vandevelde, W., Wielinga, B., 2000. Knowledge Engineering and Management: The Common KADS Methodology, first ed. The MIT Press, Cambridge, MA, ISBN: 978-0-262-19300-9.

Sevel, L., Letzen, J., Boissoneault, J., O'Shea, A., Robinson, M., Staud, R., 2016. (337) MRI based classification of chronic fatigue, fibromyalgia patients and healthy controls using machine learning algorithms: a comparison study. J. Pain 17 (4), S60.

Simard, P.Y., Steinkraus, D., Platt, J.C., 2003. Best practices for convolutional neural networks applied to visual document analysis. In: ICDAR. vol. 3.

Soheily-Khah, S., Marteau, P.-F., Béchet, N., 2017. Intrusion Detection in Network Systems Through Hybrid Supervised and Unsupervised Mining Process—A Detailed Case Study on the ISCX Benchmark Dataset.

Špačková, O., Straub, D., 2017. Long-term adaption decisions via fully and partially observable Markov decision processes. Sustain. Resilient Infrastruct. 2 (1), 37–58.

Stamate, C., et al., 2017. Deep learning Parkinson's from smartphone data. Pervasive Computing and Communications (PerCom), IEEE International Conference on 2017.

Tejpar, S., Bertagnolli, M., Bosman, F., Lenz, H.J., Garraway, L., Waldman, F., Warren, R., Bild, A., Collins-Brennan, D., Hahn, H., Harkin, D.P., 2010. Prognostic and predictive biomarkers in resected colon cancer: current status and future perspectives for integrating genomics into biomarker discovery. Oncologist 15 (4), 390–404.

Tian, Y.D., 2016. A simple analysis of AlphaGo. Acta Automat. Sin. 42 (5), 671–675.

Tieleman, T., 2008. Training restricted Boltzmann machines using approximations to the likelihood gradient. Proceedings of the 25th International Conference on Machine Learning. ACM.

Vakirlis, N., Sarilar, V., Drillon, G., Fleiss, A., Agier, N., Meyniel, J.P., Blanpain, L., Carbone, A., Devillers, H., Dubois, K., Gillet-Markowska, A., 2016. Reconstruction of ancestral chromosome architecture and gene repertoire reveals principles of genome evolution in a model yeast genus. Genome Res. 26 (7), 918–932.

Victor, D., 2015. Elon Musk and Stephen Hawking among hundreds to urge ban on military robots. New York Times.

Vu, E.L., et al., 2017. A novel electrocardiogram algorithm utilizing ST-segment instability for detection of cardiopulmonary arrest in single ventricle physiology: a retrospective study. Pediatr. Crit. Care Med. 18 (1), 44–53.

Vuokko, R., et al., 2017. Impacts of structuring the electronic health record: results of a systematic literature review from the perspective of secondary use of patient data. Int. J. Med. Inf. 97, 293–303.

Wallace, H., Tilson, H., Carlson, V.P., Valasek, T., 2014. Instrumental roles of governance in accreditation: responsibilities of public health governing entities. J. Public Health Manag. Pract. 20 (1), 61–63.

Waller, M.A., Fawcett, S.E., 2013. Click here for a data scientist: big data, predictive analytics, and theory development in the era of a maker movement supply chain. J. Bus. Logist. 34 (4), 249–252.

Wang, F.-Y., et al., 2016a. Where does AlphaGo go: from Church-Turing thesis to AlphaGo thesis and beyond. IEEE/CAA J. Autom. Sin. 3 (2), 113–120.

Wang, X., Guo, H., Wang, J., Lei, T., Liu, T., Wang, Z., Li, Y., Lee, T.H., Li, J., Tang, H., Jin, D., 2016b. Comparative genomic de-convolution of the cotton genome revealed a decaploid ancestor and widespread chromosomal fractionation. New Phytol. 209 (3), 1252–1263.

Watt, S.M., 2006. Making Computer Algebra More Symbolic (Invited) (PDF). Proc. Transgressive Computing 2006: A conference in honor or Jean Della Dora, (TC 2006), pp. 43–49.

Wayne, P.A., 2005. CLSI Protection of Laboratory Workers From Occupationally Acquired Infections; Approved Guideline—Third Edition. CLSI document M29 A3. Clinical and Laboratory Standards Institute.

Wu, M.C., Lee, S., Cai, T., Li, Y., Boehnke, M., Lin, X., 2011. Rare-variant association testing for sequencing data with the sequence kernel association test. Am. J. Hum. Genet. 89 (1), 82–93.

Yampolskiy, R.V., 2016. Artificial Intelligence Safety and Cybersecurity: A Timeline of AI Failures. https://arxiv.org/abs/1610.07997.

Yu, P., Haskell, W.B., Xu, H., 2017. Approximate Value Iteration for Risk-Aware Markov Decision Processes. arXiv preprint arXiv:1701.01290.

Zadeh, L.A., 1997. Toward a theory of fuzzy information granulation and its centrality in human reasoning and fuzzy logic. Fuzzy Sets Syst. 90 (2), 111–127.

Zayed, H., 2016a. The arab genome: health and wealth. Gene 592 (2), 239–243.

Zayed, H., 2016b. The Qatar genome project: translation of whole-genome sequencing into clinical practice. Int. J. Clin. Pract. 70 (10), 832–834.

Zhao, B., et al., 2017. A genome-wide association study to identify single-nucleotide polymorphisms for acute kidney injury. Am. J. Respir. Crit. Care Med. 195 (4), 482–490.

Ziegler, T., 2017. GITCoP: A Machine Learning Based Approach to Predicting Merge Conflicts from Repository Metadata. Diss. University of Passau.

Zmora, N., Zeevi, D., Korem, T., Segal, E., Elinav, E., 2016. Taking it personally: personalized utilization of the human microbiome in health and disease. Cell Host Microbe 19 (1), 12–20.

Zweigenbaum, P., Lavergne, T., 2016. LIMSI ICD10 coding experiments on CépiDC death certificate statements. CLEF.

Further Reading

Lakshika, E., Barlow, M., 2017. On deriving a relationship between complexity and fidelity in rule based multi-agent systems. In: Intelligent and Evolutionary Systems. The 20th Asia Pacific Symposium, IES 2016, Canberra, Australia, November 2016, Proceedings. Springer International Publishing.

Mair, V.H., 2007. The Art of War: Sun Zi's Military Methods. Columbia University Press, New York, ISBN: 978-0-231-13382-1.

R. E. Nakhleh, Consensus statement on effective communication of urgent diagnoses and significant, unexpected diagnoses in surgical pathology and cytopathology from the college of American pathologists and association of directors of anatomic and surgical pathology. Arch. Pathol. Lab. Med.—Vol 136, February 2012, 148–154.

Patrinos, G.P., Baker, D.J., Al-Mulla, F., Vasiliou, V., Cooper, D.N., 2013. Genetic tests obtainable through pharmacies: the good, the bad, and the ugly. Hum. Genomics 7, 17.

Wang, F.-Y., 2016. Complexity and intelligence: from Church-Turning thesis to AlphaGo thesis and beyonds. J. Command Contr. 2 (1), 1–4.

CHAPTER 12

Genomics England: The Future of Genomic Medical Diagnosis: Governmental Scale Clinical Sequencing and Potential Walled-Garden Impact on Global Data Sharing

Darrol J. Baker

The Golden Helix Foundation, London, United Kingdom

12.1 INTRODUCTION

Did you know the youngest person to upload and analyse a whole human genomes was a 4 year old girl from Nottingham, England... She said "30x wasn't good!"

(Analysis was on a breast cancer tumor normal pair)

In this chapter, the key features for genomic data integration within a healthcare system will be examined, using the approaches made by Genomics England and the elucidation of clinical findings by third-party providers. We will look into the approach made by Genomics England and the specific use of artificial intelligence and machine learning promoting health by collaborating companies, well-being, and fitness within a global real-time clinical setting. The particular emphasis will be given to specific clinical real-world examples and how this is likely to change the shape of clinical diagnostics medicine.

We consider that clinical genomic data findings are one of the most valuable tradable resources a country may have in future healthcare economics (Patrinos et al., 2013). The UK government has taken a ground-breaking approach by contracting out national genomic goals within the specific framework of the National Health Service England (NHS England). NHS England is pioneering a ground-breaking project to build genomics healthcare understanding aimed at future genome-stratified treatments, findings, and critical patient care in conjunction with the immediate help of recruited patient cohorts within rare diseases, metabolic, and cancer-related illnesses.

269

Human Genome Informatics. https://doi.org/10.1016/B978-0-12-809414-3.00012-7

The correct selection of key optimal genetic tests for a patient and finding some transparency in clinical decision making is hard within a complex healthcare network. There are many limitations within the diagnostic space and particular acute cases involving rare disease patients. Many factors contribute to optimal testing for clinical patient diagnosis and using proven genetic strategies to tackle these issues (Mvundura et al., 2010). Within the Magrathea framework, they have proven genetic tools and computation power that meets tough challenges discussed within the chapter on artificial intelligence in this book. Clinicians have a vast selection of tests for optimal patient understanding and UK's National Health Service; indeed, clinical healthcare service around the globe has a massive task to integrate clinical genomic medicine throughout their networks of hospitals. Over 4500 Clinical Laboratory Improvement Amendments (CLIA) and College of American Pathologists (CAP) genetic diagnostic tests can be sourced online and over 20,000 none validated or for research purposes only tests covering a variety of needs (unpublished), it is clear that a positive patient outcome is key (Lehman et al., 1998). And, although this may seem a great number, it is a mere fraction of the tests that are coming to the market and within the personalized medicine (PM) industry.

Within most healthcare services, human error is a key dichotomy to patient success. To resolve issues of losing patient data and the mismanagement of information, a mind-set change is needed; from the authors' point of view and understanding, it is a problem prevalent in all healthcare systems across the globe. In a recent ongoing case within the NHS, an incident happened, where at least 1700 patients may have been harmed by a "colossal" blunder which meant thousands of patient records were left to pile up in a warehouse (Morse, 2017). The number at risk is likely to rise as only two thirds of the 700,000 notes found had been checked (Morse, 2017). Moreover, cancer test results and child protection notes were among the documents that were missing in England (Morse, 2017).

In truth, numerous sequencing and specialized technologies can be outsourced, as discussed in previous chapters, and this can mitigate risk or indeed safe money on capital equipment purchases and, in some cases, allow for very small clinics running very large projects—relative to today's understanding of large genome projects. One would expect 10–100,000 genome projects to be the norm as prices drop and technology evolves. Many labs employ whole genome sequencing (WGS) and whole exome sequencing (WES), targeted panels, and traditional Sanger methods. Some clinics also need to consider mutation events and employ choices for copy number variation (CNV; see Chapter 2), structural variants (SV) detection for WGS, Comparative genomic hybridization (CGH), arrays and Multiplex ligation-dependent probe amplification (MLPA) as well as the standard banks of usual bloods, tissue, and culturing test. In some instances, for mitochondrial defects (mitogenomics), repeat expansions and epigenetics defects within a given diagnostic case study. The analysis of this data is a key for positive patient outcomes or indeed the ability to manage long-term patient care. These clinical sourcing services can help in easing the burden on a given healthcare unit; however, it usually comes at a high economic price to the given healthcare trust and, therefore, is limited to extreme cases or the wealthy patient. For these genomic cases, in particular rare diseases or undiagnosed cases, the British government needed a comprehensive innovative economic solution.

Finding a rare disease variant has been like finding a "needle in a haystack"; we can go one step further. It is like finding a needle in a haystack in universe of hay-stacked fields. Moreover, in most cases, the "needle" and "Haystacks" are nonrepresentative of your population. We are at such an early stage of genomic medicine that the true understanding will come with time as study volumes increase to population-sized events and technology pulls all data into an automated real-time Artificial Intelligence environment. We have seen in numerous chapters in this book that this comes down to the integration of the people, systems (political and others), technology, the ability to build leaning tools, and complex system intelligence for clinical genomic and diagnostics interpretation of findings. This is how Genomics England is building a clinical genomic integrated framework across the United Kingdom and providing a framework model copied by other countries across the globe.

12.2 UNDERSTANDING THE GENOMICS ENGLAND APPROACH TO CLINICAL GENOMIC DISCOVERY

Most medical clinicians see the Genomics England (GEL; https://www.genomicsengland.co.uk), 100,000 genomes project methodology for genome sequencing integration, as the future model for all national healthcare systems approach to clinical medical and diagnostic understanding. Due to the unique nature of the integrated healthcare system within the National Health Service in the United Kingdom, Genomics England's approach works extremely well. The UK government has a true game changing approach that combines funding models, patient buy-in, and nationwide ground routes understanding of the importance of genomics to future health needs. Most of this is still trapped within traditional in-house sequencing approaches and Genomics England has made the outsourcing model a standard process going forward. Outsourcing all the projects elements was Genomics England's plan for the beginning to fulfill the requirements for National Health Service needs and the UK governmental funders. These principal components were sequencing, interpretations, and then the feedback of findings back into the health service. The National Health Service fundamentally is at the end and indeed entry point

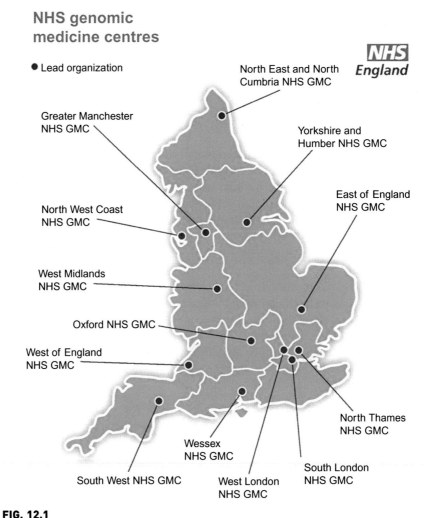

FIG. 12.1

13 Genomic Medicine Centres (GMCs) have been established by National Health Service England. These centers will lead the way in delivering the 100,000 Genomes Project. The GMCs have a track-record of providing excellence in genomic services. Eligible patients who are interested in getting involved will be referred to GMCs by their clinicians.

for clinical interpretation, understanding, and final clinical actions, albeit at a very early stage of communication with the regional National Health Service sequencing hubs (Fig. 12.1).

What this model allows Genomics England is ground-breaking clinical genomic volume due to their size. Genomics England is in fact a very lean organization and one could even argue mere paper pushes for the whole project.

Genomics England is the hub of the whole effort in the 100,000 genomes project, in effect simply controlling the contracts for all the project requirements. The first contract to be determined was sequencing. Sequencing was based on a list of vendors doing an evaluation competition, and this was a tendered competition and evaluation between all sequencing providers that applied for vendor status. To all that applied for the data, samples were sent to the providers and discs of data analysis and reporting were sent back to Genomics England for evaluation. In the end, two sequencing providers were chosen for commercial contract negotiations and, from this stage, Illumina was chosen as the final preferred bidder (Genomics England, 2016a,b,c,d). Part of the contract with Illumina states that everything needs to happen in England, which means Illumina had to make inwards investment to support this programme in England. A building is being built in conjunction with the Wellcome Trust on the Hinxton Genome Campus in Cambridge. Illumina is to decipher the genetic codes of 100,000 people in the United Kingdom to investigate the genetics of cancer and rare genetic diseases. The work will be carried out in the new state-of-the-art £27 million sequencing center that is being funded and built by the Wellcome Trust on its Genome Campus in Hinxton. In addition, the Medical Research Council (MRC) is providing £24 million to ensure that the resulting genomic data donated by participants can be properly collected, stored, and analyzed by scientists (Wynne 2014; Fig. 12.2). Illumina will be renting the building with their sequencing equipment and staffing in place to supply this contact.

Illumina also follows Genomics England's sequencing pipeline guidelines, including software, algorithms, and process of raw data into BAM, the compressed binary version of a SAM (GitHub, 2016) file that is used to represent aligned sequences, and Variant Call Format (VCF) (GitHub, 2015) specifies the format of a text file used in bioinformatics for storing gene sequence variations formats for clinical interpretations.

The next stage of the project was the integration of the sequencing information with the National Health Service in England or NHS England. NHS England contacted all the hospitals in the National Health Service (NHS) and, in 2014, NHS England canvased selected hospitals to apply to become Genome Medical Centres; there are now 13 regional centers giving geographical coverage across England. The regional centers collect samples from hospitals within their regional zone, giving an actual total number of 80 hospitals covering the whole of England for sample collection. Each center has a lead hospital being the delivery partner for the local hospitals. Of course, this is a major operation and brings into scope all aspects of medical diagnostics within the NHS: clinically certified labs, Dry/wet lab testing employed by the various agencies within the NHS diploid to make sure that sample collections, extraction, and generation of DNA are of the quality needed for whole genome sequencing. These regional centers are monitored and audited by NHS England every

FIG. 12.2
Visualization of the new £27 million Sequencing center (left), hosted on the Wellcome Trust Genome Campus. It will house the sequencing facilities of both the Genomics England's UK 100,000 Genomes Project and the Sanger Institute.

3 months to monitor performance and generally oversee the complete process and share latest findings and improvements to processes as and when they happen. Included in this is a range of trainings: both classroom and online programmes covering a broad range of understandings. A total of over 2000 researchers, clinicians, analysts, and trainees were involved in 88 applications to become part of the Clinical Interpretation Partnership (GeCIP) (Genomics England, 2016a). Over 28 teams or "domains" were established for clinical and research experts to work in as part of its GeCIP. The teams will focus not only on a vast range of disease areas, but also analytical and social sciences. Indeed, this partnership brings together over 4000 UK clinicians and scientists as well as over 500 international collaborators at the forefront of Genomic Medicine.

For Genomics England, this results in a lean and clean genome sample pipeline supply, allowing for total ownerships from patient to sample, hospital to lab, and sequence to analysis. One further issue Genomics England had was a place to hold all genomics data. Genomics England is not building anything themselves; they are looking at cloud solutions, but cloud solutions do come with data security issues and indeed the geographical location of the cloud. Luckily within the United Kingdom, the government sets up a framework where they can offer commercial companies private cloud services held within a governmentally owned and operated data center (Department of International Trade, 2016). This is the model adopted by Genomics England leading to a solution, which is of a scalable infrastructure, commercially competitive, and under governmental control: you know where the buildings, hardware, and data are with complete control of security for both onsite and online with true military grade security. This allows Genomics England to benefit from latest practices in hardware and software cloud-based data access and manipulation and computing; in fact, Genomics England only own the discs that the genotype data is housed on and rent everything else. In this system, you can maintain a continued supply of DNA samples that go to Illumina; the BAM and VCF files are called by under the given pipeline. Under the Genomics England contract, this data is stored in the data center for further phenotypic analysis and third-party user access.

The next stage of the pipeline was for Genomics England to look at the phenotype data. For Genomics England phenotype, data is needed for the interpretation of the genome and further analysis for the treatment of patients as part of the strategic contract for the various programmes within the National Health Service. The contract with the National Health Service medical centers meant that they need to collect structured phenotype data from patients based on specific data models for each given disease type (Caulfield et al., 2015). At present, this is 170 rare disease area types in the data models and further eligibility criteria are collected based on genotype/phenotype ontologies giving many layers and granularity to the data collected. The data is stored in an object relational database management system (ORDBMS) with an interface for versioning of the model. Moreover, the front-end has a social media networking aspect to it, allowing clinicians and researchers who are part of this project or have interest in a given patient case outcome in England and the wider United Kingdom the ability to discuss in detail the related aspects of the dataset: such as, its usefulness of the interpretations for a given dataset and for further steps or clinical understanding from the 4500 user base.

At present, researchers are looking for more phenotype data and clinicians as little as possible; the balance is hard to find, so the framework works by trying to strike a balance between these various groups. This dynamic tension exists in most clinical patient cases around the globe; collective understanding seems to help resolve and share in the overall understanding and delivery of key novel findings.

There is a specific data model with versioning that also allows you to write a standardized Xmachine learning file (XML) format (McDonagh, 2015) (Bernardini et al., 2003). Where the XML file is then used by the Genomics England interface, giving the regional centers three options: enter the data manually via the Genomics England graphical user interface (GUI), learn XML from their own electronic healthcare record systems (EHRS), and develop programmes accordingly or utilize the local informatists for their XML needs. This helps support all users and is a highly progressive move allowing hospitals to get their IT integrated with quarterly visits from inspection teams and general information sharing from the participating groups. Best process and practices are rapidly shared and adopted very quickly indeed across the network of clinicians and researchers. The data the hospitals can extract can reduce the burden on disease decision making and indeed are contracted and paid by Genomics England based on the samples they process based on a basic phenotype minimal datasets. There are countless other datasets that are collected in the National Health Service in England such as statistical and registry information. Genomics England does have agreements to have access to consent for patients to access their entire longitudinal medical history for correlational research beyond the initial sample collection. Therefore, everything will be in the data center that is associated with the patient (Genomics England, 2016a,b,c,d).

Genomics England has taken a ground-breaking and controversial approach not to build a traditional genomics pipeline. Instead, Genomics England decided to run another competition to build their interpretation pipeline. Genomics England used the pilot sequence data generated from the previous competition and sent this dataset to around 30 companies and academics groups, with the simple instructions to interpret these genomes to the best of your ability and work out what is wrong with the given patient and, from this simple first round, Genomics England generated a shortlist of 10 groups that were then shortened by subsequent rounds to a short list of Congenica, Omicia, Nanthealth, and WuxiNextcode for the initial contract for the first 8000 genomes to proceed. Utilizing a similar model previously selected by Genomics England, if you want to run your interpretation algorithms on this data, you will need to do this inside the Genomics England data center. With the strict rule that no data will come out of this data center, it did not matter if it is for research or clinical interpretation, everything has to run inside the Genomics England Data Centre. The companies' software was installed and was running generating clinical reports; these clinical reports were then sent back to the clinical genetics departments and medical centers for use and clinical implementation. It is this crucial stage of analysis and interpretation that is revolutionary with the large genome projects and could lead to internalization of findings.

The government has established a grant programme channeled through Genomics England to stimulate new activity in the clinical genomes space for up and coming companies and start-ups. Genomics England has funded several companies and is monitoring an effort to stimulate the genomics industry in the United Kingdom and Northern Ireland.

The feedback that Genomics England gives to the NHS England is of precise narrow scope. Genomics England is not interpreting the genome to include everything like most clinical reporting done globally; genome interpretation is resolved on the basis of the specific point of view for a given condition from the patient illness. If the patient has a specific cancer or rare disease, GEL finds what is of relevance to that particular cancer or rare disease, with current knowledge. For additional or secondary findings, and for clear variants with a very high penetrance with action-ability, and hence these lists are not as detailed, similar to reporting from commercial companies and or American 3–5 page clinical reporting's. Genomics England will only send the secondary findings information to the patient if the patient has consented to receive this information: an optional consent for the patient to receive these secondary findings or not. With a longitudinal approach to data gathering, Genomics England expects the secondary lists to increase rapidly as expertise in these given areas increases, but at this early stage the data increases are relatively small.

The current status of the Genomics England's sequencing is around 50,500 of the 100,000 whole human genomes (accessed February 2018), with a monthly increase of around 500 genomes. With the new facility opening in Hinxton, Cambridge, this per month volume will catch up with the expected 100,000 genomes in 5 years target; at current rates, this will far exceed this timeline. However, it is expected that UK Governmental funding for the GEL project will continue for many years. The genomes themselves are indeed being transferred using a dedicated "fat" fiber line from the current sequencing operation and the data center. The companies that have software installed within the data center are Congenica, Omicia, and WuxiNextcode for the first phase of clinical interpretation: three companies, three softwares, three interfaces, and one might say three sets of administration headaches. Hence, over time this is expected to be weeded down or indeed replaced. Like with previous bake-offs, Genomics England will compare and contrast performances, where some samples will go to all three in a view to understand who is best, and Genomics England will test all three companies with staged data to test each to find obvious variants and other secondary findings, but also to see if they can find more interesting patterns and understandings. And, the company has the ability to interpret the genome in an intelligent manner and more than the usual traditional approach of *Top Ten hits… with laughable clinical (are you really a clinician) interpretation* style listings. Indeed, some companies have expertise in a given area; so like all clinicians we are looking for the right match to positive result so that one would expect an evolving market place. Moreover, the importance is simply a findings base for additional clinical understanding. At the bare minimum, it is a 50% diagnostic rate and it will be intriguing to see who finally wins this battle. Or indeed, we can see a new company is emerging, as we will undoubtedly see as the tech advances.

Genomics England selected Inuvika as their technology partner for the 100,000 genomes projects to deliver secure virtual desktop environment for the 100,000 Genomes Project. Genomics England selected Inuvika's Open Virtual Desktop Enterprise (OVD) to deliver secure access to the research environment for the 100,000 Genomes Project. Consisting of a Windows virtual desktop, data center hosted applications and associated de-identified datasets; OVD publishes the users' environment, so it can be securely accessed from a standard HTmachine learning-enabled web browser from any location (Genomics England, 2016c). Genomics England is partnering with GenomOncology LLC (GO) to utilize the GO Knowledge Management System (GO KMS) as a tool for clinical reporting enablement. Genomics England will integrate the GO KMS as a key content driver to augment clinical reporting in the 100,000 Genomes Project's cancer programme, coupling Genomics England curated database with the GO KMS's data for a comprehensive clinical report comprising the most relevant drugs, prognoses, and clinical trials. Genomics England also adopts Edico Genome's DRAGEN Bio-IT Platform to increase accuracy and consistency of next-generation sequencing analysis. At the 2018 annual J.P. Morgan Healthcare Conference, Edico Genome and Genomics England announced a new partnership to strengthen the accuracy and consistency of next-generation sequencing data analysis in Genomics England's Rare Disease Pilot. (Genomics England, 2016a,b,c,d).

Due to the simple nature of these reports and ease of predictions, another criterion for these companies to consider is the ease of use from NHS England to handle. The National Health Service users will now face three differing interfaces with three differing outputs of clinical reports. In that sense, Genomics England has created a standard description based on a gene panel, so they will be curating the data via a new crowd sourcing (McDonagh, 2015) activity. To help the community to decide which of the outputs is good enough evidence based on literature. In terms of which genes are associated with which condition, and this is being used by the companies to label their outputs (Fig. 12.3), giving a common framing for the national health service employees to be familiar with, given they may switch to a national health service trust as a preference for one vendor or indeed a provider that later may not be selected for use, a common frame of reference is still in place and retraining on the software and not data legacy is the issue. Moreover, if and when other companies come onboard at a later date, they will need to conform to Genomics England and National Health Service guidelines standards. The author analogizes this to

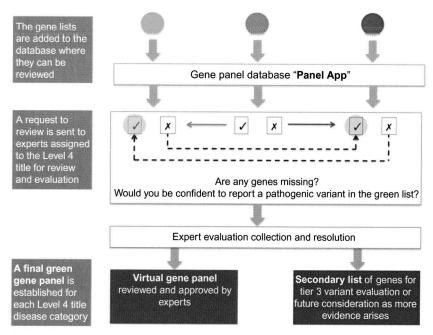

FIG. 12.3

Genomics England has developed a unique resource, the "PanelApp." It is a new crowdsourcing tool for the scientific community, allowing knowledge of rare disease genetics to be shared and evaluated. This will create comprehensive evidence-based gene panels for rare diseases. The resource is publically available for anyone who would like to view and download the gene panels. Experts can register as a reviewer to make evaluations of the gene panels.

the UK's railway gauging systems and its subsequent future roles in many aspects of global travel and transportation, which "gauging" the 4 ft. 8 in gauge had been used by imperial Roman army war chariot (or before) and relates to the approximate width of two yoked horses. And like the horse and rail tracks of the past, one can imagine many new global projects will use the tracks grounded by Genomics England as templates for their national-based clinical genome projects, gene panel development, and indeed adoption as a possible global standard. The unique size and scale of the UK's NHS and GEL excellent team do give the UK government a very strong negotiation hand.

The panel app (Fig. 12.3) for curating the gene lists is an open platform and all globally can access this app and contribute to this effort via the public login (McDonagh, 2015). There are many conditions that are needed for sign up and, depending on this eligibility of the gene that is flagged by the given gene reviewer, decisions need to be taken if the gene is definitely as prevalent for a given condition or not. In that sense, going forward, this will be a powerful diagnostics crowd-based tool for genotype/phenotype associations and diagnostic reporting within the Genomic England framework.

There are two activities that involve clinicians and researchers working together; although the whole idea of this project is for the treatment of patients within National Health Service England and the wider United Kingdom, it's all done under a novel research protocol. This research protocol is available for download from the Genomics England website. This details what ethical approval Genomics England has for genome analysis within the remit of the Genomics England contact. Fundamentally, Genomics England is returning results to the health system for clinical care and consent involves which data is available for research purposes in the same environment, which means the data does not get distributed like most genomic findings and patients can be contacted; the prospects here are vast both in clinical and commercial settings. Genomics England has taken an additional omics sample and there will be a revision of samples and analysis taken, so this is a moving area and the data and analysis are ongoing. This has a far different scope to most project looking at data for results in a single study or until funding of the project ends. This is a living, breathing, and ongoing process (Fig. 12.4) that can and will have dramatic patient impact.

Genomics England aims to extend this frame of results to include research. Genomics England has the ability to search the "Genomics England Clinical Interpretation Partnership" (National Health Service England, 2016), which is currently a network of over 2500 principal investigators (PIs). The PIs get access to the datasets in exactly the same way as the National Health Service workers do; they cannot take data out of the environment and they have to work within the Genomics England framework. This does require a change in the usual way PIs work with genomics data. Researchers traditionally are used to access data globally and download at will doing analysis at their own IT infrastructure; this is not the case within the Genomics England framework and they cannot download results anymore. Genomics England believes this is the way of the future for health data and genome analysis and, to some extent, has been so within the corporate and commercial world of genome medicine from the start. Health data stays in one place and the health system becomes able to support research efforts and activities within the IT framework of the given health body, country of countrywide consortia, and blocks such as the European Union (EU). The important part of this framework is the quality of the datasets and the protection of the nation's genomic data, patient records, and findings going forward.

The partnership has access to analysis of data and the biorepository sample allowing the researchers to apply or write for grants to various bodies to do additional work and additional data collections. It is important to note that Genomics England is only contracted for whole genome sequencing for this time, but

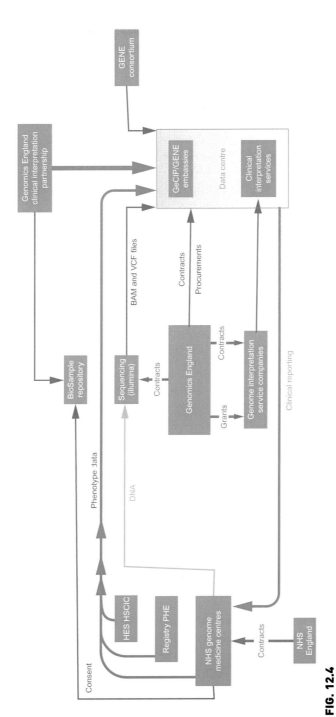

FIG. 12.4

An overview of the Genomics England process. Genomics England sits as the center hub controlling the flow of information through contract to third-party providers that run all operations and analysis within the local Genomics England framework, with no data be accessed for external use.

for further researches, it is possible to obtain grants for the biorepository samples for RNA or other epigenetics activities that is not in the remit of the Genomics England contract. However, all data and results must be within the framework of the Genomics England operation, where, again, one will expect the UK government and indeed the department of trade and investments to make significant contributions long after the initial project funding cycle ends.

There is also a set of companies in a precompetitive consortium called the Gene Consortium. They have a limited access to a limited datasets in a similar fashion and constraints as researchers; you cannot take anything away accept summary level information which is not sensitive and this does need to go through an "airlock" process to ensure that sensitive information is completely firewalled.

The consortium and clinical interpretation partnerships were advertised in 2015; the idea of the consortium was to have a domain of researchers for every known disease area, along with some crosscutting domain around analytics and generic analytics. All these groups were and are able to see the data, and even now there have been publication based on disease subsets and domain subsets for the disease areas (UK Government, 2015). They will be able to work together like similar consortia, such as Encyclopaedia of DNA Elements (ENCODE) (The ENCODE Project Consortium, 2007) or the 1000 Human Genome Consortium (The 1000 Genomes Project Consortium, 2010).

The goals of this consortium are for people, PIs, and groups taking part in this consortium. They think they will be doing basic academic research, when actually Genomics England wants them to interpret those 50 percenters that didn't get interpreted by the companies, with the simple idea that the automated processing of the genomes will be via the contracted organization. What about the areas that cannot be interpreted? They have access to all the data but can they help?; what happened when they do and use the known phenotype data to aid the interpretation of the missing areas and indeed benefit from any secondary analysis and papers that are published from this resulting new data? These all do have to be within the United Kingdom, under the established framework rules set up by the UK government for Genomics England. All the rights relating to discovery and data rights are held by Genomics England and, by direction, the National Health Service England and the UK governments Department of Health (UK Government, 2015). It is clear that to create a level playing field for anyone else can make secondary claims on analysis on the datasets and IP. However, the raw primary analysis data is held within the framework of the United Kingdom for everyone to use. In that sense, there is great space for everyone to work together to aid research and build businesses on the back end of this effort.

As a quick aside, it is important to note that although some may be horrified by the above paragraph, it is for the best of both worlds. Outside of all the great data coming from this project: genomic, drug information, statistically data, population dynamics, etc., one could spend many lifetimes on each of these areas. The fundamental key is the economics of new drug discovery within a given population. We expect clinicians and patients to find answers to difficult diagnosis and planning. It also allows commercial companies to take part, and guess what, if you find something, so long as you develop it in the United Kingdom, you can do so. It is an intriguing model one will see adopted as standard, particularly within developing and emerging economies.

At present, there are several domains Genomics England has focused on and indeed additional domains are being added. Genomics have initiated a set of domains with cancer and rare disease cross-cutting, ethics, and law. There is also variation feedback, pharmacogenomics, and health economics and epiomics. Genomics England plans to create a single environment for data held on the health service side, with an App-spaced environment where you can run Apps that deal with interpretation of the data and you can use Apps for both clinical and research purposes using a single framework model and scale the IT infrastructure based on a private or virtual private cloud; Genomics England allows companies to do this at present. The important thing here is that this is a service and you cannot take the data away and cannot download the raw data from Genomics England and utilize this on your own servers. It is a new way; a modern way of dealing with a country's natural datasets and a model that will be adopted by the wider communities going forward on their own large scale analysis projects, without the fear of losing control of your own countries' datasets to more advanced frameworks or indeed countries that have a more mature setup. This immensely helps developing countries where resources are in shortage and need is great; to aid working with partners and third-party companies to invest infrastructure in the country and for the country to benefit longitudinally in the investments made. More impressive is the ability of such small countries with rare genotypes and phenotypes to capitalize on these natural resources and indeed request back data already being used, taken, or stolen from said environments previously by researchers, biotech, and/or pharma.

This is indeed a radical change from the traditional approaches to genomics and the sharing of information within the genomic communities. Genomics England believes that the idea of data sharing where you cannot take the data away is a new process and the wide-ranging user base will accommodate this new model and adjust their work accordingly. This approach does make initially working internationally difficult, where each country has an "Information locked in a walled garden." It is for each country to deal with the retaining and indeed sharing of this information within each region. The question of how one does share information within a given shared dataset is still prevalent, even more so when information cannot leave your data center or country. Genomics England is walling all health service information that is highly sensitive and, therefore, this can only be the solution to sharing issue at present; yet, it is still trying to find and share meaningful data from patient datasets for internal and third-party resources. However, this is the only meaningful way at present to deal with healthcare genomic information in a global setting. One thing is very clear, the training of the internal base needs to be consistent and up-to-date and healthcare workers in general should have a systematic educational learning programme.

If we do talk about walled gardens in every country, then most countries are not going to allow healthcare records or any data to escape or leave their country at the individual level. But there are several groups and organizations looking into the issue of sharing information; one of them is The Global Alliance for Genomics and Health (Global Alliance) that is dealing with this problem (GA4GH, 2016; Fig. 12.5). The organization consists of over 40 countries signed up to bring together over 400 leading institutions working in healthcare, research, disease advocacy, life science, and information technology. The partners in the Global Alliance are working together to create a common framework of

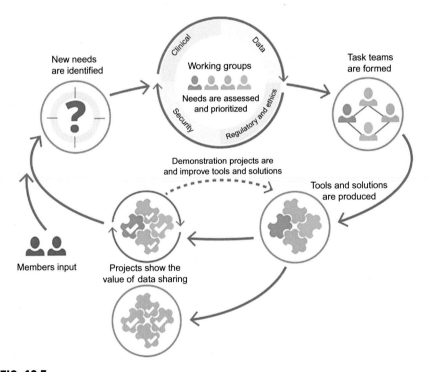

FIG. 12.5

GA4GH demonstration Projects use and improve on Work Products and demonstrate the value of data sharing through practical application of our work. These, and other large-scale projects, have broad impact on data sharing stakeholders, reach a diverse community, and may require many resources.

harmonized approaches to enable the responsible, voluntary, and secure sharing of genomic and clinical data. It is interesting to note that the Global Alliance has a complete range of sponsors from academic, instates, biotech, pharma, and governmental funding, which does show a willingness to share information and not close down genomic communication some believe inevitable. This enables the groups to share the information without breaking the privacy of the individual.

There is also a massive nationwide educational programme within the Genomics England framework and contract to nationally distribute a new set of courses both at lower and higher degree status, lectures, and materials. There is also scope for most of these efforts to be moved online; the idea is everyone within the healthcare service will be affected by the Genomics England's effort and, therefore, there needs to be training programmes and courses in place to cope with the demand at all levels of understanding, both in the technical and nontechnical aspects of the wider contract and study. Within NHS England, all will need to know some aspect of genomics and, therefore, courses are being set up at the appropriate level. Indeed, university programmes will need to evolve by updating course material to cover this undertaking. Again, GEL is leading the educational initiative and adoption of these programmes is slowly increasing on the global clinical stage.

The programme is an engine for transformation of the healthcare systems and NHS England towards clinical genome medicine. It's an engine for what is happening inside the NHS England hospitals with respect to data capture and the standardization of the information in all its facets. There are many areas and pressure points within the NHS, such as the issue of there being many patient recording charts at the end of the patient's bed from the various hospitals and no standardized form (Coiera, 2015). Hence, Genomics England is helping to galvanize a response to these long-standing standardization issues; these little niche issues are being addressed and are all coming together within the Genomics England framework project. Therefore, Genomics England can track best practices within a given NHS Trust Hospital and, via the social media aspect of the project, presents the most efficient and effective way to deal with a given problem or indeed flag a request of need for a given issue, resulting in a lean process across all aspects of clinical understanding: better results, quicker results, and better patient outcomes, while improving the worklife balance.

Finally, there is the idea that the data can be used in two ways. Genomics England currently designed for specific interpretation aimed at a given purpose. However, once you have the structure in place, there is nothing now stopping other organizations, such as the genome medicine centers generating different apps in this environment, such as pharmacogenomics applications or other such applications linked to the EHRS. If your EHRS informs you of a genome relating to a patient, you can then query that genome with apps and tools to determine if there will be adverse reactions to a given set of drugs or, indeed, if the amount needed is greater or less than the recommended dose amount (van Schaik and van Gelder, 2016). The possibilities are endless for patient care. One could even go further to suggest for certain patients; this is the first-time physicians can actually go to perfectly healthy individuals and preemptively strike at a devastating condition before symptoms present in the patient, in some instances many decades before so (Davies, 2015). It is important to note that this data is not held in the normal way of recent large cohort studies; this is open for anyone with the given access rights within the healthcare system, thereby aiding the pharmacogenetics decision making. Data are tracked and changes noted with the metadata of the file, to ensure data security and integrity. The unique makeup of the NHS within the United Kingdom makes this approach possible, though state-of-the-art in terms of genomics, and is more than 80 years in the making since the formation of the national health service on July 5th, 1948 by the Aneurin Bevan, a Welsh Labour Party politician who was the Minister for Health in the postwar Attlee government from 1945 to 1951 (National Health Service Choices, 2016).

On behalf of cancer patients, Genomics England collects tumor normal pairs and trio studies. These samples are sequenced; indeed, the long-term data and analysis from this one small aspect of the project will lead to a new era of cancer patient treatments. Moreover, the integrations of world leading cancer charities, already closely linked to this project via numerous cohort studies from previous years, provide the clinical links to NHS patients and better understanding of cancer therapeutics across the board; learnings not just in genomics, in clinical care and all aspect of patient handling. Sharing important knowledge gained from a patient outcome helps downstream patient care. The idea is to retain this understanding with the framework. This is being taken to another level with the Magrathea framework, which applies artificial intelligence, complex systems, and machine learning mathematics to aid diagnostics in the areas where cells are growing fast or indeed time is limited. The impact of this work for the future of cancer diagnosis and patient treatment will be ground-breaking. The sequencing of cancer patients at GEL for normal tissue is at the standard $30\times$ coverage and $75\times$ for the tumor sample; therefore, giving a great scale of knowledge and confidence for short-term predications and indeed for a wealth of future total patient understanding in many aspects from initial patient contact through longitudinal nature of the condition and outcomes research. And indeed, Genomics England is looking to push the sequencing in this space to $110\times$ coverage, such is the importance placed on this area of research and need for detailed understanding. The cancer programme at Genomics England is taking longer, as, one can imagine, configuring sample handling in hospitals and standardizing for rare diseases are hard enough; however, reconfiguring pathology services in terms of tumor candidates in theater is a much more difficult undertaking. FFPE is cleared that this is not going to be good enough; Genomics England has tried many pilots around the country and is needing to deal with fresh frozen samples. Therefore, the change management processes and programmes needed in theater are great and a huge undertaking. This is indeed happening, but slower than Genomics England originally expected. The focus is on the best possible practice for the best possible data, leading to the best possible patient outcomes in the end.

Genomics England will integrate the GO KMS as a key content driver to augment clinical reporting in the 100,000 Genomes Project's cancer programme, coupling Genomics England curated database with the GO KMS's data for a comprehensive clinical report comprising the most relevant drugs, prognoses, and clinical trials. The GO KMS enables Precision Medicine by allowing users to aggregate and analyze biomarker-based data within a "genomics-aware" framework that includes a diverse set of annotations including genes, pathways, drugs, alterations, transcripts, and a disease ontology. The platform leverages a large number of existing data sources including FDA, NCCN, and ASCO guidelines as well as providing exclusive API access to the expertly curated data of My Cancer Genome. In addition, the GO KMS is designed to empower researchers and clinicians alike to build and maintain their own curated knowledge repositories. GO and Genomics England will work through an initial implementation phase that will focus on extending the GO KMS to include NICE Guidelines and UK-specific clinical trials, as well as a variety of other enhancements to support clinical reporting, leading to more personalized care for NHS patients.

So, if the patient does come back into the hospital, this does allow for the addition of many position support systems, because the infrastructure has already been built to support this. Like the PAC system for images, images are available on demand wherever you are. Therefore, the genome for that given person is accessible and available where they are, for any analytical process, open to clinical interpretation, including similar venture in Holland with the genotyping cards (van Schaik and van Gelder, 2016) that are presented to pharmacy staff to prevent adverse drugs being prescribed by them or indeed the physician; also, the ChipCard in Kuwait (Fig. 12.6).

This will change the way researchers should and will think about the world, whereby researchers have been used to low coverage genomic datasets to work with, be it genotypes or whole exome data, to cover the whole genome diagnostics and this is only good enough for finding new patterns and new information for gene relationships. However, now we have got high quality-called data and genomes with very high coverage for all patients to enable the researchers to ask: Why they cannot explain the genome? Why couldn't I find out what is wrong with the patient? They now will have the complete information for this

FIG. 12.6

Kuwaiti prototype Medical ChipCard: Genotyping card based on known whole genome sequenced information.

person and will be able to build a complex real-time model, explaining what is going wrong with the given genome and pathways pertaining to a given patient. Moreover, if the model is not good enough, one can push back to the community for answers. In some respects, third-party companies are already helping with the ability to link and build better disease models; companies such as AI framework discussed in this book by the Magrathea and GEL teams are enabling all within the Genomics England datasets to link to other members within the framework through metadata approaches and links to rare disease sets, models, and associated databases of interest. The partnerships that Genomics England is making are pushing a new dynamic relationship within genome communities both locally and with other similar efforts in the pipelines from other global regions. One important factor seems to be that all are closely following the Genomic England methodology and processes to aid patient diagnosis, understanding, and long-term treatments.

Genomics England is a governmental run 5-year plan, leading to the analysis of 100,000 genomes by the year 2020. Although the project has got off to a slow process, it is gathering great speed and being integrated into full clinical care within NHS England. Beyond the 2020, Genomics England will seek further funding and indeed license access to the Genomics England framework is open to all now via the Genomics England consortium. "The Genomics England NE Consortium is open to a wide range of pharmaceutical, biotech and diagnostics sector companies in the life sciences sector. The membership fee for large companies (with a market capitalization over $1 billion) is £250,000 per company. As well as contributing financially, these companies will also have to commit a number of employees, such as scientists and bioinformatians, to the consortium. With other resources, such as the secondment of staff, Genomics England anticipates that the total investment from each company will be in the region of £500,000 in 2015. Genomics England has always been clear that the charges should not prevent small companies or start-ups with great ideas from joining the consortium. In the spirit of actively encouraging smaller companies to get involved, Genomics England has reduced the fee for companies with a market capitalization below $1 billion to £25,000 for 2015. Genomics England will also be collaborating with bioinformatics companies to help automate analysis of the genome data to improve efficiency and accuracy."

Genomics England expects to be sequencing forever and, if and when new technologies come on board, they will be ready. The move away from short base sequencing is expected in the near to medium-term future; a move to larger lengthier fragments proposed by many pore-based approaches, such as the Oxford Nanopore technology (Fig. 12.7), is expected. At present, these technologies are not suitable for clinical grade sequence analysis, yet they are very close; when they are, Genomics England will always move to the better technology, the beauty being that all datasets, framework understanding, and patient records will evolve with these new approaches and indeed new findings when made.

FIG. 12.7

Oxford Nanopore's PromethION is a standalone benchtop instrument designed for high-throughput, high sample-number analyses. Its modular design allows a new paradigm of versatile workflow where many different experiments may be run in real time, with no constraints of fixed run times. Being the same technology as the MinION portable DNA sequencing, it offers real time, long read, high fidelity DNA, and RNA sequencing.

12.3   CONCLUSIONS

It is clear to see that the "walled gardens" of genomics are here to stay; the ability of the global research community and the ability to share data in a true meaningful way to clinical diagnostics understanding is a critical issue going forward for the genome community to share information without divulging important information, be it patient consent information or indeed rare-disease findings that can have a dramatic effect on developing countries' future. In fully genomic-literate countries such as the United Kingdom with Genomics England, it is clear that this model will be adopted by other countries and indeed corporations in the short term. The process of sample for understanding and treatment will be the long-lasting legacy of this process, when genome moves to real-time analysis. It is where important steps made by the Magrathea team in artificial intelligence will link data people, systems, and technology together and add meaning to these findings into systems, processes, and pathways to patient pioneered by Genomics England. The variant-specific Apps that will be the new Google, Twitter, or Facebook of the 22nd century or may be an uber-control App such as the mentioned social media software where these genomics/medical Apps will naturally connect when our society moves progressively to a pure digital information society.

The author's usual optimism is somewhat tailored here: The walled gardening of genomic medicine is in effect happening internally with patient consent, and one cannot see how this can and will translate into a global community sharing information. At present, we are living in the golden age of data sharing; one suspects this will close like governments do with precious mineral and oil resources wealth. Business will take the lead in this partnership; one suspects offering some incentive to share longitudinal data, be it "free" or reduced healthcare or, like we see with blood donation in countries, a crude fee for sample, which in itself is a legal landmine. GEL has by far the best system in place (with all its issues) and the UK's unique size, ability, and innovative companies leading this field do tend one to believe it to be dominating the field of genomic medicine for many years to come.

Acknowledgments

The author would like to thank Professor George P. Patrinos, Professor Fahd Al-Mulla, Sebnem Daniels Baker, Ayla Baker, Hyram Baker, Yvonne Hisemen, and Bertie for their valued and continued support of this new scientific idea and research.

References

Bernardini, F., Gheorghe, M., Holcombe, M., 2003. PX systems = P systems + X machines. Nat. Comput. 2 (3), 201–213.

Caulfield, M., Davies, J., Dennys, M., Elbahy, L., Fowler, T., Hill, S., Hubbard, T., Jostins, L., Maltby, N., Mahon-Pearson, J., McVean, G., Nevin-Ridley, K., Parker, M., Parry, V., Rendon, A., Riley, L., Turnbull, C., Woods, K., 2015. The 100,000 Genomes Project Protocol.

Available from: https://www.genomicsengland.co.uk/wp-content/uploads/2015/03/GenomicEnglandProtocol_030315_v8.pdf. [(Accessed 7 October 2016)].

Coiera, E., 2015. Guide to Health Informatics, third ed. CRC Press, London. 6 March 2015.

Davies, K., 2015. The $1,000 Genome: The Revolution in DNA Sequencing and the New Era of Personalized Medicine. Simon and Schuster, London.

Department of International Trade, 2016. Information Communications Technology (ICT) in the UK: Investment Opportunities. Available from: https://www.gov.uk/government/publications/information-communications-technology-ict-in-the-uk-investment-opportunities/information-communications-technology-ict-in-the-uk-investment-opportunities. [(Accessed 7 October 2016)].

GA4GH, 2016. About the Global Alliance for Genomics and Health. Available from: http://genomicsandhealth.org/about-global-alliance. [(Accessed 7 October 2016)].

Genomics England, 2016a. Research Topics in the 100,000 Genomes Project: GeCIP Domains Are UK-Led Consortia of Researchers, Clinicians and Trainees. Available from: https://www.genomicsengland.co.uk/about-gecip/gecip-domains/. [(Accessed 7 October 2016)].

Genomics England, 2016b. Participant Information Sheets and Consent Forms. Available from: https://www.genomicsengland.co.uk/taking-part/patient-information-sheets-and-consent-forms/. [(Accessed 7 October 2016)].

Genomics England, 2016c. Genomics England Selects Inuvika as a Technology Partner for the 100,000 Genomes Project. Available from: https://www.genomicsengland.co.uk/inuvika/. [(Accessed 7 October 2016)].

Genomics England, 2016d. GenomOncology's Knowledge Management System to Analyse Cancer Samples in the 100,000 Genomes Project. September 23, Available from: https://www.genomicsengland.co.uk/genomoncology/. [(Accessed 7 October 2016)].

GitHub, 2015. VCF (Variant Call Format) Specifications. Available from: https://vcftools.github.io/specs.html. [(Accessed 7 October 2016)].

GitHub, 2016. Specifications of SAM/BAM and Related High-Throughput Sequencing File Formats. Available from: http://samtools.github.io/hts-specs/. [(Accessed 7 October 2016)].

Lehman, A.F., et al., 1998. Patterns of usual care for schizophrenia: initial results from the schizophrenia patient outcomes research team (PORT) client survey. Schizophr. Bull. 24 (1), 11.

McDonagh, E., 2015. New Rare Disease Gene Tool Launched—PanelApp. Available from: https://www.genomicsengland.co.uk/rare-disease-gene-panelapp-launched/. [(Accessed 7 October 2016)].

Morse (KCB), A., 2017. Investigation: Clinical Correspondence Handling at NHS Shared Business Services. 26 June 2017. https://www.nao.org.uk/wp-content/uploads/2017/06/Investigation-clinical-correspondence-handling-at-NHS-Shared-Business-Services.pdf.

Mvundura, M., et al., 2010. The cost-effectiveness of genetic testing strategies for lynch syndrome among newly diagnosed patients with colorectal cancer. Genet. Med. 12 (2), 93–104.

National Health Service Choices, 2016. The History of the National Health Service in England. Available from: http://www.nhs.uk/nationalhealthserviceEngland/thenhs/nhshistory/Pages/nationalhealthservicehistory1948.aspx. [(Accessed 7 October 2016)].

National Health Service England, 2016. Genomics England's Clinical Interpretation Partnerships Launched. Available from: https://www.genomicseducation.hee.nhs.uk/news/item/63-genomics-england-s-clinical-interpretation-partnerships-launched. [(Accessed 7 October 2016)].

Patrinos, G.P., Baker, D.J., Al-Mulla, F., Vasiliou, V., Cooper, D.N., 2013. Genetic tests obtainable through pharmacies: the good, the bad, and the ugly. Hum. Genomics 7, 17.

The 1000 Genomes Project Consortium, 2010. A map of human genome variation from population-scale sequencing. Nature 467, 1061–1073.

The ENCODE Project Consortium, 2007. Identification and analysis of functional elements in 1% of the human genome by the ENCODE pilot project. Nature 447, 799–816.

UK Government, 2015. Genomics Industry Study: UK Market Analysis. Available from: https://www.gov.uk/government/publications/genomics-industry-study-uk-market-analysis. [(Accessed 7 October 2016)].

van Schaik, R.H.N., van Gelder, I.H.e.T., 2016. Farmacogenetisch testen op CYP450-enzymen. Ned. Tijdschr. Geneeskd.. 160,. A9404V.

Wynne, S., 2014. Genome Campus Will Host UK National Health Service 100,000 Genomes Project. Available from: http://www.sanger.ac.uk/news/view/2014-08-01-genome-campus-will-host-uk-nhs-100-000-genomes-project. [(Accessed 7 October 2016)].

Further Reading

Asimov, 1985. Robots and Empire. Grafton Books, London.

Genomics England, 2015. Clinicians, Researchers and Industry Collaborate with the 100,000 Genomes Project. March 26, Available from: https://www.genomicsengland.co.uk/clinicians-researchers-and-industry-collaborate-with-the-100000-genomes-project/. [(Accessed 7 October 2016)].

Index

Note: Page numbers followed by *f* indicate figures and *t* indicate tables.

Printed in the United States
By Bookmasters